Transmission Electron Microscopy of Materials

TRANSMISSION ELECTRON MICROSCOPY OF MATERIALS

Gareth Thomas, Ph.D., Sc.D.

Professor, University of California
Department of Materials Science & Mineral Engineering
and Molecular and Materials Research Division
Lawrence Berkeley Laboratory

Michael J. Goringe, M.A., Ph.D.

University Lecturer in Metallurgy and
Fellow of Pembroke College, Oxford University

A Wiley-Interscience Publication

JOHN WILEY & SONS
New York • Chichester • Brisbane • Toronto

Library of Congress Cataloging in Publication Data

Thomas, Gareth.
Transmission electron microscopy of materials.

"A Wiley-Interscience publication."
Includes bibliographical references and index.
1. Electron microscopy. 2. Materials–Testing.
I. Goringe, M. J., joint author. II. Title.

TA417.23.T48 620.1'12'028 79-449
ISBN 0-471-12244-0

Printed in the United States of America

10 9 8 7 6 5 4 3 2 1

To Elizabeth and Gill

PREFACE

Transmission electron microscopy has now become an established method for the characterization of materials. Recent decades have brought innovations. In the 1950s came thin foil applications following hard on the heels of replications; in the 1960s high voltage microscopy opened up new vistas, making almost all materials amenable; and in the 1970s we have analytical microscopes by which it is at least theoretically possible to resolve and spectroscopically identify atoms. The dream of atomic resolution is being fulfilled–almost 50 years after the invention of the electron microscope itself. The demand for materials scientists and engineers with a sound training in the characterization of materials has increased tremendously, and the field of electron microscopy has outgrown the earlier textbooks. Consequently we felt that it was time to write a new book for student and researcher alike. The present text is the outgrowth of the efforts of many people, especially the students who have taken the two-quarter course sequence in the Materials Science and Mineral Engineering Department at Berkeley. Thanks are due to the many students and colleagues who have directly or indirectly contributed to the present book. We particularly thank the following: W. L. Bell (who played a major role in the early development of Chapters 2 and 3), D. R. Clarke, R. Gronsky, O. L. Krivanek, B. V. N. Rao, R. Sinclair, and R. K. Mishra. Many others contributed drawings and micrographs. We ourselves, of course, claim the responsibility for the text and for any errors that may have crept in.

The book has been organized both as an instructional and reference text. Much routine work can be done without a detailed knowledge of the dynamical theory itself; for routine work Chapters 1 to 3 and the exercises should thus be adequate. For a more rigorous background Chapters 4 and 5 should be studied. This arrangement follows the two-quarter graduate course sequence we have taught at Berkeley. A separate laboratory course gives individual practical training in the methods of specimen preparation, instrumental technique, and interpretative application of contrast and diffraction principles, usually to problems connected with students' theses projects. The importance of this practical aspect cannot be overemphasized. Weekly individual instruction, rather than group demonstrations should be given, if possible, until the researcher is confident at the instrument.

Finally, it should be reemphasized that the field of microscopy is both broad and specialized. A successful research program must therefore contain a team of personnel with expertise in the various aspects of the subject–from the very practical to the theoretical.

GARETH THOMAS
MICHAEL J. GORINGE

Berkeley, California
Oxford, England
June 1979

CONTENTS

Transmission Electron Microscopy of Materials

ONE

INTRODUCTION

1 Scope of the Book

The properties of materials are structure-sensitive. Structure is in turn determinded by composition, heat treatment, and processing. Thus it is necessary to characterize both composition and microstructure at the highest levels of resolution possible in order to understand materials behavior and to facilitate the design of new or improved materials. Such characterization requires advanced and sophisticated methods of analysis using microscopic, diffraction, and spectrographic techniques. It is in this regard that the electron microscope is such a unique instrument because it provides all the capabilities necessary for both physical and chemical analysis (Fig. 1.1). This is further illustrated in Table 1.1, which lists the techniques available as a function of the limits of spatial resolution. Of particular interest in this book is the information available by using transmission electron microscopy, the fundamental principles of which are discussed in this chapter.

Although Heidenreich[1] published a paper in 1949, based on his observations of foils of aluminum, in which the essential features of contrast from crystals were described and analyzed experimentally (a remarkable achievement of specimen preparation and electron microscopy, considering that the instrument he used was a single-condenser, 50 kV type), this paper was, unfortunately, ahead of its time and not widely read. Thus it was not really until the middle 1950s that electron microscopy of thin foils of metals began to make a significant impact upon the scientific world. In recent years the field has expanded enormously. Now instruments capable of 1.2 mV operation are standard, and a number of microscopes have been built to work at even higher voltages. The addition of microanalytical capabilities now extends electron microscopy to high resolution spectroscopy. As a result a wealth of information on the structures of a large

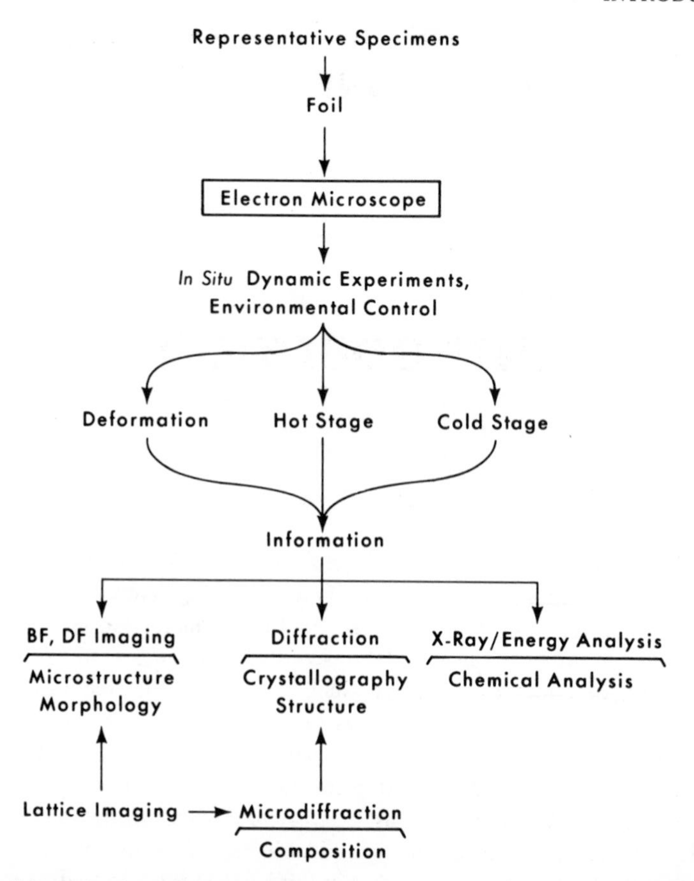

Fig. 1.1 Schematic showing the principal functions of the transmission electron microscope for characterizing materials. Courtesy MTM Association for Standards and Research; G. Thomas, *Journal of Metals*, **29**, 31 (1977).

variety of materials (metals, ceramics, organics, minerals) has become available to the scientific community, and electron microscopy has become a standard course in most university departments of metallurgy and materials science.

In spite of this growth and the appearance of several books (see the Bibliography on p. 66) since the first text[2] was published, the publication of conference proceedings, and the enormous numbers of technical articles, monographs, and brochures that have appeared, the need still exists for a comprehensive text on transmission electron microscopy (TEM). Such a text should include important recent developments, for example, high voltage electron microscopy, scanning transmission electron microscopy (STEM), special applications of many-beam diffraction effects, and the applications of TEM to a diverse group

Table 1.1 Spatial Resolution Limits in Materials Characterization

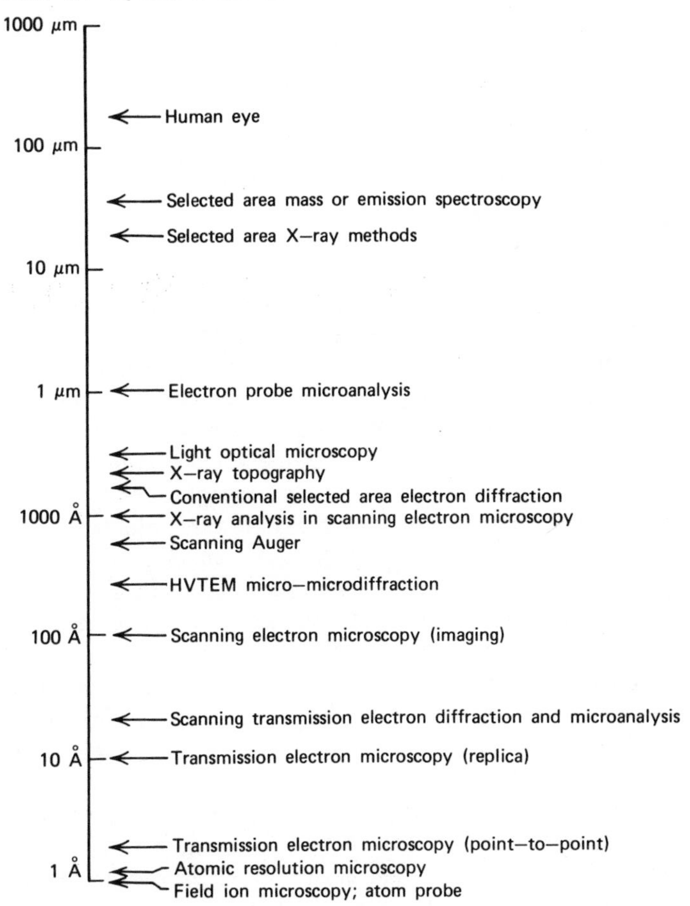

of problems and materials, and, at the same time provide practical help to the student researcher and industrial microscopist. It has been our aim to try to fill this gap by means of the present book. We have tried to indicate the situations where simple geometrical approaches are satisfactory but also their limitations. These are covered in the first three chapters. Quantitative analyses of complex materials (e.g., noncubic crystals, ceramics, minerals) require a higher level of sophistication, for example, contrast calculations and analysis. Consequently we have included two chapters that deal with the basics of the dynamical theory of electron diffraction and its applications. Throughout we have given typical examples of where electron microscopy is essential in a range of research problems representative of current trends in materials science and engineering.

Problem sets have been included also to illustrate the principles developed. Some of the most practical problems require very sophisticated techniques of analysis, and it is important that this fact be recognized. Unfortunately, such recognition is not yet universal.

2 Basis for Electron Microscopy of Materials

2.1 Wavelength and Resolution

The information that is obtained by electron microscopical methods (Fig. 1.1) is derived from the scattering processes that take place when the electron beam travels through the specimen. There are two main types of scattering: (*a*) elastic–the interaction of the electrons and the effective potential field of the nuclei–which involves no energy losses and can be coherent or incoherent (poor phase relationship); and (*b*) inelastic–the interaction of the electrons and the electrons in the specimen–which involves energy losses and absorption. It is the elastic scattering that produces a significantly informative diffraction pattern; and if the scattering centers in the specimen are arrayed in an orderly, regular manner, as in crystals, the scattering is coherent and results in spot patterns and, if the sample is a fine grained polycrystal, ring patterns. In practice inelastic scattering usually occurs as well and also produces informative regular diffraction effects, for example, Kikuchi patterns. Inelastic scattering causes absorption from which characteristic energy losses and emissions (X-ray, Auger, etc.) yield spectroscopic data. Thus the electron microscope is an extremely versatile analytical instrument (Fig. 1.1).

When an image is formed of the scattered beams, two main mechanisms of contrast arise. If the transmitted and scattered beams can be made to recombine, thus preserving their amplitudes and phases, a lattice image of the planes that are diffracting (Figs. 1.2*a* and *c*) or even structure images of the individual atoms (Fig. 1.2*b*) may be resolved directly (phase contrast). The principle is the same as that of the Abbé theory for gratings in light optics (see Figs. 1.2*a–c*). Alternatively, amplitude contrast is obtained by deliberately excluding the diffracted beams (and hence the phase relationships) from the imaging sequences by the use of suitably sized apertures, placed in the back focal plane of the objective lens (Fig. 1.3). Such an image is called a bright field image (Fig. 1.3*a*). Alternatively, a dark field image can be obtained by excluding all beams except the particular diffracted beam of interest (Figs. 1.3*b* and *c*). Examples are given in Fig. 1.4.

The basic reason for utilizing the electron microscope is its superior resolution, resulting from the very small wavelengths as compared to other forms of radia-

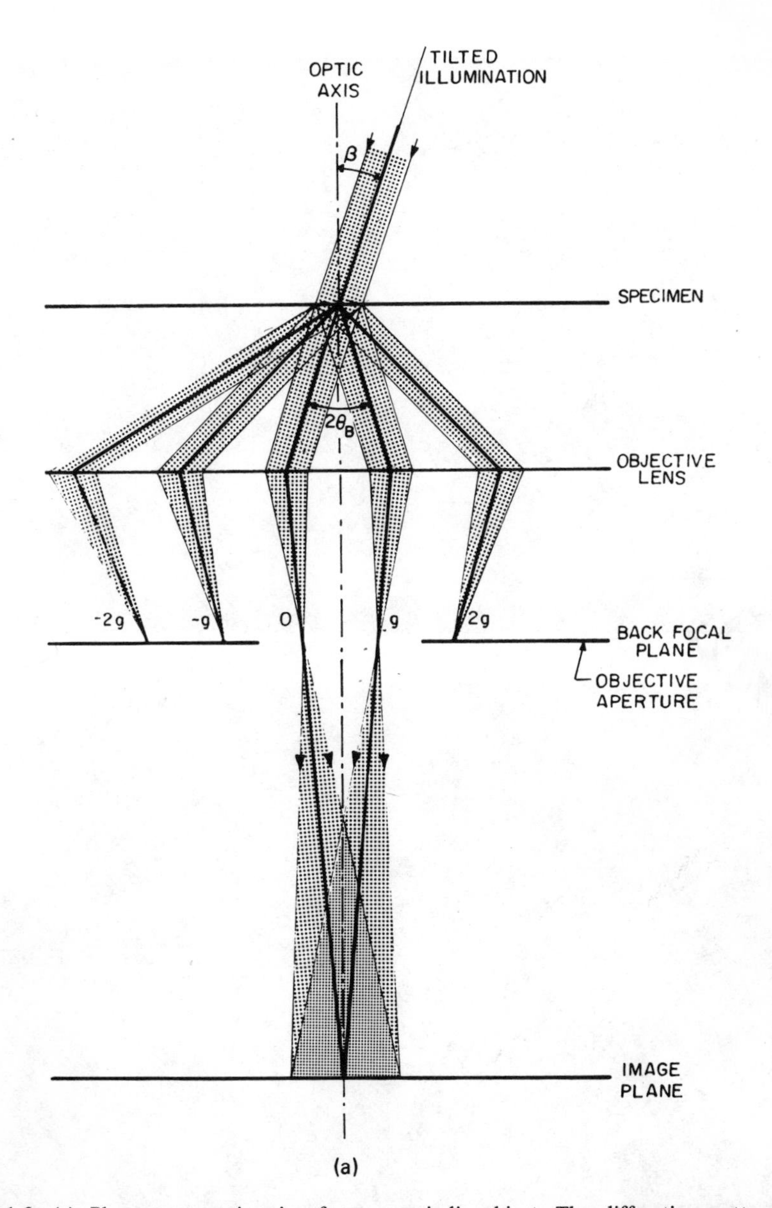

Fig. 1.2 (*a*) Phase contrast imaging from a periodic object. The diffraction pattern is formed in the back focal plane. The period $d = 1/g$ is imaged as magnified fringes if the diffracted and transmitted beams recombine at the image plane (when $2\theta < \alpha$) (see *c*). (*b*) [001] Projection of atomic resolution in a thin gold crystal. Courtesy H. Hashimoto.[6] (*c*) Lattice image of Mg_3Cd (fully ordered), showing three variants of $\{\frac{1}{2}0\frac{\bar{1}}{2}0\}$ spaced 5.4 Å and forming ordered domains. The $\{\frac{1}{2}\frac{1}{2}\bar{1}0\}$ superlattice planes of 3.7 Å are also imaged. Courtesy Pergamon Press.[4]

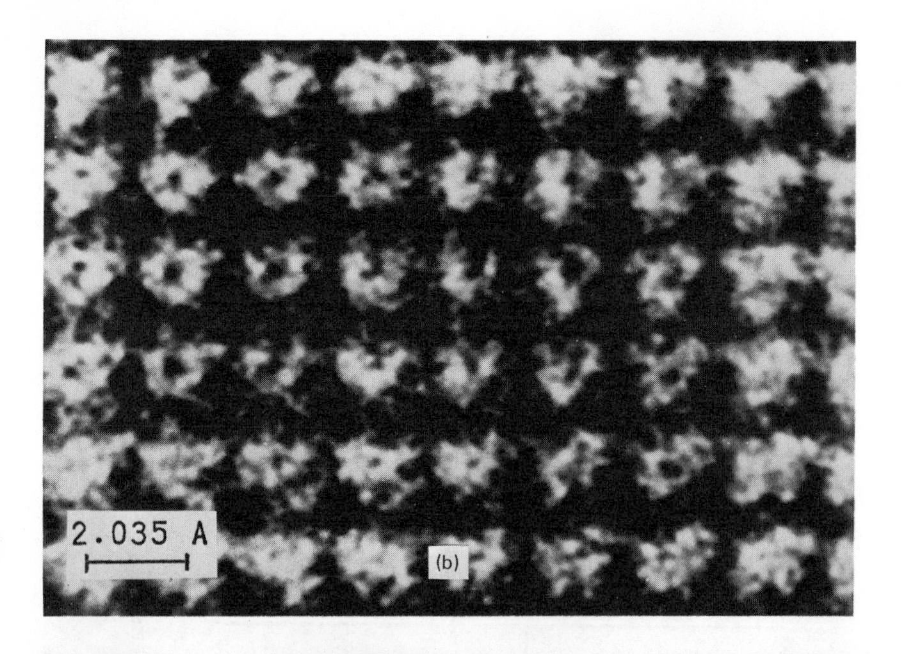

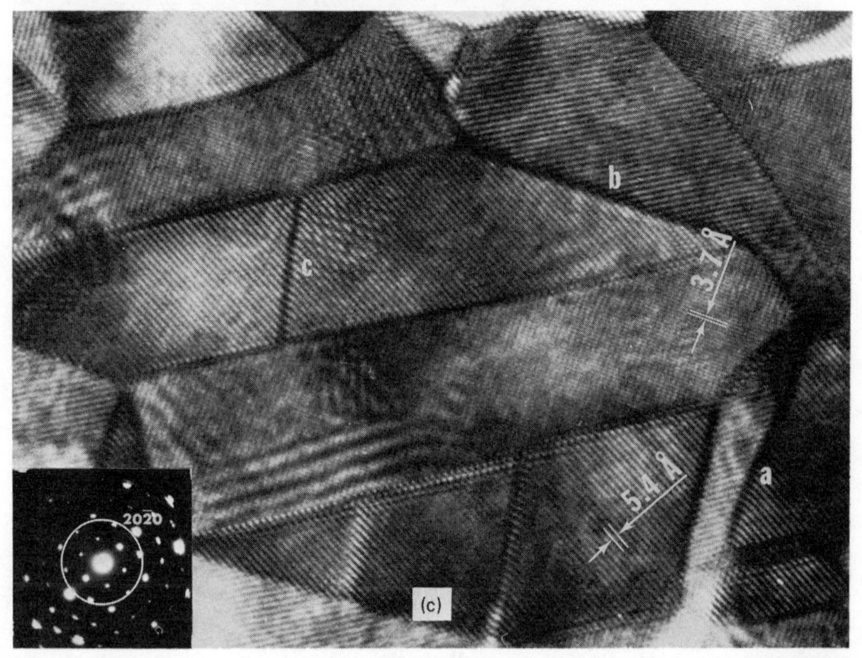

Fig. 1.2 (*Continued*)

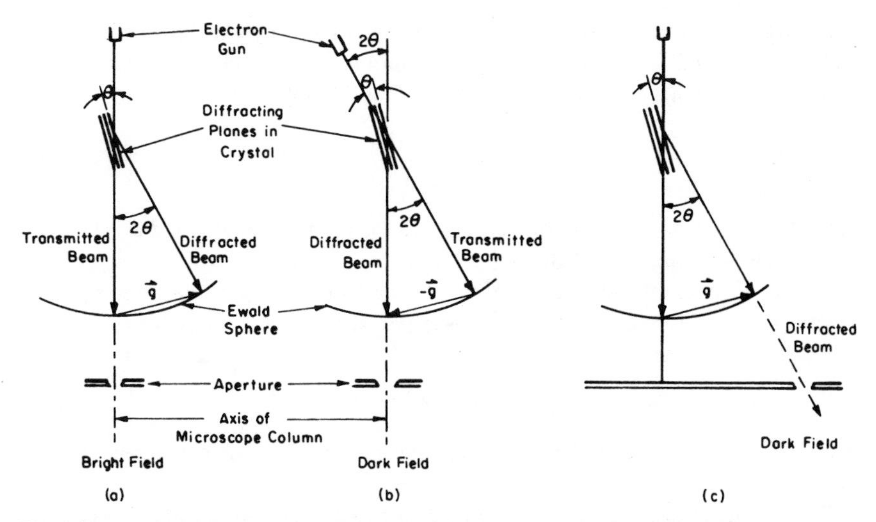

Fig. 1.3 Amplitude contrast imaging: the transmitted and diffracted beams do not recombine in the image. Objective apertures are used to stop off the diffracted beams to form a bright field image. (*a*) Dark field images are obtained by gun tilting or beam deflection (*b*) or with an off-axis aperture (*c*), all other beams being stopped off by the aperture. Courtesy Springer-Verlag.

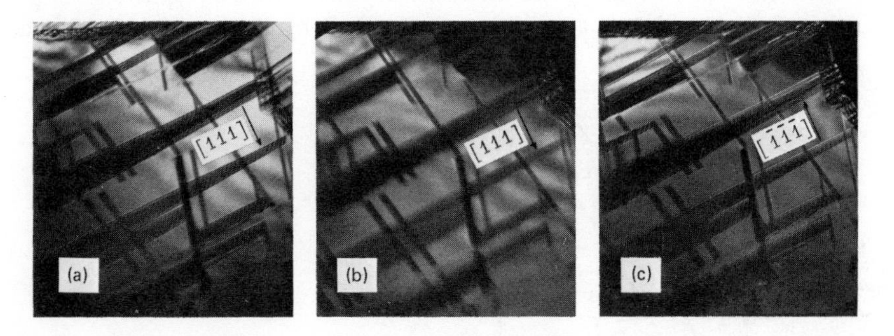

Fig. 1.4 Examples of bright and dark field images from faults in fcc cobalt. Notice the poor resolution in the off-axis aperture dark field image (*b*); see Fig. 1.3. Courtesy American Institute of Mining, Metallurgical, and Petroleum Engineers.[5]

tion (light, X-rays, neutrons). The resolution is given by the Rayleigh formula, which is derived by considering the maximum angle of electron scattering, α, which can pass through the objective lens. This formula is

$$R = \frac{0.61\lambda}{\alpha}, \quad (1.1)$$

where R is the size of the resolved object, λ is the wavelength, and α is identical to the effective aperture of the objective lens.

In the electron microscope the effective aperture is limited chiefly by spherical aberration of the imaging lens. The spherical aberration error (see the equation on p. 85 of ref. 2) is

$$\Delta S = C_s\alpha^3, \quad (1.2)$$

where C_s is the coefficient of spherical aberration of the objective lens ($\simeq$ focal length, of order 3 mm).

Thus R increases with decreasing α, whereas ΔS decreases with decreasing α. As a result, in electron optics (unlike light optics, where the aperture may be increased almost indefinitely) one arrives at an optimum aperture and minimum aberration, given by

$$\alpha_{\text{opt}} = A(\lambda^{1/4})C_s^{-1/4} \quad (1.3)$$

and

$$\Delta R_{\min} = B(\lambda^{3/4})C_s^{1/4}, \quad (1.4)$$

where A and B are constants $\simeq$ unity. This relationship is shown in Fig. 1.5.

The relativistic wavelength of electrons depends on the accelerating voltage and is given (see the equation on p. 5 of ref. 2) by the modified De Broglie

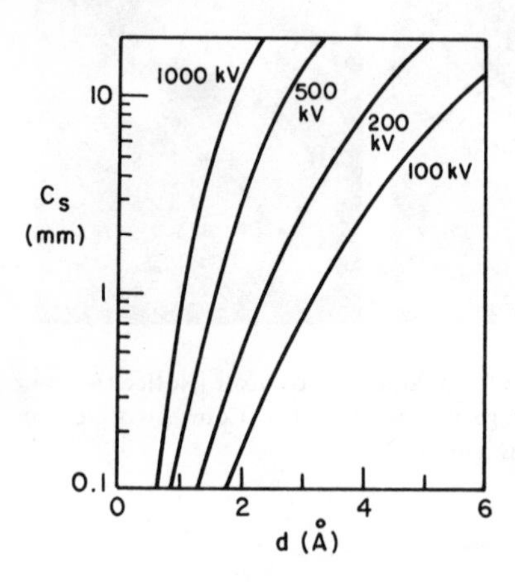

Fig. 1.5 Plot showing resolution d as a function of accelerating voltage and spherical aberration coefficient C_s.

Table 1.2 Some Properties of Electrons as a Function of Accelerating Voltage *E*

E	λ (Å)	$\chi = \lambda^{-1}$ (Å)$^{-1}$	$(v/c)^2$
80 kV	0.0418	23.95	0.2524
100 kV	0.037	27.02	0.3005
200 kV	0.0251	39.87	0.4835
500 kV	0.0142	70.36	0.7445
1 MV	0.0087	114.7	0.8856
2 MV	0.005	198.3	0.9586
10 MV	0.0012	846.8	0.9976

wavelength:

$$\lambda = \frac{h}{[2m_0eE\,\{1 + (eE/2m_0c^2)\}]^{1/2}}, \tag{1.5}$$

where h is Planck's constant, m_0 is the rest mass, e is the charge on the electron, E is the accelerating potential (V), and c is the velocity of light, that is,

$$\lambda = \frac{12.26}{E^{1/2}(1 + 0.9788 \times 10^{-6}E)^{1/2}} \quad (\text{Å}),$$

and thus decreases with increasing E. Some values pertinent to electron microscopy are given in Table 1.2.

Another advantage of the small wavelength of electrons is that the depth of field and depth of focus are very large in electron microscopes (Section 3.1.2).

Thus at 100 kV $\alpha_{opt} \simeq 6 \times 10^{-3}$ rad and $\Delta R_{min} \simeq 6.5$ Å for $C_s = 3.3$ mm. In most modern 100 kV microscopes C_s can be reduced to ~1.5 mm, giving a point-to-point resolution of about 3.5 Å. In theory this can be reduced to below 2 Å, as has been demonstrated by Hashimoto et al.[6] (see Fig. 1.2*b*) using a microscope with an objective lens with $C_s = 0.7$ mm. However, a significant gain in resolution is obtained by increasing the voltage (Fig. 1.5), and this appears to be the best way to approach atomic resolution in close-packed structures. Other factors that affect resolution are astigmatism and chromatic aberration of the imaging system, and chromatic aberration resulting from energy losses in the specimen. These errors produce poor resolution for nonaxial illumination, such as off-axis dark field imaging, which suffers from both spherical and chromatic aberration, as can be seen by comparing Figs. 1.3 and 1.4. The chromatic disk of confusion is given by

$$\Delta C = C_c\,\alpha\,\frac{\Delta E}{E}, \tag{1.6}$$

where C_c is the coefficient of chromatic aberration of the lens.

Table 1.3 Some Advantages of High Voltage Transmission Electron Microscopy

---

1. Penetration
   Light materials $\simeq$ 14 μm at 2.5 MeV
   Medium-heavy $\simeq$ 2–3 μm at 1 MeV (maximum energy for routine microscopy)
   Applications include:
   - Biological systems-natural environments, thick sections
   - Ceramics, minerals, difficult-to-thin materials
   - Nondestructive, precious samples, e.g., moon dust
   - Particle analysis (extractive metallurgy), pollution research
   - Heavy elements
   - Semiconductor devices
   - Crystals of low defect densities
   - Better statistics-larger sampling (stereomicroscopy)
   - *In situ* studies (dynamic events), environmental; gases, liquids, etc.
   - Interfaces, e.g., diffusion gradients, semiconductors, multiphase systems

2. Resolution increase
   Chromatic and spherical aberrations
   Reduced area for selected area diffraction
   Lattice defects

3. Special diffraction effects—Crystals
   Channeling-increased transmission
   Critical voltage effect
   Contrast
   - Improved magnetic contrast
   - Improved resolution of lattice defects (bright field imaging using high order systematics)

4. Displacement damage
   Simulates nuclear reactors, e.g., at 1 MeV point defect damage in iron in electron microscope in 1 min $\equiv$ 450 days in reactor
   Maximum energy should be about 1.5 MeV (1.25 MeV is threshold for uranium)

5. Ionization damage
   Decreases with increasing energy (e.g., beam-sensitive materials—biological substances, polymers, ionic crystals, etc.).

6. Higher brightness

7. Spectroscopy (X-ray or electron energy loss) superior to low voltage instruments in principle

---

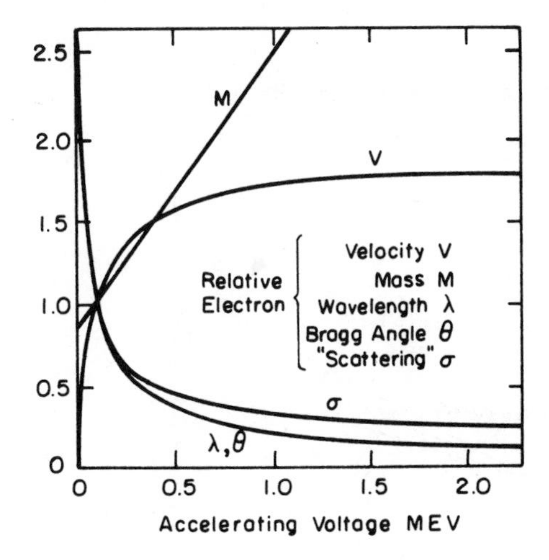

Fig. 1.6 Properties of electrons as a function of voltage, relative to those at 100 kV. Courtesy R. M. Fisher.[10]

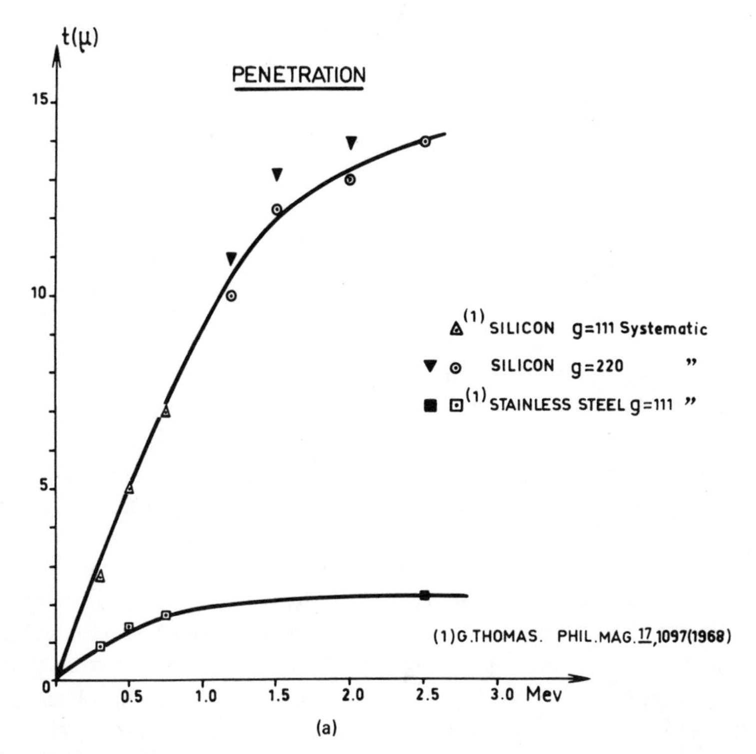

Fig. 1.7 (*a*) Experimental data on penetration in silicon and stainless steel. Courtesy *Journal of Microscopy*.[12] (*b*) Experimental data on the destruction of crystallinity (critical electron dose N_{cr}) of the amino acid 1-valine as a result of electron exposure at different voltages. Notice that even at 2.5 MeV radiation damage limits resolution to ~30 Å. The theoretical stopping power is also plotted. Courtesy Academic Press.[13]

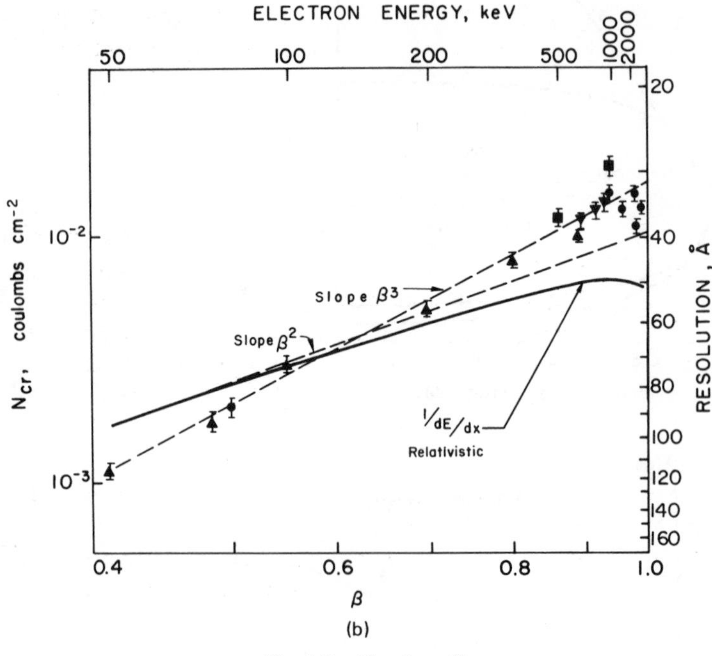

Fig. 1.7 *(Continued)*

The causes of ΔE are voltage fluctuations in the incident beam, absorption processes in the specimen and current fluctuations in the lenses. Whereas α_{opt} and ΔR_{min} change slowly with E (through λ), $\Delta E/E$ rapidly decreases with increasing E, thus decreasing ΔC, and this is one of the main advantages of high voltage operation.

High voltage electron microscopy is now well established in many laboratories in different countries. High voltages are useful for several reasons,[7-11] and some advantages are noted in Table 1.3. The effective scattering cross sections decrease with energy, as indicated in Fig. 1.6. This predicts an improvement in specimen penetration for a given level of resolution and is especially important for materials science. Experimentally, however, it is found that only in light materials (silicon, aluminum) is there a significant gain in penetration, as shown in Fig. 1.7*a*. The reduction in inelastic scattering with increasing voltage implies a reduction in ionization and other damaging processes, and this effect has been observed for several biological and polymeric solids. As shown in Fig. 1.7*b*, recent work[12,13] up to 2 MeV indicates that this damage rate decreases as $(v/c)^3$. However, knock-on damage occurs above a threshold energy that is roughly proportional to atomic number (e.g., $\simeq$ 400 keV for copper; 1.2 MeV for gold). A further gain from increasing the voltage is the rapid reduction in spherical aberration. This can be seen from eq. 1.2 and Fig. 1.5, namely, $\Delta S \simeq C_s \alpha^3$; and since $C_s \simeq \lambda^{-1}$, for a particular set of reflecting planes (see Section 2.3) $\alpha \propto \lambda$, $\Delta S \simeq \lambda^2$.

2.2 Diffraction

In crystals Bragg diffraction occurs, and eq. 1.1 can be adapted to the special case in which $\alpha = 2\theta$, where θ is the Bragg angle. Bragg's law defines constructive interference as occurring when the path difference between waves scattered by successive parallel planes of atoms of interplanar spacing d is equal to an integral number of wavelengths (phase difference is $2\pi/\lambda$ times path difference). If θ is the angle of incidence, then, as shown in Fig. 1.8, the path difference between waves 1 and 2 is $2d \sin \theta$; hence Bragg's law is

$$2d \sin \theta = n\lambda, \tag{1.7}$$

where n is the order of reflection. Equation 1.7 is essentially the same as eq. 1.1, except that now $R = d$ and the factor 0.61 is neglected.

The formation of images of lattice planes in crystals will now depend upon whether or not the recombination of a diffracted and the transmitted beam can

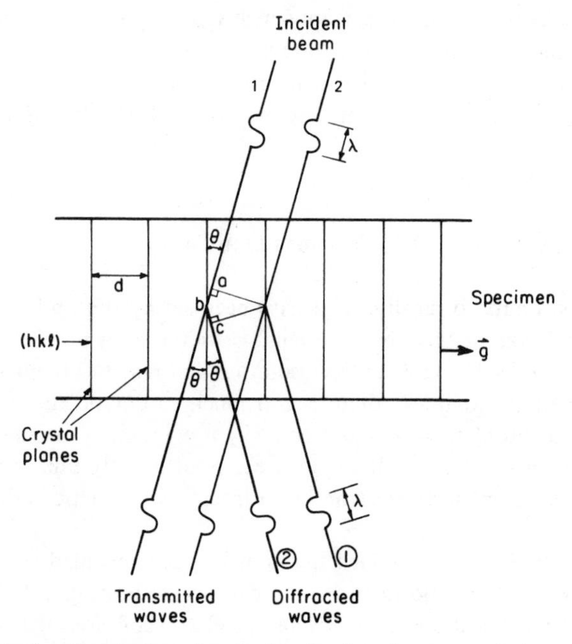

Fig. 1.8 Illustrating Bragg's law of scattering, for fast electrons and close-packed crystals, $\theta \simeq 1°$. The path difference between waves 1 and 2 is $abc = 2d \sin \theta$.

occur. Thus, if 2θ is the angle between the diffracted and transmitted beams, $2\theta \leqslant \alpha$, and fringes will be formed of spacing $x = \lambda/2\theta$ (Fig. 1.2*a*). Taking this into account, noting that for small angles Bragg's law can be written as $2d\theta = \lambda$, and adjusting the aberration equations 1.2, 1.3, and 1.6, one finds that the minimum resolvable fringe spacing $x = d$ is 2 Å or better at 100 kV. An example showing ordered domains in hexagonal Mg_3Cd is shown in Fig. 1.2*c*, and other examples are given in Chapter 3. High resolutions can be achieved only by tilting the illumination so that the diffracted and transmitted beams straddle the optic axis symmetrically, since in this orientation phase shifts from spherical and chromatic aberration cancel. The methods used for lattice imaging are described in Section 4.8. Lattice imaging was introduced by Menter[14] and has recently received considerable attention (e.g., Sinclair and Thomas[15]; See also Chapter 3, Section 12).

Figure 1.2*a* shows that, in all cases, the image is a magnified picture of the diffraction pattern (or, more strictly speaking, it is the Fourier transform of the pattern), whether the specimen is crystalline or truly amorphous. The diffraction pattern itself is formed in the back focal plane of the objective lens; and by suitable changes in magnification, utilizing projector lenses of variable focal length, it is easy to quickly obtain images and their diffraction patterns on the fluorescent screen and to record them photographically or display them on a monitor by television techniques, from which video recordings can be made. The principle of the electron microscope (Fig. 1.2*a*) can be conveniently demonstrated by setting up an optical bench, using a laser light source and a series of glass lenses. Such a system also serves for optical diffraction (Section 6, Fig. 1.39.)

2.3 Specimen Limitations

The resolution limits so far discussed have been set by the optics of the microscope, in particular of the objective lens. However, in a number of situations the specimen itself sets a limit to the resolution that may be obtained. Obviously, if the specimen is too thick, there is too much chromatic aberration for high resolution. This problem may be overcome by using thinner specimens, although some materials may be difficult to prepare in a sufficiently thin form. In others, however, there are intrinsic electron beam-specimen interactions that cannot be overcome.

The passage of electrons through the sample is accompanied by many energy transfer processes. The deposition of energy can cause heating and, depending on the scattering cross sections, more severely, inelastic collisions that lead to radiation damage. Of these collisions, ionization and bond-breaking damage will eventually destroy a specimen. Such processes are serious in nonmetallic materials and can be critical for organic and some ionic specimens.

If the degradation of the specimen occurs too rapidly, in particular before enough electrons have passed through to form an image at the required magnification, then resolution is impaired. Similarly, defect characterization is prevented if the specimen degrades before all the necessary images have been taken of the same area.

In the case of direct displacement damage (which depends upon atomic number and accelerating voltage and is not very important at 100 kV) the cross section is generally very small, so that displacement is an unlikely event during normal image-recording exposures. Therefore direct displacement is unlikely to limit the information available from the specimen; indeed, observation of the effects produced by displaced atoms[16] has become very important in understanding the void problem in materials used in nuclear reactors.

On the other hand ionization processes have very large cross sections, particularly in insulating materials (effectively zero in conductors and semiconductors). The energy deposited by the electron beam may result in breakage of chemical bonds, an effect that is obviously of paramount importance in the study of living biological materials,[17] or in certain instances in indirect displacement of atoms from their lattice sites (e.g., in alkali halides).[18] In either case the need to avoid destruction of the individual molecules or of the crystallinity of the specimen limits the dose to which the material may be exposed. As mentioned earlier, cross sections for ionization processes decrease with increasing voltage,[12,13] and hence high voltage operation may alleviate the problem. This is illustrated in Fig. 1.7*b* for the amino acid 1-valine. The radiation damage limitation on resolution is quite serious, and even at 1.5 MeV it is estimated to be ~30 Å—a twofold improvement over 100 kV, however. Although in some instances lowering the specimen temperature may also be beneficial in reducing damage,[17,18] it seems unlikely that direct molecular resolution is achievable in these materials without some kind of doping to stabilize the structure. An example from a study operating at low temperature to minimize the effects of damage is shown in Fig. 2.11. The lifetime of the anthracene crystal in the microscope is significantly increased by lowering the specimen temperature to about 20 K.

As shown in Sections 2.2 and 4.5, contamination at the specimen surfaces (which increases the amount of inelastic scattering) will also lead to a loss of resolution, and for long-exposure situations contamination may well be a limiting factor. Thus an ultrahigh, clean vacuum is important for work involving the highest resolutions.

3 The Transmission Electron Microscope

As is apparent from the preceding section, a suitable instrument is required to extract all of the information present in the specimen material being studied. This book concentrates on one of the most versatile of such devices, the transmission electron microscope (TEM). Since a number of excellent books on the design, construction, calibration, and operation of the TEM exist (see the Bibliography on p. 66), only an outline of the more important aspects for materials science applications is given here.

3.1 Components of the TEM

The first essential component of the microscope is the electron gun (see Fig. 1.9*b*), which produces a partially collimated stream of electrons of the required energy (e.g., 100 keV). In most instruments these electrons are produced initially by thermionic emission of electrons from a heated tungsten "hairpin" filament which is held at −100 kV with respect to the rest of the microscope; these electrons are subsequently accelerated via a controlling Wehnelt electrode through a hole in the anode. The electron gun thus acts as a source of electrons, of effective size 50–100 μm, diverging over a small solid angle on exit from the gun. These electrons are then focused by a set (usually two) of condenser lenses to produce the desired illumination of the specimen, ranging from the smallest possible spot (~0.1 μm in some cases, but with considerable angular spread—beam divergence) to the almost plane-parallel illumination of a large area. Recent improvements in electron gun design, necessitated by the development of scanning transmission instruments but useful also in the standard TEM, have produced effective sources which may be much smaller and brighter than those available from the tungsten hairpin. Such improvements include the use of materials of lower work function (e.g., lanthanum hexaboride) as thermionic emitters and the harnessing of the Schottky (field enhanced emission) and field emission effects. In most cases, particularly the last, ultrahigh vacua are consequently required around the filament, adding to the complexity and cost of the gun system. However, in the field emission case at least, there is some counterbalancing reduction in complexity in that the condenser system is made an integral part of the gun.

Whatever the design of the gun and condenser system, the electron beam must be capable of adjustment in both its position and its inclination. Such adjustments are provided in modern instruments by suitable sets of magnetic alignment coils, which usually also include corrections for astigmatism.

The specimen to be observed is held in a special holder and, since the microscope interior must be evacuated to a high vacuum, is introduced through an airlock. The specimen stage in which the holder fits inside the microscope is provided with the drives necessary to produce both movement and tilting of the specimen, so that different areas may be studied at controllable orientations with respect to the microscope axis and illuminating electron beam. When in position, the specimen is usually surrounded by a cooled anticontamination shield.

The objective lens focuses on the specimen and, for high resolution, has a very short focal length (1 to 3 mm). Thus it is inevitably of the "immersion" type, and some of the upper part of the objective lens also acts as part of the condenser system, modifying the electron beam before it strikes the specimen. Except in very high resolution or convergent beam studies this effect is usually

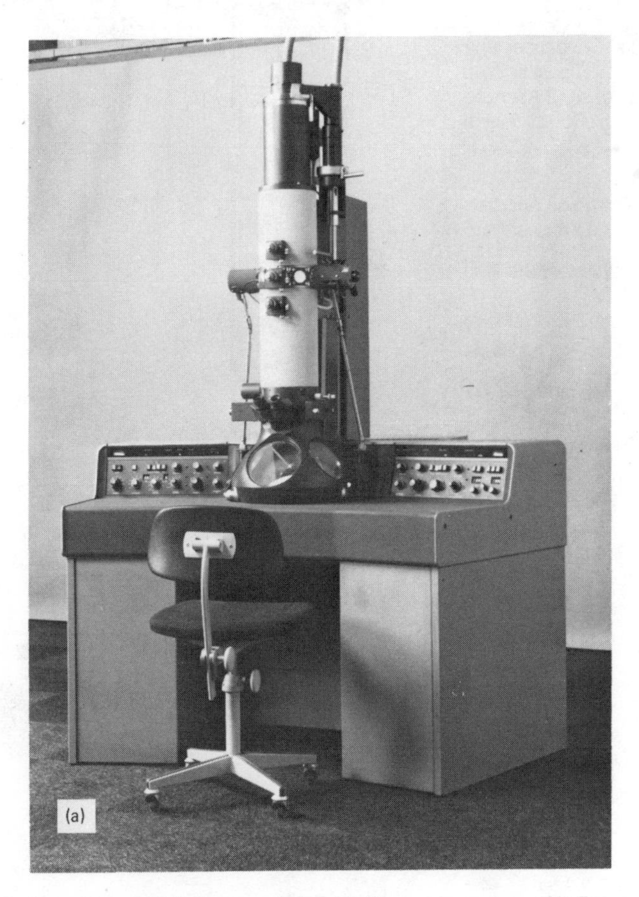

Fig. 1.9 (*a*) A modern 100 kV transmission electron microscope. (*b*) Cutaway to show principal features. Courtesy Philips Electronic Instruments.

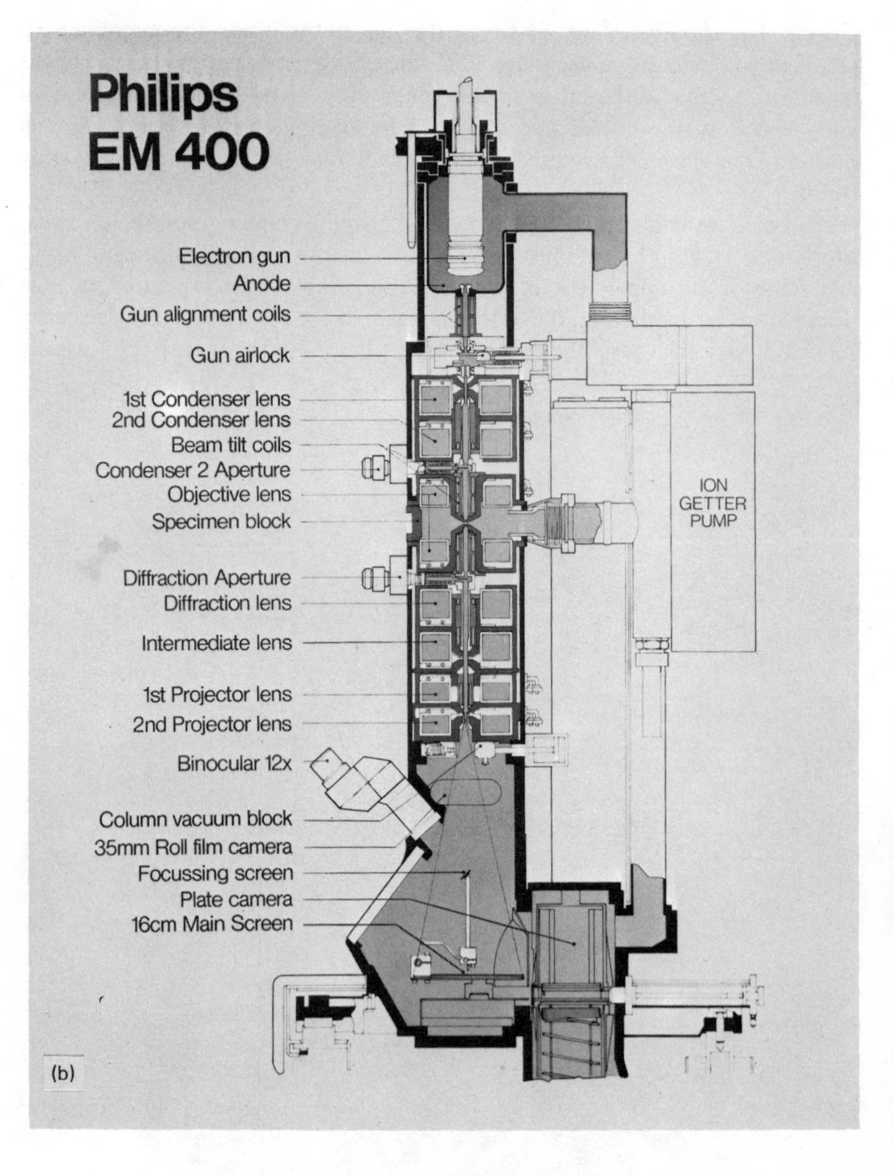

Fig. 1.9 (*Continued*)

ignored. The diffraction pattern is formed initially in the back focal plane of the objective, and a magnified image in its image plane (see Fig. 1.2*a*). The projector lenses (of which there may be two, three, or four) transfer either the diffraction pattern or the image onto the viewing screen with varying degrees of magnification to produce a variable camera length or variable magnification image, respectively.

Movable sets of apertures are provided in three locations: (*a*) in the condenser system to collimate the electron beam and modify its intensity; (*b*) in the objective back focal plane to select electrons at a particular inclination for subsequent magnification (e.g., for bright field, dark field, or lattice imaging operation); and (*c*) in the objective image plane to select electrons from a particular area of the specimen, for selected area diffraction. Usually a set of apertures consists of three differently sized circular holes in otherwise impenetrable material, any of which may be centered on the area of interest, as well as withdrawn completely. An alternative system is to provide a set of four mutually perpendicular blades which may be moved to change both size and position of the "hole."

In most modern electron microscopes all the lenses (with the exception of the integral electrostatic condenser system of the field emission source) are of the magnetic type. These lenses are energized by highly stabilized (a few parts per million in the case of the objective lens) direct current sources, the current through the lens in addition being variable for focusing. The high voltage supply must be similarly stable. In the case of instruments operating at 100 kV suitable voltages are produced by standard electronic techniques (e.g., transformer voltage multiplication), but at high voltages (greater than ~600 kV) methods such as Cockcroft-Walton voltage multiplication are required. In all cases stabilization is achieved by means of large negative feedback loops.

The interior of the whole microscope column is kept under vacuum by means of a number of sets of pumps, often arranged to evacuate regions where gases may be introduced (e.g., photographic film cassettes, specimen region), or where a particularly clean environment is required (e.g., electron gun, specimen region).

Most of the features outlined above may be seen in Fig. 1.9. The change of scale required when the electron energy is raised to 1 MeV is noted in Fig. 1.10.

3.1.1 *Electron Lenses.* The lenses in modern electron microscopes are designed so that resolutions close to the atomic level are possible. The three most important aberrations considered in lens design are related to the "shape" of the magnetic field (comparable to the shape of glass lenses in optical microscopes), that is, the variation of the magnetic field vector with position in the lens pole piece. These three defects are spherical aberration, chromatic aberration, and astigmatism. The first two defects result in the blurring of images, as given by eqs. 1.2 and 1.6. In practice, lenses are designed to minimize these defects by careful control of pole piece design, and voltage and current stabilization.

Astigmatism arises from defects in the machining of the lens pole pieces and, unlike spherical and chromatic aberration, can be corrected by the application of a compensating magnetic field. Astigmatism is also induced when magnetic materials are imaged and can be a problem if the pole pieces become contaminated (e.g., by losing a specimen in the microscope). The procedures to be adopted for correcting astigmatism are given in Section 4.8.

Fig. 1.10 (*a*) A modern high voltage electron microscope. (*b*) Cutaway to show principal features. Courtesy Kratos/AEI.

An important aspect of magnetic lenses is the rotation that occurs as the strength of the lens is changed. Thus images rotate with changing magnification unless the lenses are specially compensated. These rotations must be calibrated as described in Section 4.2.

3.1.2 *Depths of Field and Focus.* As a result of the small electron wavelength and small angular apertures of imaging lenses the depth of field and focus

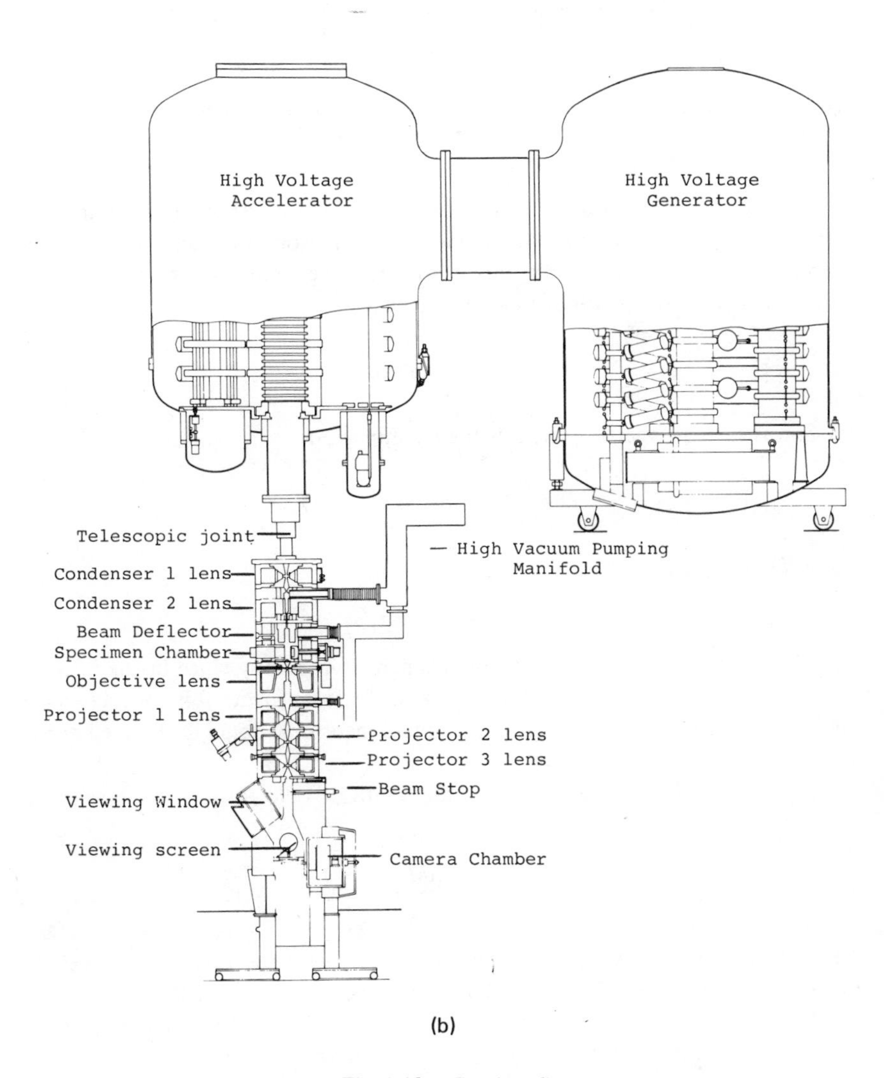

Fig. 1.10 (*Continued*)

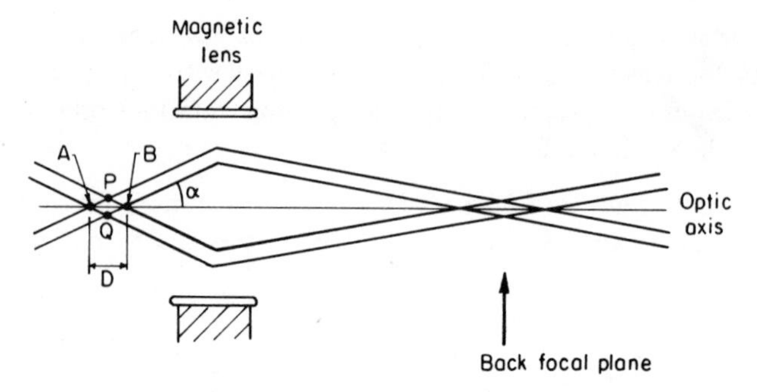

Fig. 1.11 Separation of object points P and Q is the resolution limit d. Rays from these points cross the optic axis at A and B and are equally sharp. $D = A - B$ is the depth of field.

(related by magnification) in an electron microscope is extremely large in comparison to light microscopy. It is one of the reasons that scanning electron microscopy is so useful for topographic studies, for example, fracture surfaces.

From Fig. 1.11 the depth of field D is given as

$$\frac{d}{2} = \frac{D}{2} \tan \alpha,$$

where α is the lens aperture and d is the resolution, or

$$D = \frac{d}{\alpha} \quad (\text{since } \alpha \simeq \tan \alpha).$$

For $d = 1$ nm and $\alpha \simeq 2 \times 10^{-3}$ rad,

$$D = 500 \text{ nm},$$

which is much larger than is possible by light microscopy since in the latter α may be 70°. Thus for foils $\sim\frac{1}{2}\,\mu$m thick all features in the foil are focused at once, and a projection of the in-focus three-dimensional detail is obtained at the screen or photographic negative.

The depth of focus D_f is related to the depth of field through the magnification M:

$$D_f = \frac{dM^2}{\alpha}.$$

For all practical purposes D_f is infinite, and this is the reason why electron microscopy is so useful for studying not only specimens in transmission but also rough surfaces (e.g., in fractography), and why it is not necessary to refocus when exposing plates or film not located at the viewing screen.

3.1.3 *Resolution.* Ultimately the resolution of the electron microscope is determined by the spherical aberration coefficient C_s of the objective lens and the electron wavelength, as given by eqs. 1.3 and 1.4. This is true for both TEM and STEM instruments (Section 5). As shown in Fig. 1.5, even if C_s is maintained constant or if $C_s\lambda$ is constant, as the accelerating voltage is increased the resolution must also increase (through decreasing λ), so that, at 500 kV or higher, atomic resolution even in close-packed structures (lattice constants ~4 Å) becomes theoretically possible. The optimum voltage is determined by the specimen stability and knock-on damage (proportional to atomic number) and by the voltage stabilization. For most metals, therefore, 500 kV appears to be optimum. At this voltage the theoretical point-to-point resolution is ~1.5 Å for $C_s \sim 1$ mm (Fig. 1.5).

3.2 Operating Functions of the TEM

Figure 1.12 illustrates the principal results of electron scattering by a sample and, as a result, the principal sources of information that can be obtained (see also Fig. 1.1). The operating modes are as follows: transmission electron microscopy (TEM), scanning electron microscopy (SEM), scanning transmission electron microscopy (STEM), and microanalysis (by X-ray and/or energy loss analysis or Auger analysis of surfaces).

In the TEM mode, which is the one most often encountered, the microscope is operated (*a*) to form images by bright field, dark field, or lattice image phase-contrast modes and (*b*) to form diffraction patterns by using selected area apertures and focusing the intermediate lens on the diffraction pattern formed in the back focal plane of the objective lens. Simple ray diagrams to illustrate these two modes are shown in Fig. 1.13. In the following sections the specially important

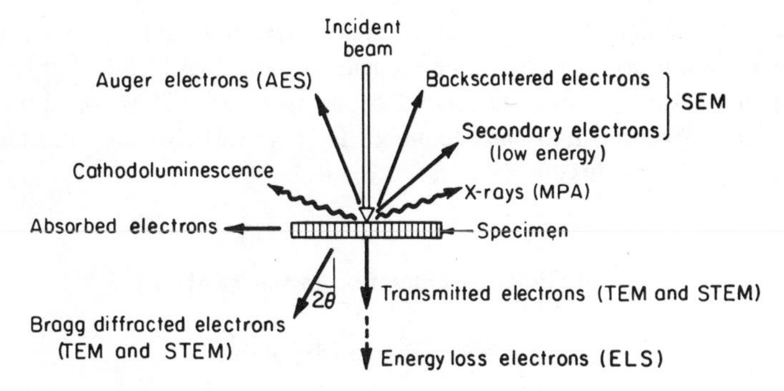

Fig. 1.12 Schematic showing electrons and electromagnetic waves emitted from a specimen as a result of elastic and inelastic scattering of the incident electron waves (see also Fig. 1.1).

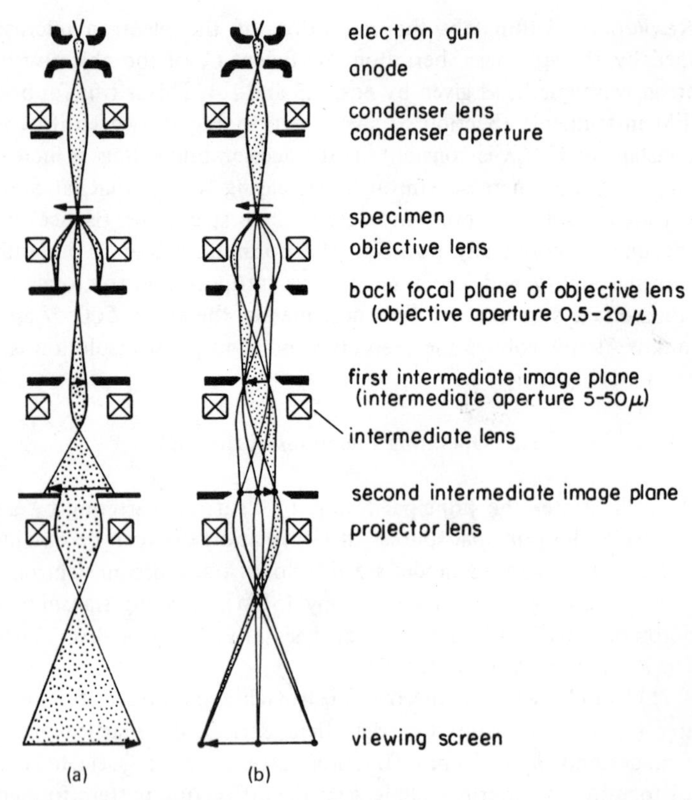

Fig. 1.13 Schematic ray diagram for a three-lens imaging microscope operated (*a*) for imaging and (*b*) for selected area diffraction. Notice the inversion relation between the image and its corresponding diffraction pattern.

methods of dark field imaging, selected area diffraction, and lattice imaging are discussed. An important point to be recognized from Fig. 1.13 is that the image and the diffraction pattern are related by an inversion. Thus, when analyzing images, such inversions must be accounted for (as must also any rotations), and procedures for doing this are given in Section 4.2.

4 Practical Operation of the TEM

4.1 Selected Area Diffraction

As the collimated beam of electrons passes through the crystalline specimen, it is scattered according to Bragg's law (Fig. 1.8). The beams that are scattered at small angles ($\lesssim 1$ to 2°) to the transmitted beam are focused by the objective lens to form a diffraction pattern at its back focal plane (Fig. 1.13). When the intermediate and projector lens system is properly focused, a magnified image of the back focal plane of the objective lens will be projected on the viewing screen. An intermediate aperture may be inserted at the first intermediate image plane to limit the field from which the diffracted information is obtained. The intermediate selected area diffraction (SAD) aperture makes it possible to obtain diffraction patterns from small portions of the specimen. This technique is very useful, since a direct correlation can be readily made between the morphological and crystallographic information of very small areas. It is also necessary for establishing the diffracting and contrast conditions in the image. The technique is of particular importance when more than one phase is present in the specimen.

Most electron microscopes have fixed positions of the intermediate aperture, which in turn fixes the position of the first intermediate image plane. To correctly obtain a diffraction pattern, it is first necessary to focus the intermediate lens on the first intermediate image plane. This is most easily done by removing the objective aperture and varying the intermediate lens current until the halo surrounding the image of the intermediate aperture disappears. Then the objective aperture is replaced, and the image focused by varying the objective lens current. This procedure focuses both the objective and the intermediate lens on the first intermediate image plane. If care is not taken to correctly focus the objective lens at the first intermediate image plane, errors will be made in relating the diffraction pattern to the area imaged inside the intermediate aperture. Finally, the strength of the intermediate lens is reduced until the image of the back focal plane of the objective lens is focused on the second intermediate image plane, resulting in a magnified image of the diffraction pattern being projected on the viewing screen. Ray diagrams illustrating the modes of operation for imaging and selected area diffraction are shown in Fig. 1.13.

The main source of errors in selected area diffraction are a consequence of (*a*) spherical aberration of the objective lens and (*b*) incorrect focusing of the objective lens. These factors must be considered in correlating images and diffraction patterns.

Spherical aberration causes displacement of the image in the plane of the SAD aperture. By tracing this displacement back to the specimen (Fig. 1.14), it is seen that the transmitted information and the diffracted information are obtained from slightly different areas, and the value of the displacement is given by eq. 1.2, that is,

$$y = C_s \alpha^3 .$$

The error is therefore a function only of C_s, the spherical aberration coefficient, and α, the scattering angle. Since the error is independent of the aperture size, its relative importance becomes greater as the aperture size is reduced; also, because the displacement is proportional to α^3, the error is greater with higher order

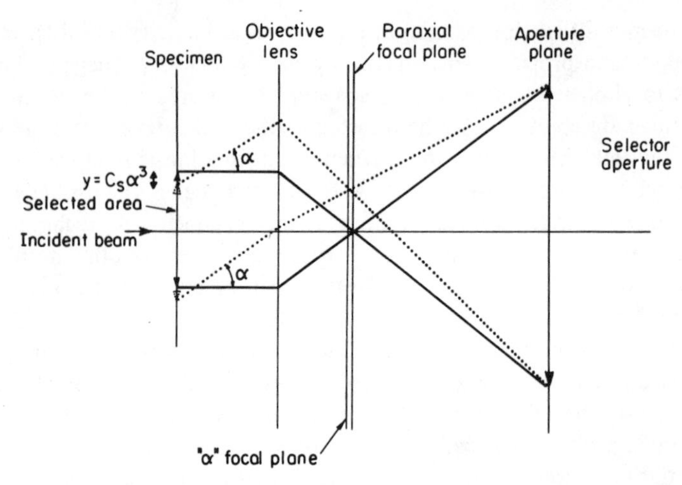

Fig. 1.14 Formation of the first intermediate image by the objective lens, illustrating the effects of spherical aberration.

beams. Although the spherical aberration coefficient is a function of the objective lens quality, it is approximately proportional to λ^{-1} (see Fig. 1.5), making it advantageous to use higher accelerating voltages. Thus, if $C_s \propto \lambda^{-1}$ and $\alpha \propto \lambda$, then $y \propto \lambda^2$.

The other major source of error is improper focusing of the objective lens. If the objective lens is defocused by a distance D, the SAD aperture relates to the plane of focus AB, as shown in Fig. 1.15. The area from which the diffracted

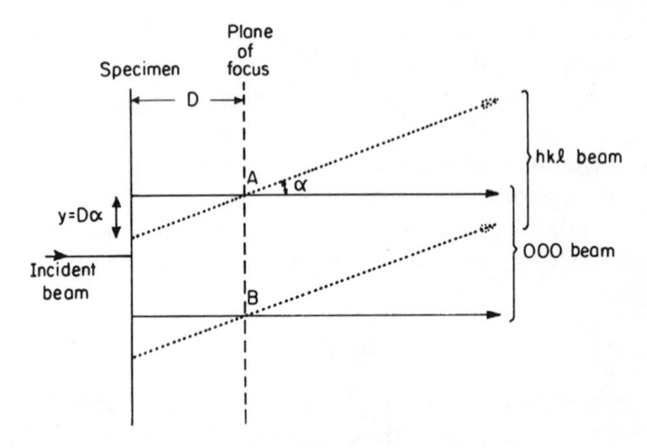

Fig. 1.15 Illustrating errors in selected area diffraction arising from incorrect focusing.

beam originates is shifted parallel to the diffraction vector by the amount

$$y = D\alpha,$$

where D, the distance between the plane of focus of the objective lens and the specimen, is positive for overfocusing and negative for underfocusing, and α is again the scattering angle.

When the errors from both spherical aberration and defocusing are considered together, the total displacement at the specimen is given by

$$y = C_s\alpha^3 + D\alpha. \tag{1.8}$$

It can be seen from this equation that for small values of α the amount of the displacement may be minimized by operating at a slightly (negative) defocused condition. However, at large values of α the α^3 term dominates. To illustrate the effect of errors on image displacement, typical values of C_s, D, and α can be used, as follows:

$$C_s = 1.6 \text{ mm}, \qquad D \simeq \pm 2\ \mu\text{m}, \qquad \lambda = 0.037 \text{ Å (at 100 kV)}.$$

For a crystal with lattice parameter ~4 Å, from Bragg's law a typical value of α is

$$\alpha \sim \frac{\lambda}{d} = 1.5 \times 10^{-2}.$$

This gives a typical value of the displacement as

$$y = 1.6 \times 10^3(1.5 \times 10^{-2})^3 \pm 2(1.5 \times 10^{-2})$$

$$= 5.4 \times 10^{-3} \pm 3.0 \times 10^{-2} = -2.5 \times 10^{-2}\ \mu\text{m} \quad \text{or} \quad 3.5 \times 10^{-2}\ \mu\text{m}.$$

If the objective lens has a magnification of 25X, this uncertainty projected to the first intermediate image plane is approximately 0.8 μm. For an accelerating voltage of 1 mV (i.e., λ = 0.0087 Å), considering the same values of C_s, D, and d, the displacement is found to be

$$y = (1.6 \times 10^3)(4.4 \times 10^{-3})^3 \pm 2(4.4 \times 10^{-3})$$

$$= 1.4 \times 10^{-4} \pm 8.8 \times 10^{-3} = -8.7 \times 10^{-3}\ \mu\text{m} \quad \text{or} \quad 8.9 \times 10^{-3}\ \mu\text{m}.$$

Again, when this value is projected to the first intermediate image plane, the uncertainty becomes 0.2 μm. This shows that at higher accelerating voltages the SAD error can be significantly decreased. Clearly it is also essential that the microscope be properly aligned and focused. Normally the smallest SAD aperture used is 5 μm.

4.2 Calibration of Rotation and Inversions

To be able to successfully correlate the images and their corresponding diffraction patterns, is it necessary to know not only precisely from where the diffraction pattern originates, but also how it physically relates to the orientation of the specimen. As the electron beam is focused through the magnifying projector lenses, the image must go through a series of cross over points (Fig. 1.13). Also, as the strength of the magnetic lens is varied, the image experiences a rotation (angular shift). Therefore, to effectively use microscopy and diffraction information, the geometrical relationships between the image and the diffraction pattern must be known.

The simplest method to determine the image rotation relative to the diffraction pattern is to use a specimen of MoO_3 crystals on a carbon support grid. Such specimens can easily be obtained by heating a strip of molybdenum and catching the oxide smoke on the carbon support film. The MoO_3 crystals tend to grow as long plates with straight edges running parallel to ⟨100⟩. If the image of the MoO_3 crystal is photographed for a series of intermediate lens settings and its diffraction pattern is superimposed on each image, the rotation can easily be measured (e.g., Fig. 1.16).

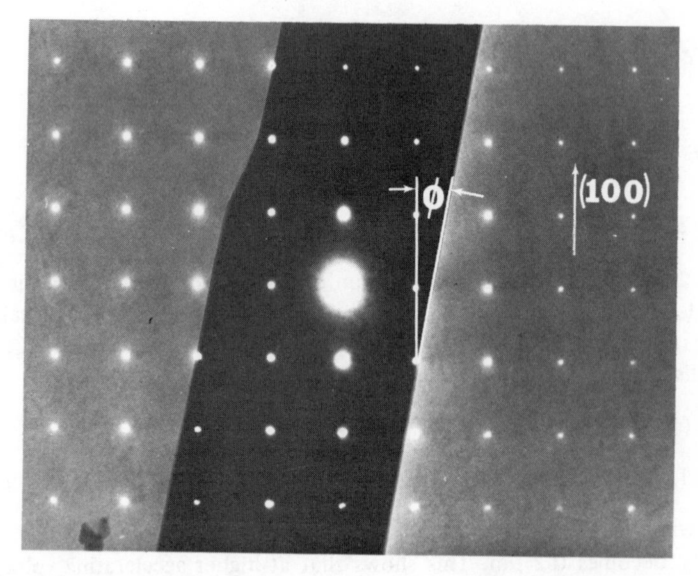

Fig. 1.16 Single crystal of MoO_3 with its selected area diffraction pattern superimposed (100 kV). The rotation ϕ ($\simeq 11°$) is the angle between the edge of the crystal and the [100] row of spots (×7750).

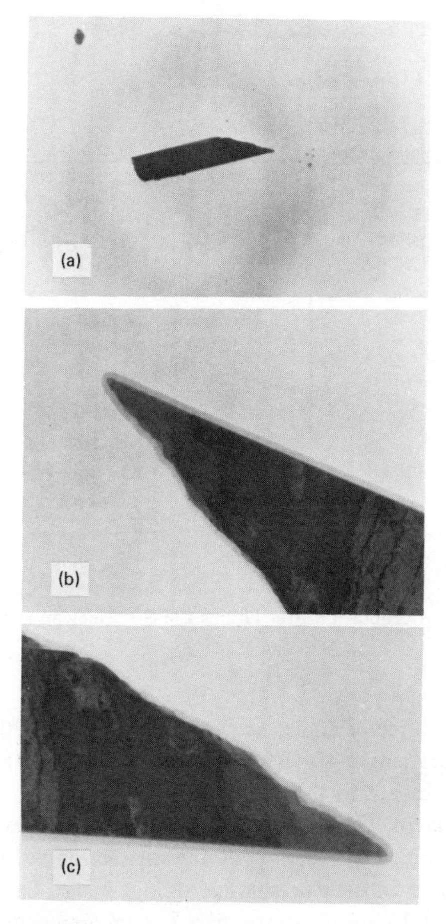

Fig. 1.17 The image inversions of a MoO_3 single crystal, taken with a Philips EM 301 equipped with a high resolution stage: (*a*) low magnification (×1300–×6000), (*b*) intermediate magnification (×7250–×3000) and (*c*) high magnification (×36,000–×360,000).

It can be seen from the ray diagrams of Fig. 1.13 that there is an inversion between the image and the diffraction pattern. This effect is present in all microscopes; however, microscopes with more advanced imaging systems may present a more complicated situation. Figure 1.17 shows the image inversion of a MoO_3 specimen taken with a Philips EM 301 for three different ranges of magnification. Figure 1.18 shows an example of a magnification-rotation calibration for a Philips EM 301 microscope, illustrating the inversion change with magnification. Since the Philips is equipped with a diffraction lens, care must be

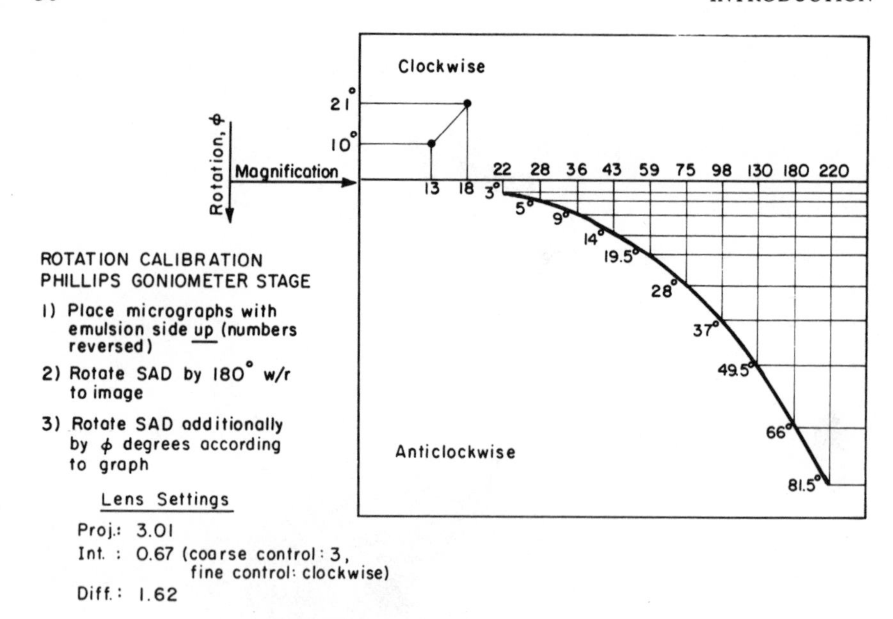

Fig. 1.18 The rotation calibration of a Philips EM 301 electron microscope equipped with a goniometer stage.

taken to focus it correctly since, as shown by Fig. 1.19, the image will invert when the diffraction lens is either under- or overfocused. If the direction of the foil normal, with respect to the electron beam, is to be known, any inversions must be accounted for. One of the most frequent exercises in electron microscopy involves relating the direction of the diffraction vector to structural features, such as defects, in the electron microscope image. Calibrations such as the one shown in Fig. 1.20 are thus essential. The recommended steps are given in the following list, with reference to Fig. 1.20:

1. Place image and diffraction pattern negatives emulsion side up (Fig. 1.20*a*) on a viewing light box or table.
2. Rotate the diffraction pattern with respect to the image, allowing for rotation-inversion calibrations (Figs. 1.20*b* and *c*).
3. When properly oriented, mark the direction of the diffraction vector (or other crystallographic data) on the nonemulsion side of the image (Fig. 1.20*c*).
4. Make a contact print so that a permanent record can be kept of the crystallographic relations. Afterward the marks can be erased.

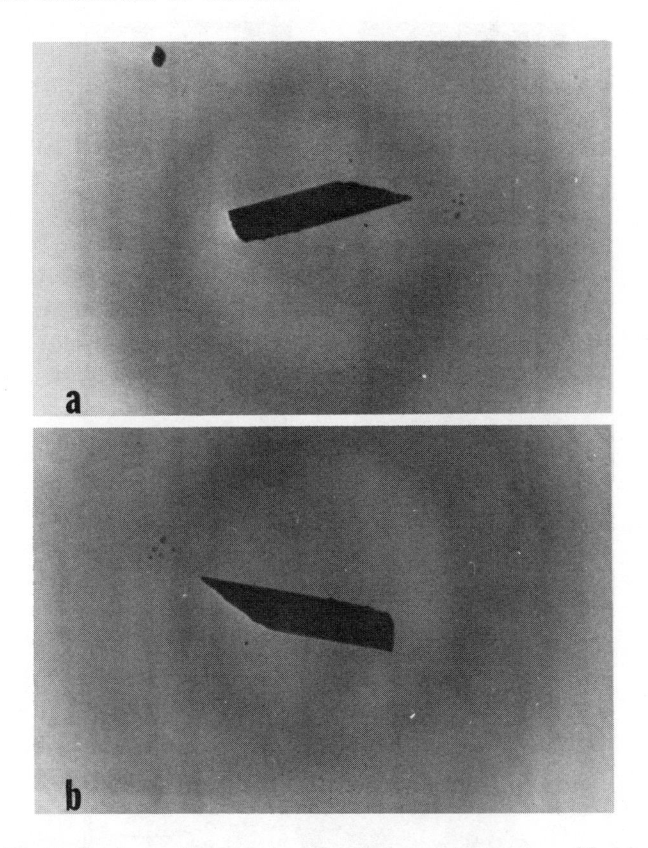

Fig. 1.19 Illustrating image inversion in a four-lens imaging system with (*a*) underfocus and (*b*) overfocus of the "diffraction" lens. Specimen is a MoO_3 single crystal.

5. Plot the data obtained on a stereographic projection, thus facilitating such analyses as true orientation and trace analysis.

4.3 Magnification Calibrations

Accurate magnification calibrations are required if quantitative measurements are to be made from electron micrographs. A wide variety of methods are available for image magnification calibrations;[19] however, the methods most commonly used are diffraction grating replicas for low and intermediate magnification and direct lattice imaging of known crystal lattice plane spacings for high magnifications. Permanent calibrations are accurate to within ±5% and can be improved to ±2% for an *in situ* calibration.

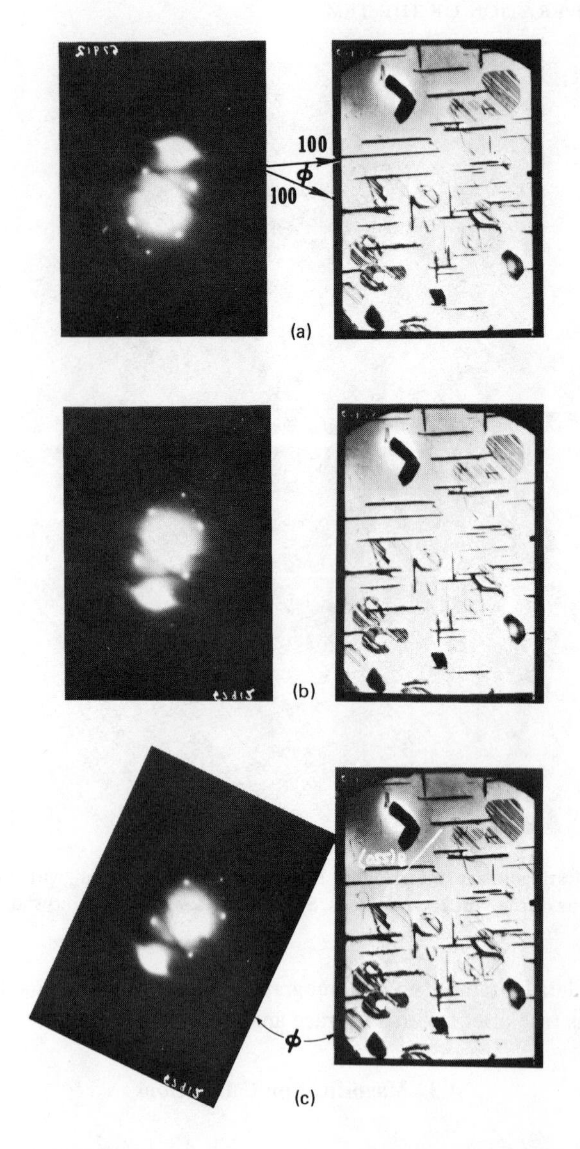

Fig. 1.20 An Al-4wt% Cu alloy aged to produce θ' precipitates (exhibiting $\{100\}$ habit planes) illustrates the procedure to correctly account for electron microscope inversions and rotations. (*a*) EM negatives are placed emulsion side up. Notice the angle ϕ between [100] in the image and diffraction pattern. (*b*) The diffraction pattern is rotated 180°. (*c*) The diffraction pattern is rotated a further ϕ degrees. The rotation in this case is ~23°, which is the angle between the trace of the θ' on (100) in the image and the corresponding 200 diffraction spot in the diffraction pattern. (Siemens Elmiskop IA operating at 100 kV and at an magnification of $\times$14,000.) Mark, in ink, the direction of the diffraction vector on the nonemulsion side of the EM image. Notice the positions of the identification numbers on the image and pattern.

Standard diffraction grating replica specimens are available in either carbon or silicon monoxide, the latter material being more stable under an electron beam. The number of lines per millimeter available are 617, 1134, and 2160, suitable for the magnification range 100X to 200,000X. At low magnifications the distance between lines may be measured, and at high magnifications the distance between two recognizable points may be used. For added accuracy the replica may be mounted with the specimen, or latex balls of a known size may be placed on the specimen.

Direct lattice imaging (Section 4.8 and Chapter 3, Section 10) may be used to calibrate magnifications greater than 200,000X. From a lattice image of a known standard specimen (gold is convenient), the magnification may be determined by measuring the interplanar spacing on the microscope plate and comparing this value with the known *d*-spacing of the material. To minimize errors the material used to calibrate the microscope should be similar in lattice spacing to the material being investigated.

4.4 Camera Constant

Calibration of the diffraction pattern magnification is made through knowledge of the camera length (i.e., $\lambda L = rd$), as discussed in Chapter 2, Section 2. An accurate determination of λL not only is important for indexing diffraction patterns but also is essential if more than one phase is to be identified. With proper care, errors associated with the calibration should generally be less than ±1%, and added accuracy may be obtained by performing an *in situ* calibration. Commonly used materials are thallous chloride and gold, gold being more stable at high beam currents, because thallous chloride may begin to evaporate. Both thallous chloride and gold may be evaporated directly on the microscope specimen in order to perform an *in situ* calibration. The specimen may also be partially masked if calibrant-free areas are desired.

4.5 Contamination

Contamination of the specimen surface in the electron beam, which is due to the interaction of the electrons with residual gases of the microscope vacuum, can lead to errors in diffraction pattern analysis if it is not accounted for. Hydrocarbons or water molecules are cracked and adsorbed on the specimen surface, building up a contamination layer of increasing thickness. The size and the thickness of the layer depend on variables such as beam intensity, diameter, and energy, quality of the vacuum, and composition of the residual gases, as well as specimen temperature and exposure time. A calibration of contamination versus beam exposure can be done by measuring the thicknesses of various contaminated spots after shadowing them at an angle.[20] If the elastic constants are known, the thickness can also be calculated from the elastic bending of the foil under the contaminating layer.[21]

Contamination may also cause serious problems in interpretation of fine detail in the image. This is discussed in Chapter 2.

4.6 The Dark Field Technique

When the image is formed by a diffracted beam (Fig. 1.3*b*), the imaging technique is called the dark field technique. The regions contributing to the diffracted beam intensity are revealed in a dark field image (Fig 1.4*b*), and thus this method is very useful in the characterization of complex microstructures.

The simplest way to obtain a dark field image from a crystalline specimen is to position the objective aperture around a diffracted beam and then operate the microscope in the normal manner (Fig. 1.4*c*). However, since the diffracted beam is inclined to the objective axis in this case, spherical aberration of the objective lens limits the image resolution. Because of this aberration all image points will be elongated by an amount $3C_s \cdot \lambda^2 \cdot \Delta\alpha \cdot |\mathbf{g}|^2$, where C_s is the spherical aberration coefficient; λ, the wavelength of the electrons; $\Delta\alpha$, the divergence of the diffracted beam; and **g**, the reciprocal lattice vector (see Chapter 2, Section 1) corresponding to the reflection used (Fig. 1.4). With a beam focused to give a reasonable image intensity on the screen in a 100 kV microscope, this elongation may be ~50 Å for d (=1/|**g**|) = 2 Å and C_s = 2 mm. This affects the resolution of the dark field image seriously and makes it rarely useful, except for preliminary sampling of the specimen.

This effect of spherical aberration on the dark field image can be overcome by tilting the illumination system so that the chosen diffracted beam is aligned accurately along the objective axis of the microscope. Modern microscopes are equipped with electromagnetic beam deflection systems which make the operation simpler and quicker than the mechanical gun tilting method used in some older models.

To take a beam-tilt dark field image of an area of the specimen oriented for a two-beam (000 and **g** beams) imaging condition, the incident beam is tilted in such a way that the diffraction condition remains unchanged when the reflected beam is along the optic axis (Fig. 1.21). To ensure that this is the case, *the transmitted beam must be moved to bring the weakly excited* −**g** *reflection toward the center of the viewing screen.* As this beam approaches the optic axis, it becomes strongly excited and this is the reflection to use for the dark field image. In the reverse situation, that is, when the transmitted beam is tilted so as to bring the strongly excited spot **g** toward the center of the screen, the original two-beam condition is disturbed; in fact, the reflection 3**g** becomes excited. This is illustrated schematically in Fig. 1.21*c*. It must be emphasized here that, after the transmitted beam has been tilted correctly to obtain the dark field image, the direction of **g** is reversed with respect to the bright field situation (see the direction of the arrow in Fig. 1.21). This is an important point to realize when analyzing defect geometries and the like.

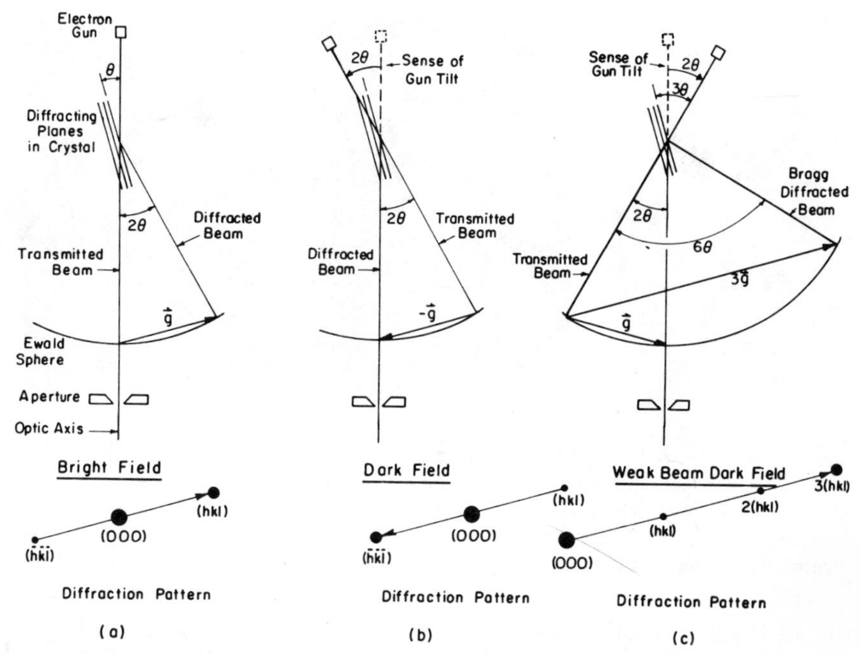

Fig. 1.21 Ray diagrams and Ewald sphere constructions for (*a*) conventional two-beam bright field, (*b*) high resolution two-beam dark field, and (*c*) (g, 3g) weak beam dark field. To go from (*a*) to (*b*), the beam (gun) must be tilted by an angle 2θ in the sense shown by the arrow in (*b*). This brings the $\overline{hkl}$ spot in (*a*) toward the optic axis. Instead, if the gun is tilted to bring the *hkl* to the center as in (*c*), the diffraction condition changes and the beam corresponding to 3g will be excited. The image formed by g in (*c*) corresponds to a (g, 3g) weak beam case. Note the sense of g as indicated in the figures.

The experimental procedure required to obtain a beam-tilt dark field image, then, is as follows:

1. Align the microscope so that the transmitted electron beam is accurately along the optic axis. This is done by following the standard alignment procedures described in the microscope manuals.
2. Orient the specimen for the desired reflecting conditions.
3. Tilt the incident beam, using the electromagnetic tilt controls, while the microscope is set in the diffraction mode. The tilting mode must be such that the incident beam is moved toward the reflected beam desired for use in the dark field imaging mode (rather than the reverse situation where the reflected beam is moved toward the incident beam position).
4. Align the reflected beam accurately along the optic axis of the microscope. Use this beam, and take the image as in the bright field case.

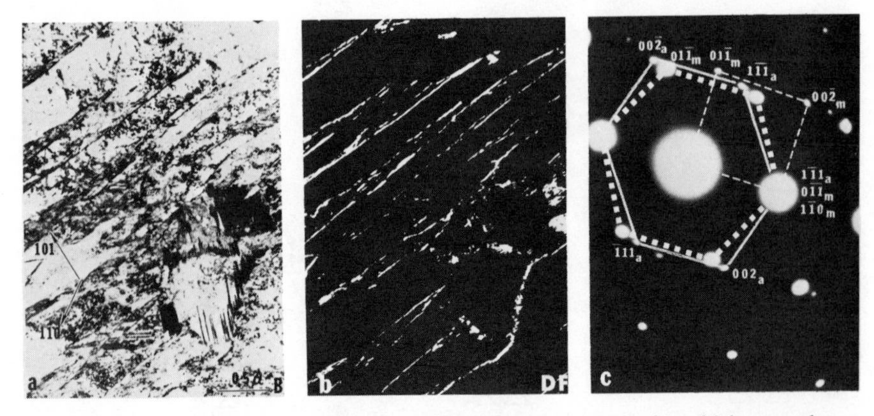

Fig. 1.22 (*a*) Bright field, (*b*) dark field, and (*c*) diffraction pattern from high strength martensitic steel. The dark field image (*b*) is obtained from the 002 austenite reflection and exposes the retained austenite phase as bright regions. The bright field image alone is not sufficient to identify the austenite regions in such cases. When the diffraction pattern is complex, as in (*c*), dark field imaging using individual spots greatly facilitates the phase identification. *a* represents austenite and *m* martensite. Courtesy B. V. N. Rao.

One of the main applications of the dark field technique is in facilitating the interpretation of selected area diffraction patterns from multiphase specimens, twinned specimens, and so on. Figure 1.22 illustrates the power of the technique in the identification of retained austenite in high strength steels. Also, much useful contrast information due to the anomalous absorption effects discussed in Chapters 4 and 5, such as the top-bottom effect in the case of stacking fault images and depth oscillations in the contrast from small particles, can be obtained easily from dark field images. The usefulness of a special dark field technique to study defects is discussed in the following section.

4.7 High Resolution Techniques for the Study of Lattice Defects*

Two important features of interest in the study of lattice defects are the geometry and the nature of the individual defects. In most cases such studies (of, e.g., dissociation of dislocations, small defect clusters, high density defects whose strain fields do not overlap) require examination of the images at high resolutions, and it often becomes necessary to use the microscopes at resolutions approaching their highest capabilities. Apart from instrumental optical errors, the resolution is also limited by factors such as diffraction contrast. As discussed in Chapters 4 and 5, the image width of a defect (such as a dislocation) is approx-

*This section requires application of contrast and diffraction knowledge as described in later chapters.

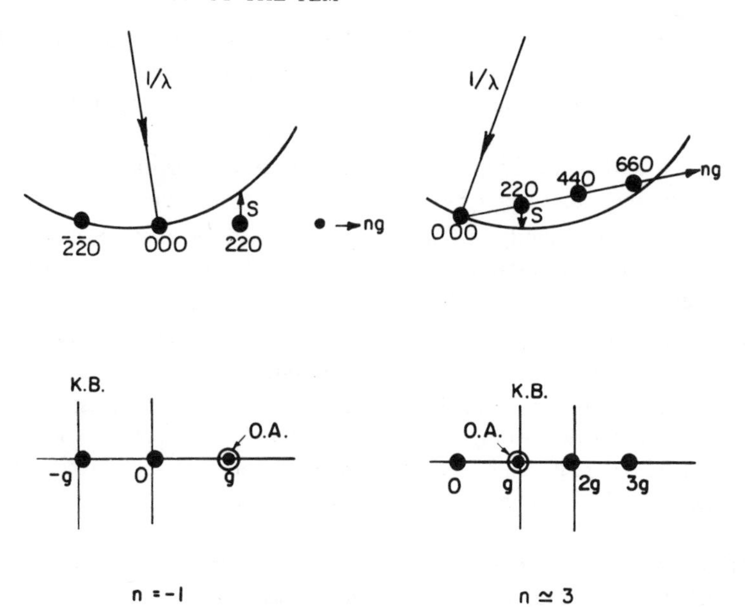

Fig. 1.23 Top: Two possible diffraction conditions in copper crystal to obtain $s(g = 200) \simeq 2 \times 10^{-2}$ (Å)$^{-1}$ for 100 kV electrons. Below: Sketch of the corresponding diffraction pattern, showing position of the 200 Kikuchi band (K.B.) and objective aperture (O.A.).

imately $\xi_g/3$, where ξ_g is the extinction distance for the reflection g. The value of ξ_g, in turn, varies inversely as the energy difference ΔE between the two branches of the dispersion surface for the particular reflection. Thus the images can be made narrower if the extinction distance (or, rather, the effective extinction distance) can be made as small as possible. There are two main ways of doing this.

1. If the distance from the reciprocal lattice point to the reflecting sphere is s (see Fig. 1.23 and also Fig. 2.6*a*), then, since increasing s is proportional to increasing E, s is made larger (but with enough intensity to preserve contrast), ΔE (effective) is made larger, and so ξ_g is made smaller. This is effected by tilting the specimen so that the reciprocal lattice point is tilted away from the Ewald sphere. This tilt is limited by the need to provide enough contrast to record the image—hence the name "weak beam method."
2. The other way to make ξ_g small is to use high order reflections, so that the upper branches of the dispersion surfaces are used; this also increases E and hence decreases ξ_g^{eff}. Other techniques, such as the multiple beam diffraction method,[22] have also been applied to achieve the same effect, although these are less frequently used.

4.7.1 *Weak Beam Method: Dark Field Imaging.* In high resolution studies of dislocations it is desirable that the image be related to the real dislocation geometry. For this purpose the images should have the following properties:

1. Each dislocation should give rise to a single peak in the image, and the peak should be very close to the position of the dislocation core.
2. The standard invisibility criterion should apply to the image peaks from each dislocation core separately. This is important in the analysis of dissociated dislocations.
3. The half width of the image should approach the instrument capabilities and hence resolution not be limited by diffraction contrast.
4. The image should be insensitive to small changes in foil thickness, dislocation depth, lattice orientation, and so on, so that lack of precise knowledge of these parameters does not limit interpretation of the image. Accurate determination of these parameters is generally difficult.

Provided that certain diffraction conditions are satisfied, weak beam images satisfy all these requirements. The technique has been successfully, and widely applied to the study of defects in crystalline solids (e.g., refs. 23–25).

To obtain weak beam images, the incident beam is tilted off the optical axis and a diffracted beam that is far off the reflecting position and therefore very weak is used to form the image. The experimental precautions relevant to all high resolution studies are applicable to the weak beam imaging mode. The dark field reflection must be aligned axially in the objective lens to minimize the effect of abberations, and the image must be free from astigmatism and movement. Since the dark field intensity is weak, the images require rather long exposure times (often $\simeq 30$ sec). This imposes severe requirements on the specimen and stage stability. Also, the low image intensity makes focusing difficult. However, in most modern microscopes equipped with electromagnetic beam deflectors, the appropriate focal conditions can be set using the bright field image. If the bright field and the weak beam channels are perfectly aligned, experimentally, it is found that the weak beam image remains in focus for the same focal conditions as the bright field image. (Of course the diffracting conditions for both modes are not the same, but this does not affect the focus.)

The diffraction conditions used for imaging are determined by the required accuracy of the analysis. For defects the optimum conditions of the microscope capabilities are met as follows:

(i) $|s_g| \gtrsim 2 \times 10^{-2}$ Å^{-1} (for $\mathbf{g} \cdot \mathbf{b} \lesssim 2$).

(ii) $|w_g| = |s_g \xi_g| \gtrsim 5$ so as to reduce the variation in contrast width with depth of the defect in the foil.

(iii) No other reflections (systematic or nonsystematic) should be strongly excited.

The diffracting conditions will vary for different materials and reflections for a particular value of $|s_g|$. From geometrical considerations, $s_g = (n - 1)g^2 \lambda/2$ when the reflection ng is strongly excited (n need not be an integer). For 100 kV electrons in copper with $g = \langle 220 \rangle$, $s_g = \pm 2.2 \times 10^{-2}$ Å^{-1} for $n = -1$ or $+3$. But for the same conditions in silicon, the same value of s_g is obtained when $n \simeq -3$ or 5. The Ewald sphere construction and the schematic of the diffraction condition for copper are shown in Fig. 1.23.

The experimental procedure to obtain a weak beam image of a defect is as follows:

1. Choose the reflection **g** to be used to form the image. In the case of defects, standard contrast analysis often dictates the choice of **g** for imaging.
2. Determine what the Kikuchi pattern will look like when conditions (i) and (ii) above (for s_g and w_g) are satisfied for this reflection (e.g., compare with a Kikuchi map).
3. Align the microscope for high resolution imaging as pointed out earlier, using the beam deflectors and so on.
4. Orient the specimen, using the specimen stage tilt controls, to obtain the diffraction condition as determined in step 2 above, with **g** on the optic axis. Make sure that no other reflections are strongly excited.
5. Attempt to focus the weak beam dark field image in this setting. If this is too difficult, use the bright field image as discussed above. Sometimes it may be necessary to take through focal series pictures to be sure of optimum focus.
6. Expose the plate. Be generous with the exposure time; an overexposure at least will show whether there is anything to be seen.

4.7.2 *High Order Reflection Method: Bright Field Imaging.* An alternative method, utilizing the long extinction distance associated with high order reflections, which is especially applicable at high voltages was introduced by Bell and Thomas in 1971.[26] In this case some of the disadvantages of the high resolution dark field mode are easily overcome. The specimen is oriented so that a higher order reflection of the systematic row is in strong diffracting condition. The image is formed using the transmitted beam. Since the background intensity is very high, focusing is easy and the exposure times required are short, minimizing the risk of losing resolution because of mechanical or other instabilities.

This technique is very easy at higher voltages because of the flatter Ewald sphere as λ^{-1} increases (Table 1.1), but it is by no means restricted to high voltages. The specimen can be oriented to excite a higher order reflection at 100 kV

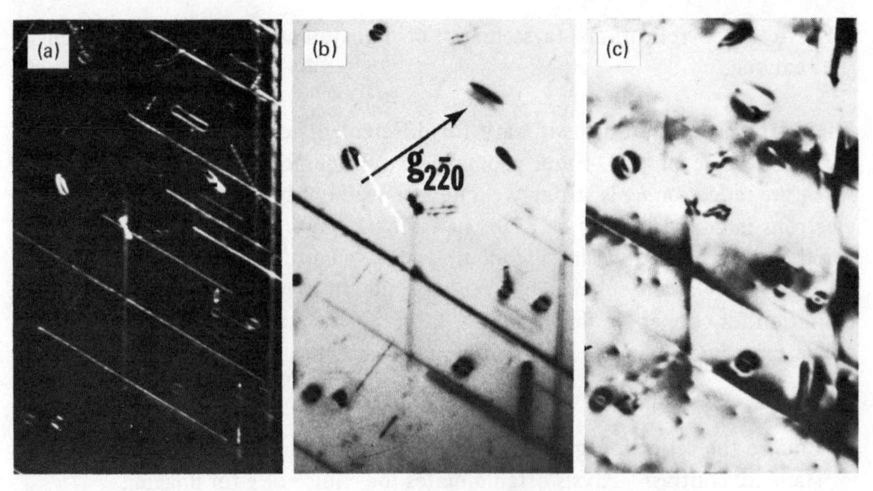

Fig. 1.24 (*a*) (g, 3g) weak beam dark field, (*b*) (0, 3g) high order bright field, and (*c*) (0, g) two-beam conventional bright field image of an area in B^+ implanted silicon. The image widths in (*a*) and (*b*) are markedly different from those in (*c*). The detailed geometry of the defects become clear in the high resolution images, but is not as apparent in (*c*).

for most materials. It must be pointed out, however, that with this technique image detail and image resolution are improved at the cost of contrast. Therefore the defect will no longer be visible, because of undetectable contrast, if too high a reflection is used. *This puts a practical limit on the possible order of the reflection* that can be used, depending on the specimen orientation and the specific defect under consideration. Figure 1.24 shows an area imaged in high order bright field mode, weak beam dark field mode, and conventional bright field mode and clearly illustrates the power of the two special high resolution techniques discussed.

4.8 Lattice Imaging

The proper conditions for producing lattice images which best represent the structure of the specimen must be chosen by comparing images obtained at the microscope with computed images based upon the dynamical theory (Chapters 4 and 5). To accomplish this, it is necessary that a number of experimental parameters be known and/or controlled with precision, namely, (*a*) specimen thickness, (*b*) specimen orientation, (*c*) accelerating voltage, (*d*) objective aperture size, (*e*) objective lens aberration coefficients, and (*f*) focusing condition. Dynamical calculations reveal, for instance[27] that, at 100 kV, two-dimensional or structural image contrast can be directly interpreted in terms of the projected charge density of the object only when the specimen is no thicker than

~50 Å and the objective lens is set at ~500 Å underfocus. The correct focus condition is obtained by offsetting the errors due to spherical aberration and objective lens defocus (i.e., underfocus–lens weaker than exact focus). This condition, known as the Scherzer focus,[28] is given by the following equation:

$$\Delta f_s = 2.5 \left(\frac{C_s \lambda}{2\pi} \right)^{1/2}. \tag{1.9}$$

This relationship and similar ones are discussed in more detail in Chapter 5, Section 9. Other examples of microscopy performed in an out-of-focus mode occur in the study of magnetic materials, as in Lorentz microscopy (see Chapter 3, Section 11). These restrictions as to specimen thickness and lens setting, as well as the desired levels of resolution, require optimum performance from both microscope and operator. The following summary is meant to highlight some of the experimental techniques for obtaining lattice images with emphasis on convenience and reproducibility of results.

4.8.1 *Preliminaries.* The microscope must first be prepared by rigorous alignment and cleaning procedures. These are normally outlined in individual operators' manuals and should be followed closely and performed frequently. (Axial alignment methods are also described in Section 4.6 and 4.7.) During cleaning, special attention should be paid to the areas of the gun, to prolong filament life, and to the specimen cartridge, to reduce contamination under high flux lattice imaging conditions.

When focusing lattice images, a high brightness source, high screen phosphor efficiency, and high magnification capabilities are most helpful for direct visualization of fine-contrast features. However, these features may require minor modifications in commercially supplied instruments.[29,30] Additional significant improvements in resolution can be achieved through the use of higher accelerating voltages.[30]

Special care must be taken in specimen preparation, or at least in the selection of appropriate areas for study, to ensure that the foil is thin enough for lattice imaging. It may also be necessary to tailor preparation methods for the purpose of obtaining proper specimen orientation,[31] for example, if a high resolution tilting stage is not available.

Although all of the above preliminary procedures may not be absolutely essential for obtaining lattice images, they are certainly valuable in ensuring convenient and reproducible high resolution operation of the microscope. Every lattice imaging session therefore should be preceded by a survey of the condition of the instrument, allowing time for necessary maintenance and the attainment of appropriate levels of column vacuum and electronic stability.

4.8.2 *Image Formation: Procedures.* In the diffraction mode (Fig. 1.25):

1. *Tilt and/or translate the specimen into the proper orientation for lattice imaging.* The orientation for structural imaging (having two-dimensional detail) is a symmetrical one and must be accurate to the highest order reflection visible on the screen. For materials having unit cell dimensions in the 5 Å range (i.e., metals and alloys), a systematic orientation may be necessary, with the planes to be imaged at the exact Bragg condition.
2. *Center the pattern on the optic axis.* Symmetrical patterns will already have been centered if microscope alignment is maintained. For tilted beam imaging, the beam deflectors (switched on as they would be in conventional dark field operation) are manipulated so that both the transmitted beam and the strongly diffracted beam are at the same distance from the optic axis along the systematic row. Accuracy is essential and may be facilitated by electronically switching the deflector coils rapidly between "dark field" (high angle) and "bright field" (low angle) positions. In this manner the normal position of the transmitted beam (locating the optic axis) may be compared with the tilted configuration; both patterns appear simultaneously during rapid switching because of residual fluorescence of the viewing screen.
3. *Insert an objective aperture.* For structural imaging the aperture must be large enough to admit on the order of 20 reflections, but should not exceed the resolution limit (~3.5 Å) set by the spherical aberration coefficient of the objective lens. Similarly, for lattice fringe imaging the aperture should be just large enough to enclose the two strong beams of interest. In any case the objective aperture, as the diffraction pattern, should be perfectly centered about the optic axis.

   A lattice image may also be formed by combining only diffracted beams (two or more) within the objective aperture. This "dark field" technique has the advantage of removing the diffuse scattering at the low angle region of the transmitted beam, which contributes to background noise. A major difficulty of the technique, however, is the lower intensity of illumination, which hinders focusing. Figure 1.25 compares the various diffraction and objective aperture geometries for these techniques. The situation in Fig. 1.25*a* is achieved with high voltages and/or crystals of large lattice parameter. At 100 kV and for close packed crystals the normal situation shown in Figs. 1.25*b* and *c*, i.e. lattice images only can be formed.

In the image mode:

4. *Increase the magnification of the area to be imaged.* Through small changes in objective lens current the various specimen images corresponding to the individual beams within the objective aperture are made to shift past one

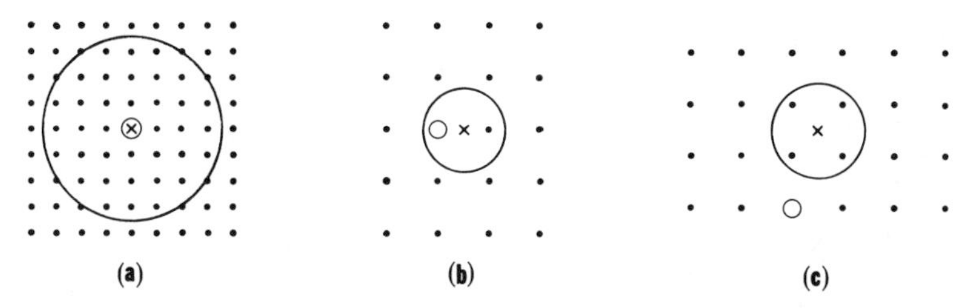

Fig. 1.25 Schematic of diffraction pattern/objective aperture configurations for (*a*) structural image, (*b*) two-beam tilted-illumination image, and (*c*) dark field lattice image. The open circle denotes the transmitted beam; the optic axis is located at x. Compare with Fig. 1.2*c*.

another. For the simple two-beam case these are a bright image (corresponding to the diffracted or dark field beam) and a dark image (corresponding to the transmitted or bright field beam), as shown in Fig. 1.26. Their simultaneous appearance can serve to verify that specific foil areas (e.g., the second-phase particles in Fig. 1.26) are suitably oriented. Small corrections in tilt can be performed rapidly, using changes in focus to observe image shifts.

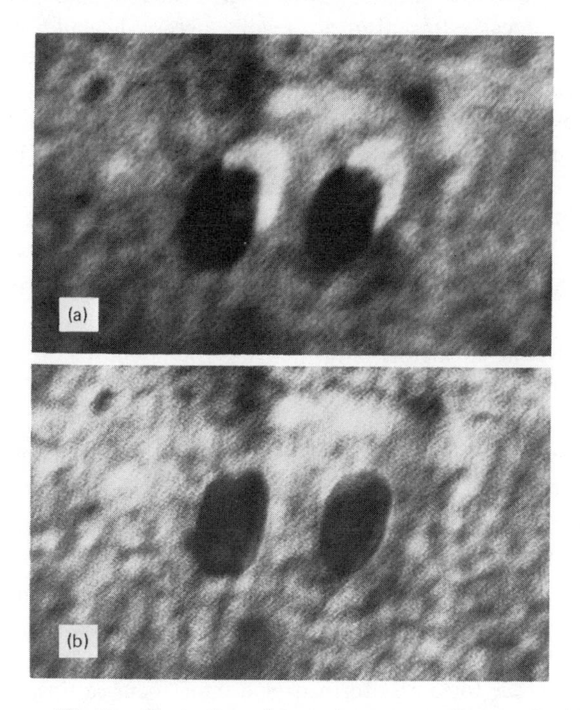

Fig. 1.26 Low magnification illustration of a two-beam lattice image (tilted illumination) in a defocused (*a*) and a focused (*b*) state. Exact focus is determined by absolute superposition of the bright field (dark) and dark field (light) images.

The proper orientation has been achieved when the image shift with focus is symmetrical about the regions of interest. After any specimen tilting the diffraction pattern should again be checked for proper centering of the objective aperture.

5. *Focus and correct astigmatism.* Astigmatism is most easily corrected when viewing a small perforation in the foil, in the same manner as when using a holey carbon film. At working magnification for lattice imaging it may be possible to view only a portion of the perforation at any one time; however, as long as there is enough curvature to include two orthogonal tangents, the image can be stigmated. In this case the stigmators are adjusted until the out-of-focus Fresnel fringe at the periphery of the hole has uniform width everywhere. This method may also be employed at the edge of a foil, again provided that there is at least a quarter-arc curvature in the field of view.

   Unfortunately it is not always possible to employ the Fresnel fringe method. Successful lattice imaging frequently requires that astigmatism be corrected *on the area to be imaged.* This is particularly true in more complex microstructures, where diffraction conditions, structure factors, or foil thickness may undergo rapid variations over small distances, and the corresponding changes in local astigmatism are readily observed. Under these conditions the clarity of the background phase contrast is the best indication of proper astigmatism correction.

   When viewing a thin foil at high magnification, a fine-scale background contrast should be apparent near the condition of best focus. If this is not the case, either the specimen is too thick and offers little chance of lattice resolution, or the image is severely astigmatic. For the latter condition, one should:

   (i) Recheck the focus for best setting while observing the background contrast at high magnification.

   (ii) Adjust the stigmators until the background contrast is sharp and has no apparent directionality or elongation.

   (iii) Repeat (i) and (ii) until the background contrast is observed to go through a minimum with variations in focus, but is unquestionably sharp otherwise.

   At this point the image is properly corrected for astigmatism, and the minimum in background contrast represents an "exact focus" setting.

6. *Photograph the image.* If fringes are not visible on the screen, or if structural imaging is being done, a through-focus series should be made. A photograph is taken, beginning with the exact focus condition, at each incremental change in focus for both overfocus and underfocus settings. Hence, for 2 Å lattice fringes at 500,000X magnification, a typical series might range from

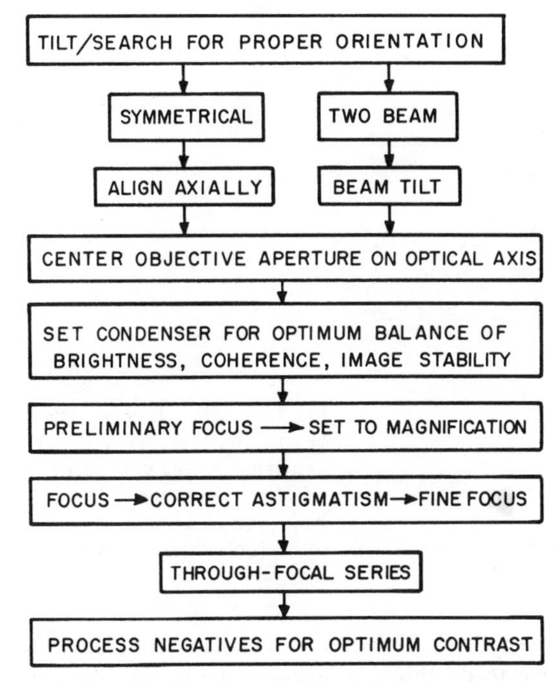

Fig. 1.27 Schematic of lattice imaging procedures.

-500 Å to +500 Å in 100 Å increments. For larger interplanar distances the extent of defocus can be increased and the number of photographs in the through-focus series decreased.

Exposure times should be kept to a minimum, and particular attention should be paid to alternative processing procedures for enhancing negative contrast and/or speed when necessary (see film manufacturer's instructions). Finally, each lattice image (or series) should be accompanied by a double exposure of the diffraction conditions and aperture shadow to be used in subsequent analyses. A summary of these procedures is presented in Fig. 1.27.

4.9 Convergent Beam Diffraction

An alternative method of decreasing the size of the area giving rise to the diffraction pattern is to illuminate only a very small area by focusing the electron

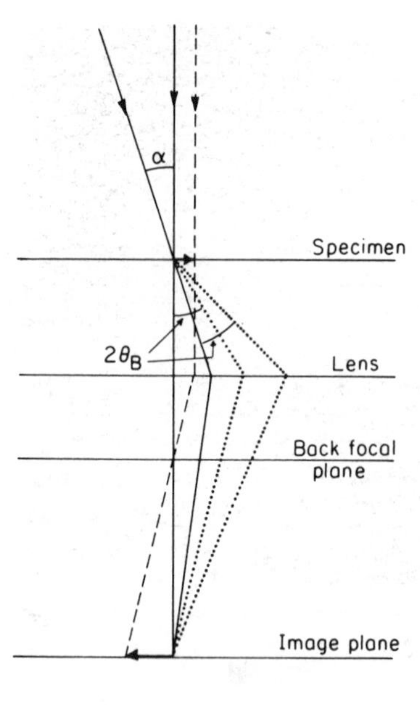

Fig. 1.28 Ray diagram illustrating convergent beam diffraction, drawn asymmetrically with one extremum of the convergence cone on the optic axis (for convenience). The diffraction disk in the back focal plane will not overlap provided that the total angular convergence α is less than $2\theta_B$ (θ_B = Bragg angle).

source onto the specimen. However, if reasonable intensities are to be retained, it is not possible to do this and maintain the parallel illumination condition; some convergence (range of incident beam directions) is introduced (see Fig. 1.28). As a result the diffraction "spots" become "disks," as illustrated in Fig. 1.29. Provided that the total convergence angle α (Fig. 1.28), which is usually controlled by the size of the condenser aperture, is less than twice the Bragg angle, this does not present too many problems since the disks do not overlap.

Apart from the small size of the area studied (which is a function of a number of microscope lens parameters), the principal advantage of the convergent beam technique is the range of information available in one picture. Each diffraction disk is produced from exactly the same small area, without any of the shifts introduced by imaging lens aberrations, and each disk contains the intensity diffracted by this same area of crystal for a range of incident orientations. This means that "rocking curves" (see Chapters 4 and 5) may be directly compared with many-beam calculations with a minimum of intensity standardizations, and a number of the underlying diffraction parameters simultaneously measured. An example of such an observation is given in Chapter 2 (see Fig. 2.30).

The main reason why the convergent beam technique is not more widely used is that, when such a small region of the specimen is illuminated, contamination is extremely rapid, and the smaller the spot the worse the problem. This be-

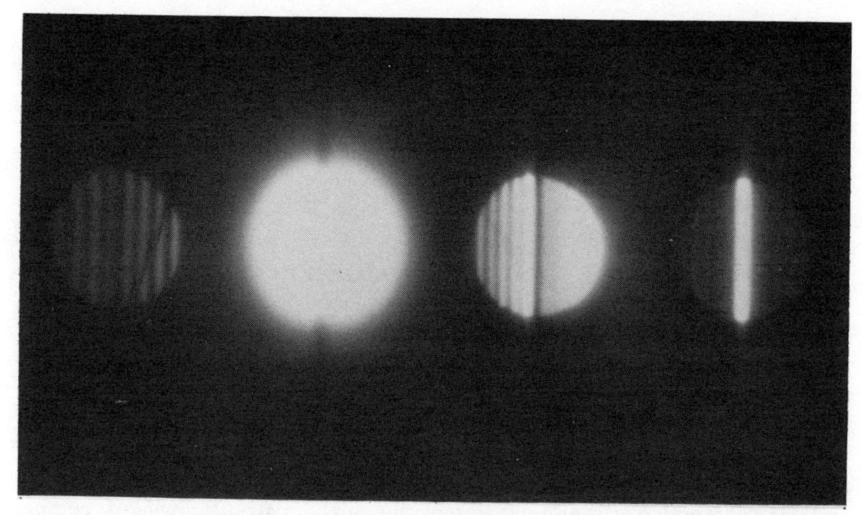

Fig. 1.29 A convergent beam diffraction pattern from a copper specimen at 200°C. The diffraction spots displayed are $\bar{1}\bar{1}\bar{1}$, 000, 111, and 222, respectively, with the last reflection exactly satisfied; 1000 kV, specimen thickness ~100 nm. Courtesy D. Imeson and A. Moodie.

comes very apparent when, in order to check the area which has been studied, the microscope is returned to normal imaging, and what is seen is a stalagmite of dirt. However, methods of minimizing the contamination exist, such as heating or cooling the specimen or improving the standard cooled anticontamination devices, enabling convergent beam techniques to be used in many microscopes. The minimum spot size and range of convergence angles available depend on the particular microscope and are usually optimized only in a nonstandard operating mode. Ultrahigh vacuum operation may be required to minimize the contamination problem, especially if spectroscopy is to be carried out on small areas (e.g., STEM operation; see Section 5).

4.10 Stereoscopic Techniques

Although in the 20-odd years since the advent of transmission electron microscopy more than 99% of all micrographs taken have been single exposures, it is worth remembering that single-exposure microscopy is to stereomicroscopy as cyclopic is to normal binocular vision. The view is superficially the same, but something is lost.

In normal vision each eye sees an object from a slightly different vantage point, and perception of depth is achieved when the brain fuses the two images. In electron microscopy the same effect is achieved by viewing simultaneously two micrographs taken in slightly different directions in a stereoscopic viewer. [If smaller pictures spaced at the interocular separation (~65 mm) are used, many people are able, with practice, to dispense with the viewer.]

The change in the apparent separation ($x^1 - x$) of two objects located at different depths, ΔZ, is known as parallax. In electron microscopy the parallax p is related to the tilt angle θ by the relation[32]

$$\Delta Z = \frac{p}{2M \sin (\theta/2)}, \qquad (1.10)$$

where M is the total magnification of the image seen in the viewer. The maximum parallax that can be accommodated comfortably by the human eye is 3 to 5 mm. Therefore to obtain maximum depth perception it is necessary to adjust the tilt angle and total magnification to give the optimum parallax. To do this, however, an estimate of the foil thickness must be made. The relation between θ and M as a function of foil thickness has been given in graphical form by Hudson and Makin[33] and is a useful aid in selecting appropriate tilt angles. A comprehensive theoretical treatment of electron stereo microscopy with estimates of the errors involved has been published by Nankivell.[34]

In practice two precautions must be taken to obtain the high quality stereo pairs necessary for quantitative analysis:

1. Since in eq. 1.10 ΔZ refers to vertical height separations, its application to foil thickness measurement is valid only when the sample is tilted by equal amounts from the foil position perpendicular to the electron beam.
2. When the microstructural features of interest are imaged by diffraction contrast, it is essential to maintain these conditions constant to avoid changes in the appearance or position of the feature (e.g., the apparent position of a dislocation depends on $\mathbf{g} \cdot \mathbf{b} \cdot s$; therefore $\mathbf{g}$ and s must not change). In practice this is readily achieved in electron microscopes equipped with biaxial tilting stages by monitoring Kikuchi or spot patterns and tilting the specimen so as to maintain $\mathbf{g}$ and s constant. In fact, this procedure is recommended for obtaining stereo pairs under any conditions since the Kikuchi pattern corresponding to each of the stereo micrograph pairs contains all the information necessary for accurately determining the tilt angle and tilt axis.[35]

The three main uses of stereo microscopy in TEM are as follows:

1. As a qualitative aid in visualizing the depth distribution of structural features in a foil.
2. For determining foil thickness as a precursor to further quantitative analysis.
3. For quantitative assessments of the spatial distribution of defects in a foil.

In addition several levels of sophistication may be employed in stereo analysis. In a purely qualitative way the perception of depth is often very useful for deciphering complex or overlapping structures, or for observing crystallographic features and relationships (e.g., in readily identifying sets of loop or precipitate habit planes). This is particularly true in high voltage work, where the use of thicker specimens often leads to greater overlap.

Foil thicknesses are readily measured if features lying on, and intersecting, each surface can be distinguished, simply by measuring the parallax with a scale. Frequently in quantitative work the precise position of the surface is established by depositing islands of metal or other distinguishable features on the foil.

The determination of depth or spatial distributions of defect clusters or other features requires careful stereomicroscopy. For example, the contrast exhibited by small vacancy or interstitial clusters (the so-called black-white contrast) is characterized by a vector **l** drawn from the black to the white lobe. The scalar product $\mathbf{g} \cdot \mathbf{l}$ is of opposite sign for vacancies and interstitials, but it also reverses regularly with increasing distance from the foil surface. Thus, in order to determine the nature of a defect, its position in the foil must be accurately established. For applications such as this one more expensive stereoscopic viewers, of the type used in aerial surveying, are employed. By optical means a "floating spot" is incorporated in these instruments, and the apparent position of the spot in the foil can be changed with a micrometer drive. This allows depth differences to be measured with good precision.

A further degree of sophistication has been introduced by Thomas and Cuddy,[36] who have extended the floating spot concept and installed a floating grid. Moreover, the viewer has also been equipped with displacement transducers[37] to convert x–y coordinate points to electrical signals and to connect all the outputs to a minicomputer. In this manner the spatial coordinates and size distributions of point defect clusters in irradiated foils are obtained with a minimum of tedium, and the output in the form of numerical data or histograms is displa directly at the computer readout.

Bell[38] has developed a new extension of stereomicroscopy that is begin find a number of applications in materials science. Dubbed "$2\frac{1}{2}$-D micr the technique takes advantage of the dark field image shifts that o defocusing of the objective lens to create an artificial parallax a stereo effect.

The image position is shifted parallel to the operating recipr **g** by an amount given by

$$y = \Delta D \lambda g,$$

where ΔD is the change in focus and λ the electron w features are present in the foil which give rise to d

allax arising from defocusing will be

$$p = M\,\Delta y = M\,\Delta D \lambda\,\Delta g$$

In practice, two dark field micrographs are taken with equal amounts of over- and underfocus and the pictures are viewed in a stereo viewer in the normal way. The features characterized by the different g values then appear to be located at different depths. The power of the technique lies in its ability to "unscramble" information in a complex microstructure, and Bell and co-workers have used it to delineate regions of different lattice rotation in heavily deformed material, to distinguish different precipitate phases, and to differentiate between small vacancy and interstitial clusters. A very promising application is the analysis of dislocation densities in heavily deformed or martensitic materials.

5 The Scanning Transmission Electron Microscope

Scanning electron microscopy (SEM) as such has been deliberately omitted in order to keep the scope of the book within manageable limits. However, in the last few years there has been increasing interest in the use of scanned electron beams in *transmission* microscopes (STEM). In the following sections these developments are outlined, preceded by a very brief description of the scanning principle. Fuller expositions may be found in the texts listed in the bibiliography at the end of the chapter.

5.1 Basic Features of Electron Scanning Systems

The basic components of a typical electron scanning system are shown in Fig. 1.30. The system consists of (*a*) an electron gun; (*b*) a number of condenser lenses, the purpose of which is to produce a highly demagnified electron spot

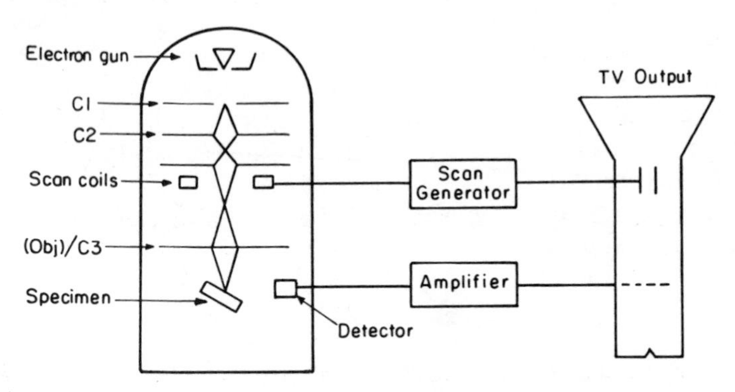

Fig. 1.30 Schematic outline of a scanning electron microscope system.

on the specimen (three are shown here, consistent with that function in a conventional electron gun; a field emission gun system would normally include the functions of some of these lenses); (*c*) scanning coils to scan the spot across the specimen on a suitable raster of small size (e.g., a television-style raster); (*d*) a scan generator driving both the coils and one or more TV output devices; and (*e*) a detector/amplifier system that modulates the brightness of the TV output device.

Thus, provided that there is a constant delay time in the detection stage (which is taken into account in the display stage), there is a one-to-one correspondence between a point on the TV output screen and a point on the specimen. Resolution in the case of scanning systems is governed by (*a*) the size of the electron spot or "probe" on the specimen surface, (*b*) the extent to which this spot spreads as it penetrates into the specimen, and (*c*) the total current in the spot which, together with production and detection efficiencies, determines the length of time the spot must stay in one place for the signal-to-noise ratio to be acceptably high. The magnification is purely geometrical—no magnifying lenses need be used.

To a large extent what is detected gives the instrument its name: scanning electron microscope, X-ray microanalyzer, and so on. The various particles and radiation that are presently employed are shown schematically in Fig. 1.12. A standard SEM detects either backscattered electrons or secondary electrons, or possibly both by switchable energy filtering (see Fig. 1.31); by insulating the specimen from the rest of the microscope, a "specimen current" image may be formed using the electrons absorbed by the specimen. Image constrast may then be modified by using amplifiers of different slopes and by mixing the various signals. A scanning Auger microscope detects the characteristic Auger electrons and is thus an *analytical* device as well as an image-forming system. The specimen also emits a wide spectrum of electromagnetic waves (see Fig. 1.12), ranging from optical wavelengths (the cathodoluminescence used in the light-emission

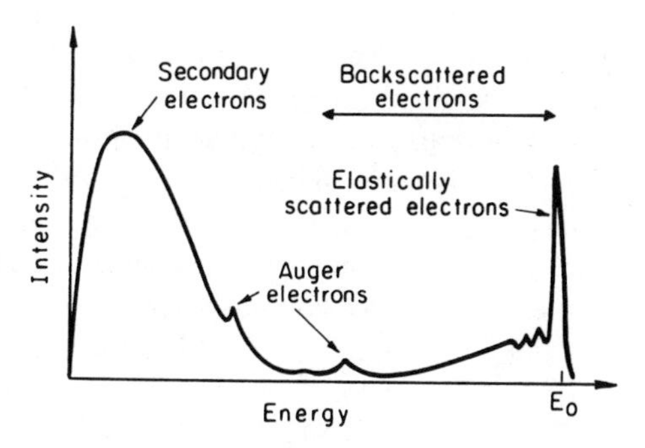

Fig. 1.31 The energy spectrum of the electrons emitted from the specimen.

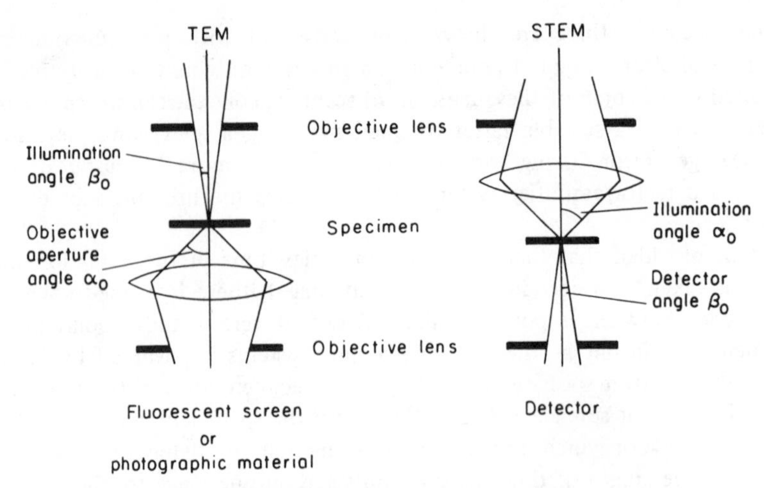

Fig. 1.32 Illustrating the reciprocity between TEM and STEM.

microscope) to the characteristic X-rays measured in the X-ray microanalyzer. If the specimen is sufficiently thin, some of the electrons are transmitted with little energy loss, and it is these that are detected in the STEM.

By comparing Figs. 1.30 and 1.13, it may be seen how simple it is, in principle, to construct a hybrid instrument capable of all the scanning functions mentioned above, as well as transmission. Addition of a scanning system and detectors above the specimen in Fig. 1.13 yields a SEM, while addition of scanning and detectors anywhere below the specimen produces a STEM system. Thus hybrid systems, as opposed to dedicated instruments, may be made extremely versatile, although they are inevitably subject to design compromise in regard to either ultimate performance or ease of change of mode, or perhaps both.

5.2 Imaging in the STEM

Scanning a fine probe across the specimen on a raster and modulating the brightness of the similarly scanned image output device with the intensity of the transmitted electrons obviously allows a magnified image to be formed. As will be discussed further in Chapter 5, there is a reciprocity relationship between the STEM and TEM images. Roughly speaking, the two systems are inverted, the functions of illuminating and collecting apertures being interchanged as shown schematically by the use of the angles lettered α and β in Fig. 1.32. Thus, under conditions of similar α and β, the STEM and TEM images are found to be similar, as illustrated in Fig. 1.33. The principal advantage claimed for the STEM mode is for penetration and resolution in thick specimens, where, because the lenses

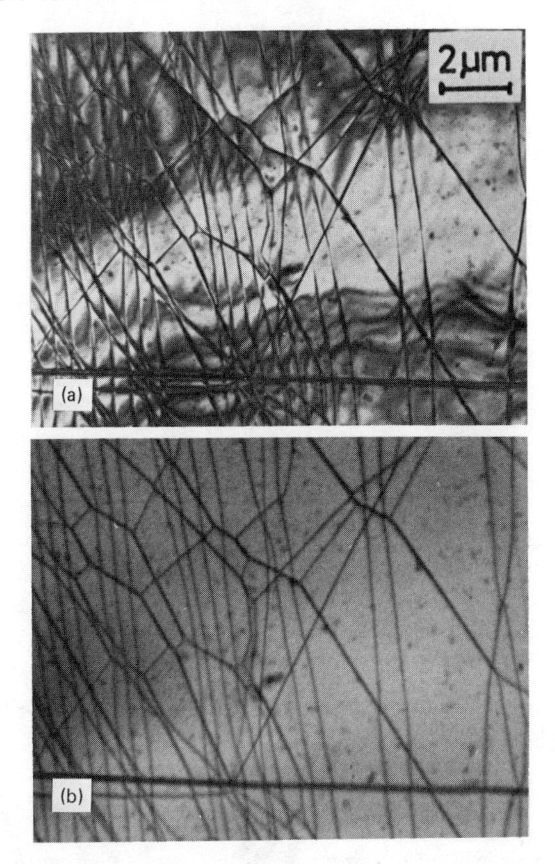

Fig. 1.33 (*a*) TEM and (*b*) STEM images of the same area of a specimen of MoS_2, showing dislocations both taken at 80 kV. Courtesy *Journal of Microscopy*.[45]

after the specimen, if present at all, merely guide the electrons into the collector as indicated in Fig. 1.34, rather than forming an image, chromatic aberration effects are avoided. Such an improvement in resolution in thick specimens has yet to be conclusively demonstrated, however. The main advantages of the STEM system in practice are in microdiffraction and microanalysis, as will be discussed in the subsequent Sections.

Like TEM, the STEM system may be operated in bright and dark field modes. Again as in the case of TEM, the dark field image may be formed either by deflecting the diffracted beam (this time below the specimen) into a centrally placed objective aperture, or by displacing the aperture to allow through only a diffracted beam, as illustrated in Fig. 1.35.

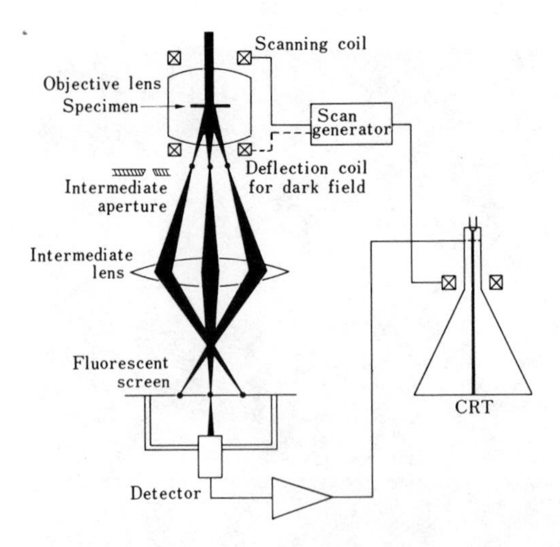

Fig. 1.34 Ray diagram of formation of STEM image and micro-microdiffraction pattern. Courtesy *JEOL News*, 15E-1.

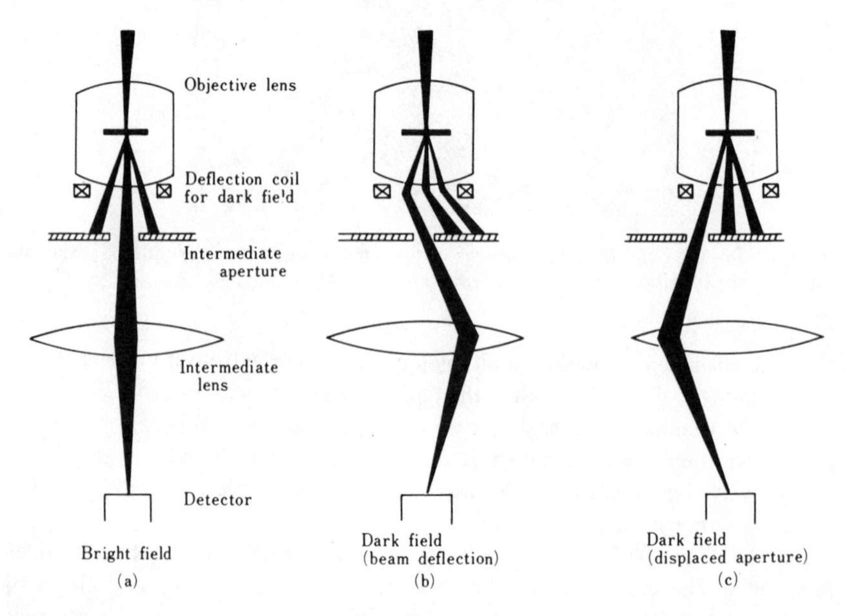

Fig. 1.35 Ray diagrams of formation of bright and dark field images in STEM. Courtesy *JEOL News*, 15E-1.

5.3 Microdiffraction in the STEM

As noted in Section 4.1, because of optical errors, especially spherical aberration, there is a limitation on the minimum size of the area giving rise to the diffraction pattern in the selected area diffraction mode. With the very fine probe available from the condenser lenses of a scanning system the area is selected as the total illuminated area. Because beam convergence and minimum spot size are interrelated, the selected area is usually larger than the smallest attainable, for example, 0.2 μm diameter at a convergence of 10^{-4} rad. The ray diagram for this mode is identical to that for selected area diffraction, shown in Fig. 1.13, except that the aperture is not required.

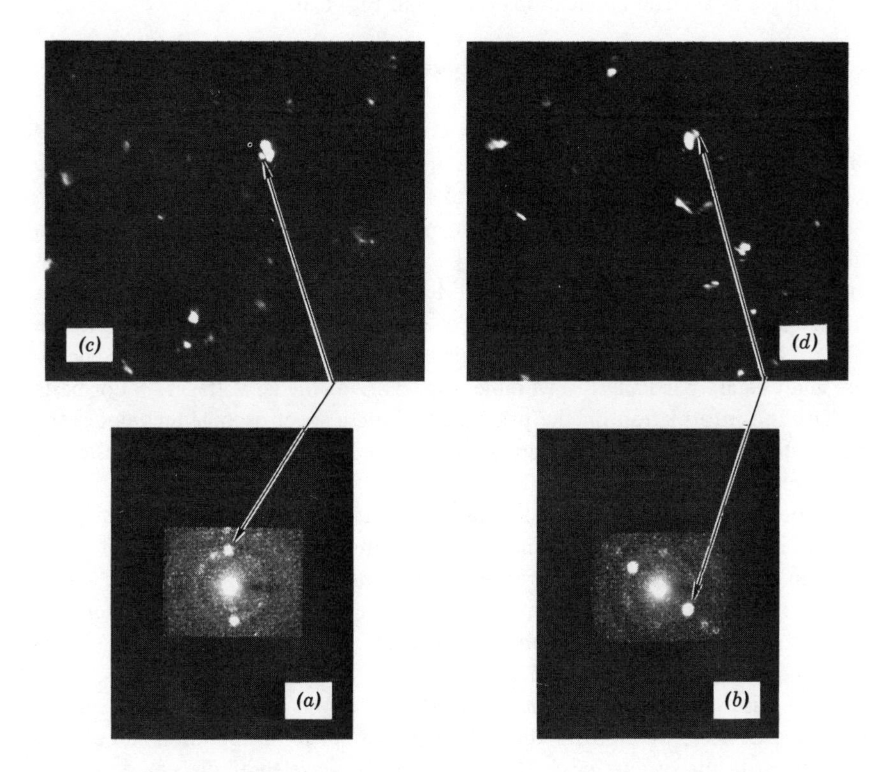

Fig. 1.36 (*a*, *b*) Rocking beam microdiffraction patterns from crystallites of germanium ~50 Å in diameter, indicated by the arrows in the corresponding dark field micrographs (*c*, *d*). The dark field images were taken in the conventional TEM mode, using the arrowed reflections, which arose in the microdiffraction pattern from the small crystallites indicated. Courtesy R. H. Geiss, IBM Research Laboratory, San Jose.

Even smaller areas may be used to form diffraction patterns in a scanning transmission mode as indicated in Fig. 1.34, such areas being easily related to the corresponding STEM image. The technique usually adopted is to stop the electron probe at the requisite point in its raster and then either (*a*) apply scanning fields to the "dark field" deflection coils denoted by dashed lines in Fig. 1.34, or (*b*) record the diffraction pattern on the fluorescent screen or photoplate, as illustrated in Fig. 1.34. In this case a compromise must again be made between probe size and convergence, typically 20 nm diameter at a convergence of 10^{-3} rad.

Alternatively, diffraction information may be displayed in the convergent beam mode already described in Section 4.9 or in the rocking beam mode.[39,40] An example of such an application is shown in Fig. 1.36.

5.4 Microanalysis in the STEM

Elemental chemical analysis is possible whenever irradiation results in quantized absorption or emission processes which are characteristic of the elements in the specimen. The well known K, L, M, and so on X-ray emission is produced when radiation such as incident electrons has enough energy to eject an electron from one of the inner shells, resulting in X-ray emission of characteristic wavelength as another electron falls into the vacant level. Any such process, if detectable, offers a means for chemical analysis. For electron microscopes operating in the 100 to 1000 kV range both characteristic X-ray emission and electron energy loss analysis are particularly favorable, especially at higher voltages.[41] For basic background information on the principles of these spectroscopic techniques the reader should consult the appropriate books listed in the Bibliography at the end of this chapter.

As noted above, the addition of suitable detectors enables the STEM system to be used as a microanalyzer, an implication of the *transmission* mode being that the thin foil, in which the electron beam spreads sideways hardly at all, allows extremely small areas to be sampled. Typically an area 20 nm in diameter may be analyzed; this may be the same area as that giving rise to the scanning transmission diffraction pattern noted above.

In the case of *X-ray microanalysis* the detectors may be solid state, energy dispersive devices, which offer the advantages of speed of operation and ease of attachment to the system, or crystal spectrometer, wavelength dispersive devices, which are usually preferred for quantitative studies. In all cases, of course, care must be taken with specimen mounting and electron and X-ray collimation to prevent spuriously produced X-rays from entering the detector system. An example of X-ray microanalysis in the STEM mode is shown in Fig. 1.37.

Elements that may be satisfactorily measured using their characteristic X-ray emissions are limited to those of atomic number Z, greater than about 11; alternative techniques must be employed for elements of smaller atomic number. One such technique is *electron energy loss spectrometry* (ELS).[41] In this case the detector to be added to the STEM system is an electron energy spectrometer; this can distinguish between low Z elements by their characteristic K-shell losses, which occur in the range of a few hundred electron volts, separated by several tens of volts from each other. Thus a device of moderate resolution is sufficient provided that adequate signal is available above background. An example of such an analysis in a (Be-Si-O-N) ceramic is shown in Fig. 1.38, in which it can be seen that the light elements are detected and qualitatively analyzed by ELS but are not detected by X-ray analysis. Thus ELS is particularly promising for ceramics and minerals.[42,43]

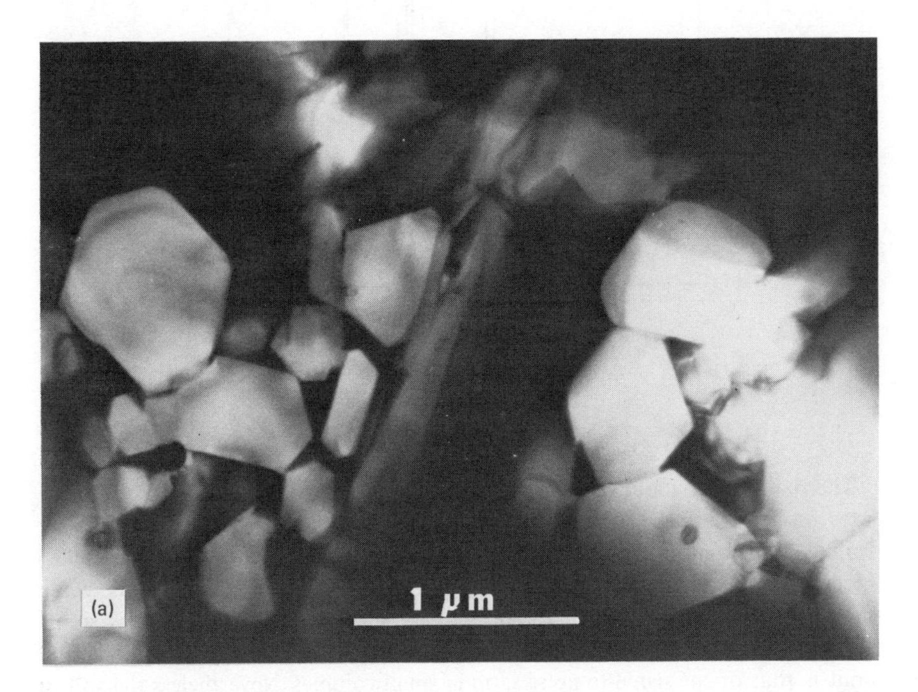

Fig. 1.37 (*a*) Image of an yttria-fluxed Si_3N_4 polycrystal, showing faceted silicon nitride grains (light contrast) with $\{10\bar{1}0\}$ habit amidst the yttrium-silicon-oxynitride phase (dark regions). (*b*) STEM X-ray spectrum from one of the yttrium-silicon oxynitride grains, ~3000 Å diameter, shown in (*a*). Other spectra show some variation from region to region, but the very high impurity content is common to all areas. Courtesy D. R. Clarke.

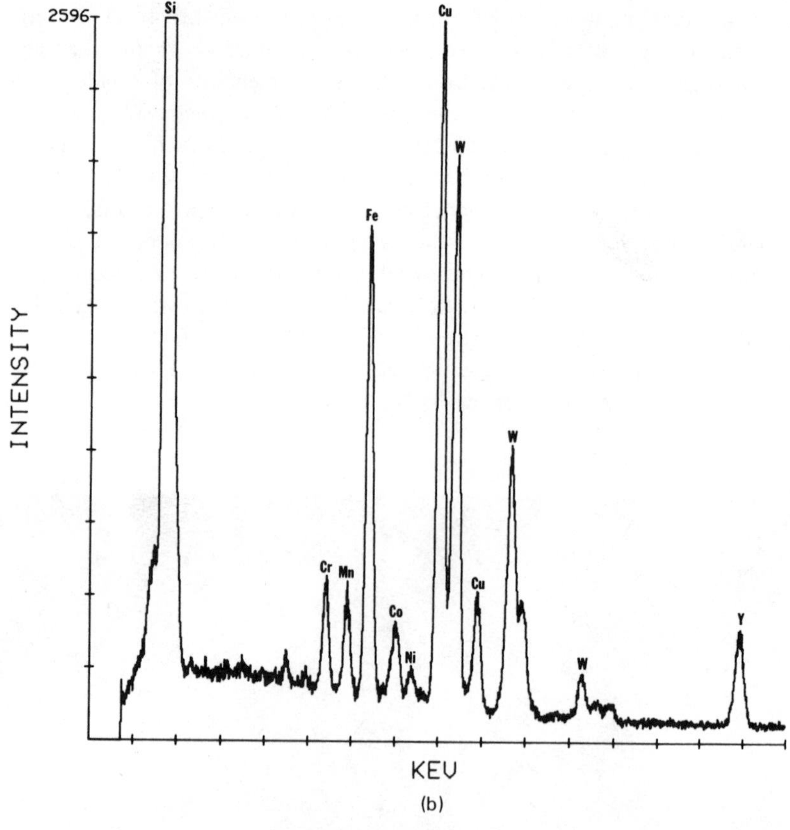

Fig. 1.37 (*Continued*)

The main point about X-ray analytical limits is that for finer and finer probe sizes one is eventually limited by the X-ray signal from the volume irradiated. Unlike the situation in conventional microprobe analysis of thick (or bulk) specimens, the dispersion of electrons in a thin foil is quite small; thus, while the spatial resolution is high (~100 Å), the spectroscopic resolution is somewhat lower ($\sim 10^{-3}$), depending upon atomic number. Thus the X-ray detection limit is that of the signal-to-noise ratio in small volumes. Nevertheless the ability to chemically analyze small volumes by STEM methods is perhaps the most attractive feature of these instruments. Some further examples are given in Chapter 3, Sections 10 and 11.

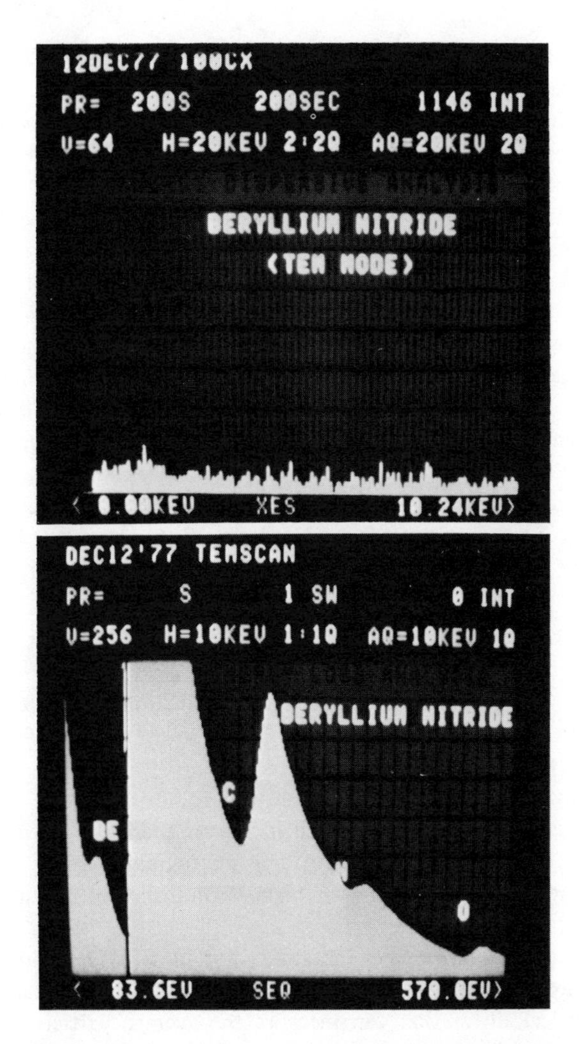

Fig. 1.38 STEM X-ray spectrum (upper) and electron loss spectrum (lower) from a Be-Si-O-N ceramic, illustrating the improved detection of light elements in the latter case. The carbon loss comes from a thin layer of carbon evaporated onto the foil to minimize charging.

6 Optical Diffraction and Electron Microscopy

An optical diffractometer is the simplest and most efficient device for performing two-dimensional Fourier transforms yet invented. As such, it finds many applications in the analysis of electron micrographs.

The diffractometer usually consists of a laser as a source of coherent light, a collimator-pinhole assembly that illuminates a micrograph negative with a "clean" light beam, and one or more lenses that form the diffraction patterns of the negative. A simple diffractometer is shown in Fig. 1.39. With a diffraction lens of sufficient quality such a design will give very adequate diffraction patterns.

Without the electron microscopy negative (image or diffraction pattern), the light would converge to be focused at a single point at the center of the viewing screen. The inserted negative modulates the amplitude of the transmitted passing light wave and causes diffraction away from the center. If the negative is an electron microscope image containing a single periodicity, the laser beam will be diffracted into a pair of "Bragg" spots, each a distance D from the screen center, where

$$D = \frac{\lambda L}{d}.$$

Here λ is the wavelength of the light, L the diffraction camera length, and d the spacing of the periodic modulation (analogous to the formation of the electron diffraction pattern).

When many different periodicities are present in the image, the diffraction pattern shows a pair of spots for each one of them. In fact, it is a Fourier transform of the illuminated area of the micrograph. This is very useful for analyzing both low resolution and high resolution electron micrographs (or diffraction patterns).

For *low magnification images* the ability to identify the periodicities present serves in analyses of material texture. For instance, when there are second-phase particles, optical diffraction can measure the average particle size and shape, directional ordering of elongated particles, and so on (Fig. 1.40).

For *lattice images* the periodicities revealed by optical diffraction are more or less directly the periodicities of the original specimen examined in the electron microscope. The optical diffraction pattern therefore rather resembles the electron diffraction pattern. The correspondence is not absolutely exact, however, because the optical diffraction pattern is the Fourier transform of the image (i.e., intensity), whereas the electron diffraction pattern is a Fourier transform of the amplitude of the electron wave. This means that even under ideal circumstances the optical diffraction pattern is obtained by autocorrelation of the electron one. The autocorrelation changes the intensities of the

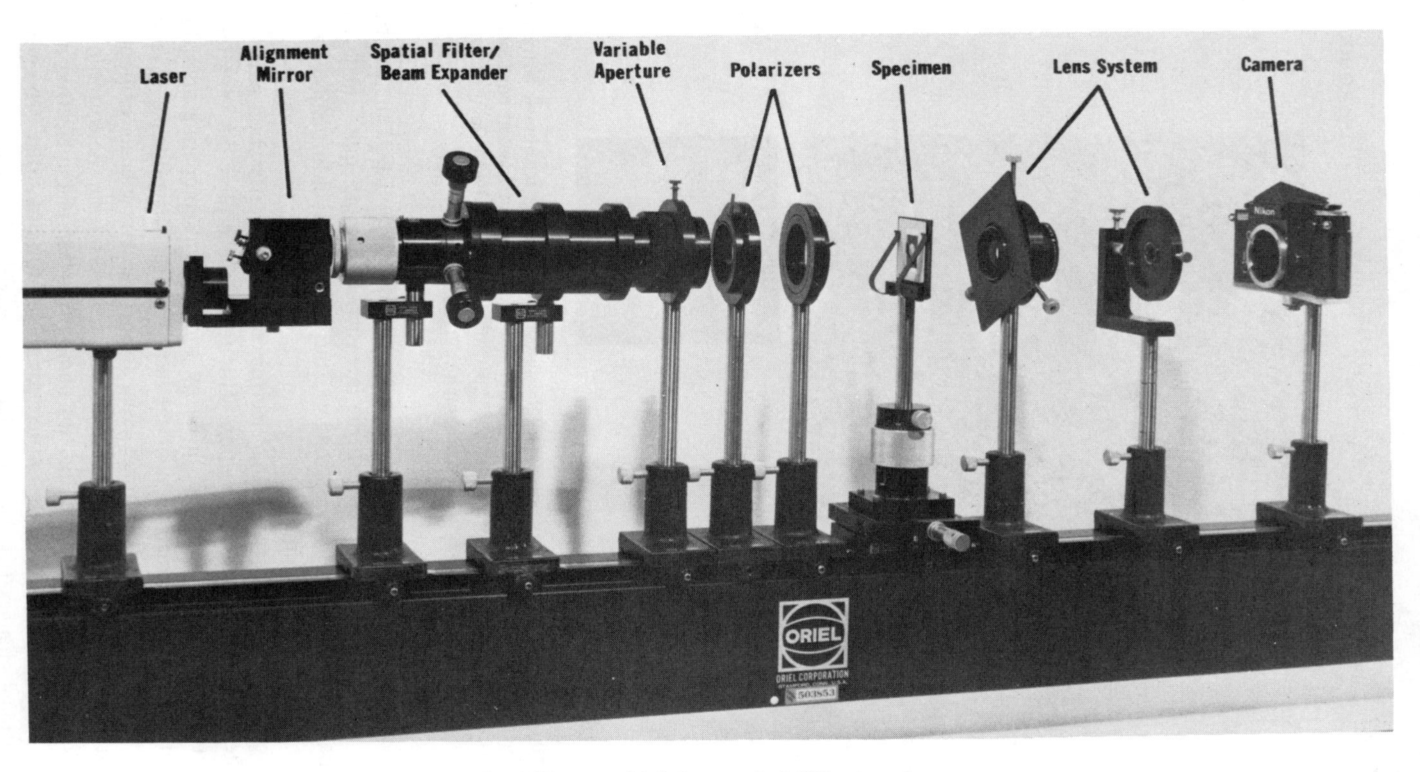

Fig. 1.39 A multiple lens optical diffractometer.

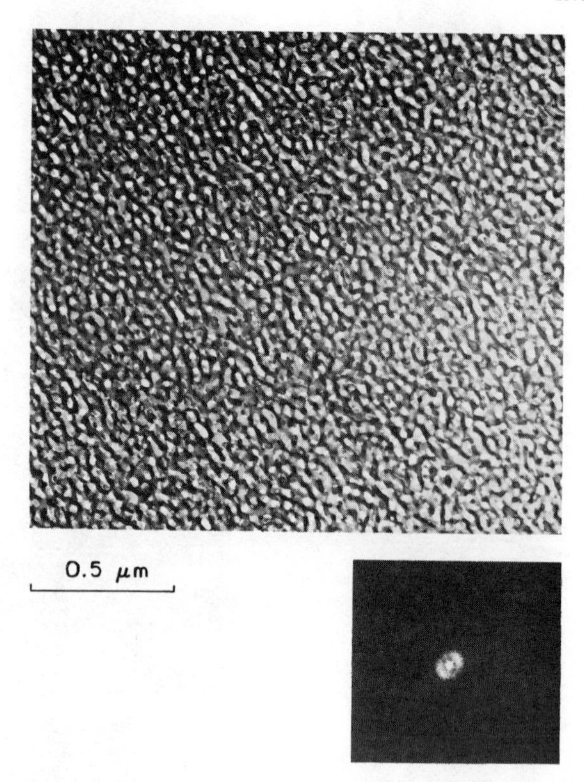

Fig. 1.40 Analysis of particle size, shape, and directional ordering by optical diffraction. The dimensions of the pattern relate to the average size of the particles; in particular, its arced nature shows up their directionality. Specimen: step-aged Fe-Cr-Co-V alloy after thermomagnetic treatment of 630°C. Micrograph courtesy Y. Belli.

original spots and introduces some new ones. It does not change the positions of the original spots, however, and the intensity changes are quite small if the specimen does not scatter too strongly. Thus geometrical information can be obtained without compromise.

The correspondence between the electron and optical diffraction patterns is exploited by *optical microdiffraction*. This technique probes the variation of the lattice parameters of the specimen by forming and comparing optical diffraction patterns from small areas of the lattice image. The principal advantage of the technique is that it is much easier to form a narrow beam of light that illuminates an area of only a few lattice fringes than to form the corresponding 10 to 20 Å electron probe required for microdiffraction in the electron microscope. Thus, for instance, the lattice parameter variation that accompanies compositional changes (e.g., in spinodal decomposition or in polytypism) can

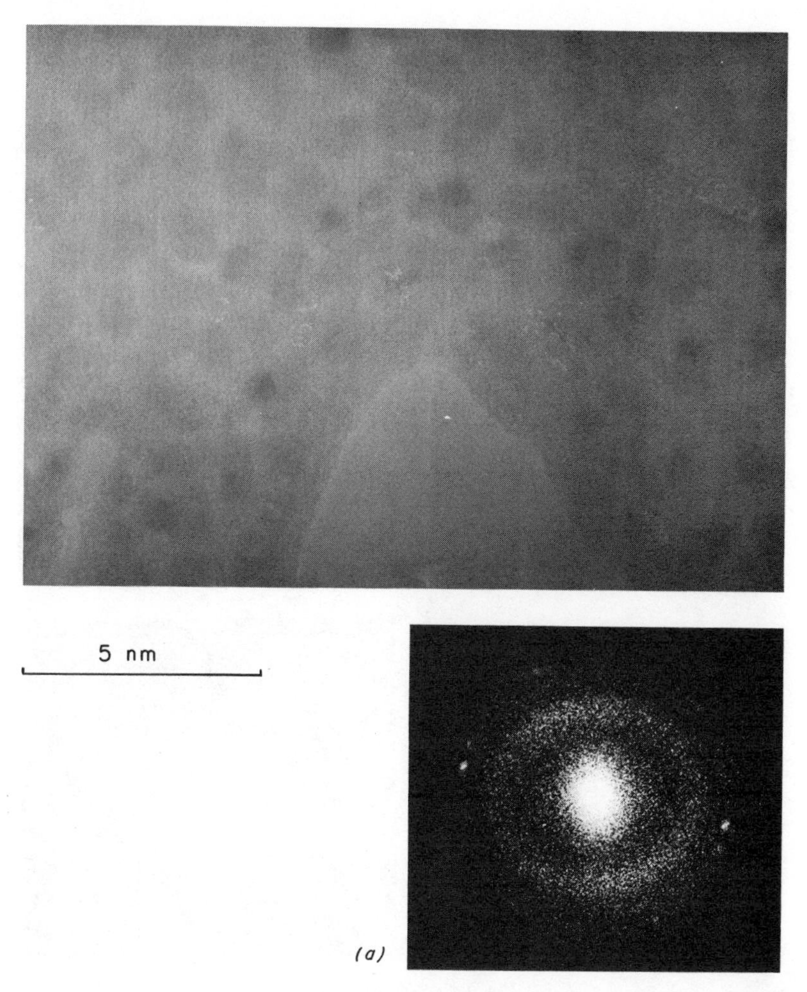

Fig. 1.41 Micrographs and their corresponding optical diffractograms of an amorphous carbon film (also containing small germanium crystallites which gave rise to the spots in diffractograms) at (*a*) Gaussian focus and (*b*) Scherzer defocus. The Gaussian focus diffractogram indicates a small residue of astigmatism, while the broader ring of the Schnerzer defocus diffractogram confirms the increase in range of spatial frequencies transferred at that defocus. The diffractograms should be compared with the contrast transfer functions of Fig. 5.36. Courtesy O. L. Krivanek.

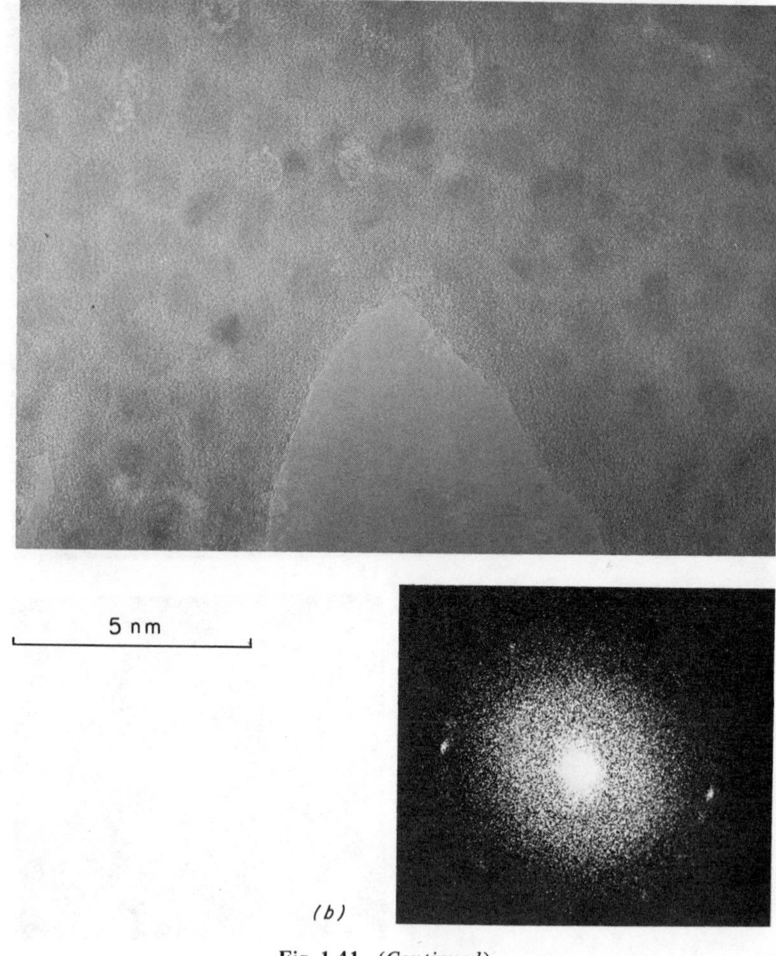

Fig. 1.41 (*Continued*)

be studied as described in Chapter 3 (Section 3.11.2, Figs. 3.47 to 3.50). The presence of a second phase and its lattice fringes can be detected in different areas of the micrograph.

Another very useful application is in *determining the microscope operating conditions by optical diffraction analysis* of the micrograph of a thin amorphous film. A thin phase specimen contains many periodicities (spatial frequencies).[44] The electron microscope "filters" these spatial frequencies–it images only some of them. The actual frequencies imaged depend on many microscope parameters: defocus, astigmatism, the spherical and chromatic aberration coefficients of the objective lens, the spread of the energies and angles of illumination, and the angle of the illuminating beam with respect to the optic axis. The selective imaging of the spatial frequencies is described by the *contrast transfer function* (CTF), which is discussed in more detail in Chapter 5, Section 9.1.

When a high resolution micrograph of an amorphous film is placed in the optical diffractometer, its diffraction pattern shows directly the square of the CTF (see Fig. 1.41). Since the CTF is influenced by all the parameters mentioned above, even a cursory glance at the diffraction pattern gives one a good idea of how well the electron microscope has performed, and precise measurement of the radii and intensities of the characteristic rings makes possible exact evaluation of all the parameters. The defocus and spherical aberration coefficients, for instance, can be determined to a precision of ±50 Å and ±0.1 mm, respectively.

The final application mentioned here is in filtering noise out of electron images. Sometimes one knows that the interesting part of the image has a certain periodicity whereas the noise is aperiodic. Such a case is ideally suited to *optical noise filtering*, in which the formation of an optical diffraction pattern is the first of three steps. In the second step the diffraction pattern is passed through an opaque screen with holes that replace the diffraction viewing screen. The holes allow the interesting information to pass through, while the aperiodic noise is screened out. In the third step the diffraction pattern is retransformed to an image that looks just like the original, but with the noise removed.

7 Specimen Preparation

Specimens for the electron microscope must obviously fit into the holder mentioned in Section 3.1 and hence are usually limited to a disk of maximum external diameter approximately 3 mm and thickness about 0.5 mm. The actual area of interest must be a few tens of nanometers thick and, since it will be studied at, say, 20,000X magnification, need only be on the order of microns or tens of microns in extent to be adequately usable. Therefore small pieces of material held between conventional grids or suitably sized disks are ideal.

The following methods have been successfully used for studies of materials:

1. Replication of the surface of a bulk specimen by means of a thin, electron-transparent layer which may be stripped off.
2. Cleavage of the material until thin enough.
3. Microtomy (direct cutting of thin sections).
4. Collection of small particles which may be either naturally occurring or produced by grinding.

5. Evaporation of the material to deposit a thin film on a suitable substrate.
6. Similar deposition from chemical reaction in liquid or vapor phase.
7. Thinning from bulk with minimal disturbance to preexisting structures in the material, by chemical, electrochemical, or ion-sputtering techniques.

These various methods, including their fields of application and their limitations, are discussed in a number of books (see the Bibliography). By far the most important in the study of materials is the seventh method, and Appendix A gives details of a number of recent developments and a wide range of references.

8 Bibliography

In the appended bibliography we have selected a number of books in the "mainstream" of transmission electron microscopy in which details of many of the topics touched on in this chapter may be found. However, for an appreciation of the geometry of defects and orientation relationships, interpretation of diffraction patterns, and many other problems, a number of additional skills are required, such as stereographic projection analysis, stereoscopic techniques, and, of course, basic crystallography. Publications on these topics are also included, and, to assist the reader to ascertain some of the basic skills required, some questions in these areas are included in the exercises.

Books (in alphabetical order)

Crystallography and Crystal Defects, A. Kelly and G. W. Groves, Addison-Wesley, Reading, MA, 1970.

Diffraction Physics, J. M. Cowley, North Holland/Elsevier, Amsterdam, 1975.

Durchstrahlungs–Elektronenmikroskopie Fester Stoffe, E. Hornbogen, Verlag Chemie, Weinheim, 1971.

Einfürung in die Elektronenmikroskopie, M. von Heimendahl, F. Vieweg, Braunschweig, 1970.

Electron Microscopy and Structure of Materials, G. Thomas, Ed., University of California Press, 1972. Berkeley, Los Angeles, London (See also the 1963 book, edited by G. Thomas and J. Washburn.)

Electron Microscopy in Material Science, U. Valdrè, Ed., Academic Press, New York, 1971.

Electron Microscopy in Mineralogy, H-R. Wenk et al., Eds., Springer Verlag, Berlin, 1976.

Electron Microscopy of Thin Crystals, P. B. Hirsch, A. Howie, R. B. Nicholson, D. W. Pashley, and M. J. Whelan, Butterworths, London, 1965. (revised edition published by Krieger, New York, 1977).

Electron Optics and Electon Microscopy, P. W. Hawkes, Taylor and Francis, Ltd., London, 1972.

Fundamentals of Transmission Electron Microscopy, R. D. Heidenreich, Wiley-Interscience, New York, 1964.

Interpretation of Electron Diffraction Patterns, K. W. Andrews, D. J. Dyson, and S. R. Keown, Hilger and Watts, London, 1967.

Manual on Electron Metallography Techniques, ASTM Technical Publication 547, American Society for Testing Materials, Philadelphia, 1973.

Modern Diffraction and Imaging Techniques in Materials Science, S. Amelinckx et al., Eds., North Holland Press, Amsterdam, 1970. 2nd Ed. (rev.) 1978.

Modern Metallographic Techniques and Their Applications, V. A. Phillips, Wiley-Interscience, New York, 1971.

Physical Aspects of Electron Microscopy and Microbeam Analysis, B. M. Siegel and D. R. Beaman, Eds., John Wiley, New York, 1975.

Practical Electron Microscopy in Materials Science (4 vols), J. W. Edington, Macmillan, London, 1975.

Practical Methods in Electron Microscopy (several volumes), A. Glauert, Ed., North Holland, Amsterdam, 1972 onward.

Scanning Electron Microscopy Applications to Materials and Device Science, P. R. Thornton, Chapman and Hall, London, 1968.

Stereographic Projection and Applications, O. Johari and G. Thomas, Wiley-Interscience, New York, 1969.

Transmission Electron Microscopy of Metals, G. Thomas, John Wiley, New York, 1962.

X-ray Microanalysis in the Electron Microscope, J. A. Chandler, in series entitled *Practical Methods in Electron Microscopy*, Audrey M. Glauert, Ed., North Holland, New York, 1977.

Conference Proceedings

Electron Microscopy and Analysis Group (EMAG) (of the Institute of Physics, London), published approximately biennially, latest 1977.

Electron Microscopy Society of America (EMSA), published annually by Claitors, Baton Rouge, LA.

European Regional Conferences, published every 4 years, latest 1976.

High Voltage Electron Microscopy, published approximately triennially, latest 1978.

International Congresses, published every 4 years, latest 1978.

Scanning Electron Microscopy, O. Johari et al., Eds., published annually by IIT Research Institute, Chicago.

Exercises

Below are set out examples of the content of Chapter 1, some of which will serve as an "orientation quiz." The purpose of the latter is to orient the student with regard to the background (e.g., basic crystallography) needed to pursue electron microscopy. Many of these topics will be discussed in more detail in Chapter 2, (e.g., the solution to question 1.6) but if the present exercises prove difficult it may be necessary to consult the references suggested.

1.1 Why is the image of a specimen inverted with respect to the specimen, but the diffraction pattern is not? Is this relationship independent of the number of lenses after the objective lens, or not?

1.2 In selected area diffraction (SAD) using a conventional 100 kV TEM, why is it not useful to use apertures with effective diameters of less than about 0.5 μm to select areas for diffraction? In what ways can this limitation on SAD be improved?

1.3 Derive the simple expression (equation 1.10) relating foil thickness t to parallax for a foil tilted by equal amounts ($\theta/2$) from its initial position perpendicular to the electron beam.

1.4 Calculate the optimum tilt angle for taking a stereo pair if the electron optical magnification is 20,000, an X2 stereo viewer is to be employed, and the foil thickness is ~5000 Å (take $p = 4$ mm).

1.5 (*a*) Define structure factor, F. Calculate the structure factor for NaCl.

(*b*) Draw the reciprocal lattice for NaCl. Can you index all rel-points? Include only points for which $F \neq 0$.

(Reference: any text on diffraction, for example, Cullity, *Elements of X-ray Diffraction*, Addison-Wesley.)

1.6 Make a scale drawing to illustrate the reciprocal lattice-reflecting sphere concept for Bragg's law, using Cu K_α X-rays and 100 kV electrons for an aluminum single crystal, and show why the Laue method requires polychromatic radiation in the X-ray and neutron diffraction cases. See Chap. 2.

(Reference: Barrett and Massalski, *Structure of Metals*, McGraw-Hill.)

1.7 Plot the standard (114) cubic stereographic projection. Show the position of the $(1\bar{1}0)$ pole after twinning on (111). Check your answer by using the twin matrix algebraic calculation.

(References: Johari-Thomas, *Stereographic Projection and Its Application*, Wiley, 1969.)

1.8 An electron probe of 10 nm diameter is formed in a 100 kV STEM. The convergence semiangle α of the probe is 5×10^{-3} rad. If it is assumed that the electron source brightness is $B = 10^{10}$ A^{-2} sr^{-1}, and the aberrations of the probe-forming lens are neglected, how many electrons pass through the probe per second?

1.9 (*a*) With the probe of Exercise 1.8, what is the minimum concentration of nickel impurity in a 300 Å thick copper foil that can be detected by X-ray microanalysis in a STEM system which can collect 5000 Cu K_α counts before contamination prevents further analysis? (Assume that nickel becomes detectable when the number of counts in the Ni K_α peaks exceeds the statistical variation in the background by a factor of 5, and that the background strength is 2% of the main Cu K_α peak.)

How does the minimum detectable concentration change when (*b*) the probe diameter is increased to 100 nm or (*c*) the specimen thickness is increased to 120 nm, assuming that the beam current density and convergence are unaltered?

1.10 Show by simple ray diagram sketches why spherical and chromatic aberrations are critical problems in high resolution electron microscopes (but are not so serious in optical microscopes). What can be done to minimize these aberations? Why is chromatic aberration more significant for light materials (low Z) than heavy ones (high Z)?

1.11 (*a*) From the expressions needed in Exercise 1.10, namely,

$$\alpha_{opt} = \lambda^{1/4} C_s^{-1/4}$$

and

$$\Delta r_{min} = 0.61 \lambda^{3/4} C_s^{1/4}$$

complete the following table:

Voltage (kV)	λ (pm)	α_{opt}		Δr_{min} (Å)	
		$C_s = 1$ mm	$C_s = 3$ mm	$C_s = 1$ mm	$C_s = 3$ mm
100	3.7				
500	1.42				
1000	0.87				

and plot the data graphically. Compare your results for Δr to those for d in Fig. 1.5.

(*b*) If you wished to resolve the Na and Cl atoms in NaCl, which conditions would you choose (orientation of crystal, electron optical), and why?

(*c*) For the apertures in the table (i.e., α_{opt}) how many beams will be scattered into each aperture (compute $n \cdot 2\theta$ from Bragg's law) for this crystal (NaCl) if it is oriented in [001] at the three voltages given?

References

1. Heidenreich, R. D., *J. Appl. Phys.*, **20**, 993 (1949).
2. Thomas, G., *Transmission Electron Microscopy of Metals*, John Wiley, New York, 1962.
3. Thomas, G., *Modern Diffraction and Imaging Techniques in Materials Science* (Eds. S. Amelinckx et al.), North Holland, Amsterdam, 1970, p. 159. (New Ed. available 1979)

4. Sinclair, R. and Dutkiewicz, J., *Acta Met.*, **25**, 235 (1977).
5. Thomas, G., *Trans AIME*, **233**, 1608 (1965).
6. Hashimoto H., Endoh H., Tanji T., Ono A., and Watanabe, E., *J. Phys. Soc. Jap.*, **42**, 1073 (1977).
7. Dupouy, G., *Adv. Opt. Electron Microsc.*, **2,** 167 (1968).
8. Cosslett, V. E., *Modern Diffraction and Imaging Techniques in Materials Science* (Eds. S. Amelinckx et al.), North Holland Press, Amsterdam, 1970, p. 341. See also Howie, A., *ibid.*, p. 295.
9. Bell, W. L. and Thomas G., *Electron Microscopy and Structure of Materials* (Ed. G. Thomas), University of California Press, 1972, p. 23. See also Fisher, R. M., *ibid.*, p. 60.
10. Fisher, R. M., *Proceedings of the 26th Conference of the Electron Microscopy Society of America*, Claitors, Baton Rouge, LA, 1968, p. 324.
11. Humphreys, C. J., *Phil. Mag.*, **25,** 1459 (1972).
12. Thomas, G. and Lacaze, J. C., *J. Microsc.*, **97**, 301 (1973).
13. Howitt D. G., Glaeser, R. M., and Thomas, G., *J. Ultrastructur. Res.*, **55,** 457 (1976).
14. Menter, J. W., *Adv. Phys.*, 7, 299 (1958).
15. Sinclair, R. and Thomas, G., *Met. Trans.*, **9A,** 373 (1978).
16. Goringe, M. J., *J. Microsc.*, **16,** 169 (1973).
17. Glaeser, R. M. *J. Microsc.*, **112,** 127 (1978).
18. Hobbs, L. W., Hughes, A. E., and Pooley, D. M., *Proc. R. Soc.*, **A332,** 167 (1973).
19. Edington, J. W., *Practical Electron Microscopy in Materials Science, Monograph 1: The Calibration and Operation of the Electron Microscope*, Macmillan (Philips Technical Library), 1974.
20. Knox, W. A., *Proceedings of the 32nd Conference of the Electron Microscopy Society of America*, Claitors, Baton Rouge, LA, 1974, p. 560.
21. Hirsch, P. B., et al., *Electron Microscopy of Thin Crystals*, Butterworths, London, 1965.
22. Hashimoto, H., Kumao, A., and Endoh, H., *Proceedings of the 8th International Congress of Electron Microscopy*, Canberra, Australia, 1974, p. 244.
23. Cockayne, D. J., *Z. Naturforsch.* **27,** 452, 1972; *J. Phys.*, **35C7,** 141 (1974).
24. Jenkins, M. L., *Phil. Mag.*, **29,** 813 (1974).
25. Karnthaler, H. P. and Wintner, E., *Phil. Mag.*, **32,** 81 (1975).
26. Bell, W. L. and Thomas, G., *Electron Microscopy and Structure of Materials*, University of California Press, 1972, p. 23.
27. Lynch, D. F., Moodie, A. F., and O'Keefe, M. A., *Acta Crystallogr.*, **A31,** 300 (1975).
28. Scherzer, O., *J. Appl. Phys.*, **20,** 20 (1949).
29. Phillips, V. A. and Hugo, J. A., *Micron*, **3,** 212 (1972).
30. Johansen, B. V., *Micron*, **4,** 121 (1973).
31. Bourret, A., Desseaux, J., and Renault, A., *Acta Crystallogr.*, **A31,** 746 (1975).
32. Heidenreich, R. D. and Matheson, L. A., *J. Appl. Phys.*, **15,** 123 (1944).
33. Hudson, B. and Makin, M. J., *J. Phys.* E: *Scient., Instrum.*, **3,** 311 (1970).
34. Nankivell, J. F., *Optik*, **20,** 171 (1963).

35. Thomas. G., *Modern Diffraction and Imaging Techniques in Materials Science* (Ed. S. Amelinckx et al.), North Holland, Amsterdam, 2nd Ed., 1978, p. 399.
36. Thomas, L. E. and Cuddy, L. J., *Proceedings of the Conference of the Electron Microscopy Society of America*, Claitors, Baton Rouge, LA, 1970, p. 38.
37. Thomas, L. E. and Lentz, S., *Proceedings of the Conference of the Electron Microscopy Society of America*, Claitors, Baton Rouge, LA, 1974, p. 362.
38. Bell, W. L., *J. Appl. Phys.*, **47**, 1626 (1976).
39. Geiss, R. H., *Appl. Phys. Lett.*, **27**, 174 (1975).
40. Woolf, R. J., Joy, D. C., and Titchmarsh, J. M., *Proceedings of 5th European Congress on Electron Microscopy*, Institute of Physics, London, 1972, p. 498.
41. Isaacson, M. and Johnson, D., *J. Ultramicrosc.*, **1**, 33 (1975).
42. Several authors in *Developments in Electron Microscopy and Microanalysis 1977*, Conference Series 36, Institute of Physics, London, 1977, pp. 141–154.
43. Several authors in *High Voltage Electron Microscopy 1977*, Japanese Society of Electron Microscopy, Tokyo, 1977, pp. 225–234.
44. Krivanek, O. L., *Optik*, **45**, 97 (1976).
45. Joy, D. C. *J. Microsc.*, **103**, 1 (1975).

TWO

GEOMETRY OF ELECTRON DIFFRACTION

1 Bragg's Law and the Reciprocal Lattice

1.1 Geometry

From eq. 1.7 and Table 1.1 it may be seen that for fast electrons the Bragg angles are very small. For example, for a cubic crystal with $a_0 = 4$ Å, the Bragg angle θ for the 200 reflection at 100 kV = $0.037/4 \simeq 10^{-2}$ rad or $\simeq 0.5°$ $[d_{200} = a_0/(h^2 + k^2 + l^2)^{1/2} = 2$ Å$]$. This means that Bragg diffraction will occur only for planes that lie within a few degrees of the incident beam, that is, planes whose poles lie close to 90° to the incident beam. If $\mathbf{g}_1 \ldots \mathbf{g}_n$ are vectors perpendicular to the reflecting planes $\{hkl\}_1 \ldots \{hkl\}_n$ and $[uvw]$ is the direction of the beam, then to a first approximation

$$[\mathbf{g}_1] \cdot [uvw] = [\mathbf{g}_n] \cdot [uvw] = 0. \tag{2.1}$$

Thus the diffraction pattern will contain the vectors g lying normal to the beam. In terms of the stereographic projection the vectors g or the poles of the planes $\{hkl\}$ may be determined by drawing a great circle at 90° to $[uvw]$. This great circle will contain the poles satisfying condition 2.1. Conversely, if one wishes to select a diffraction vector $\mathbf{g} = uvw$, the great circle of $[uvw]$ contains all the orientations that contain that particular g. This is illustrated for cubic crystals by the standard projection shown in Fig. 2.1.

The vectors g normal to the reflecting planes are called reciprocal lattice vectors. They have magnitude equal to the reciprocal of the d-spacings of the particular hkl planes involved and terminate at reciprocal lattice points (rel-points);

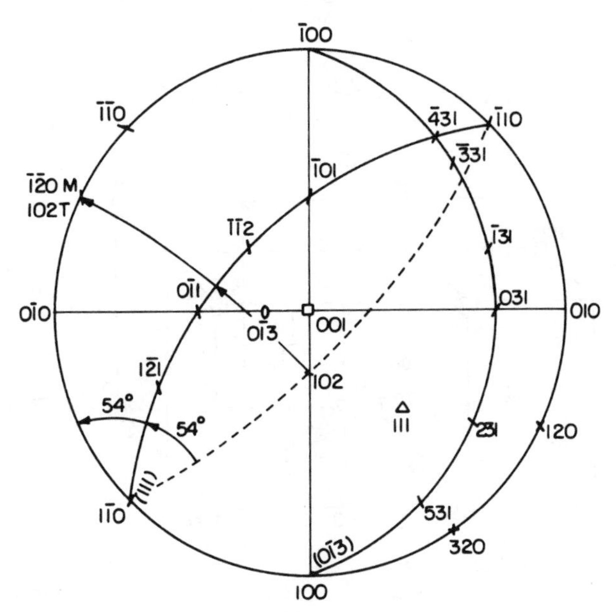

Fig. 2.1 Standard cubic [001] stereographic projection showing great circles for poles 90° from [111] and $[0\bar{1}3]$, respectively. The dotted circle contains poles that will twin by reflection to the basic [001] circle for (111) twinning; for example, 102 (twin) will then superimpose on $\bar{1}20$ (matrix).

see Fig. 2.2. Thus

$$|\mathbf{g}| = \frac{1}{d} = \frac{\sqrt{h^2 + k^2 + l^2}}{a_0} \quad \text{for cubic crystals.} \tag{2.2}$$

If **a***, **b***, **c*** are the basis vectors of the reciprocal lattice, then, relative to a point chosen as the origin, every other reciprocal lattice point can be reached by a reciprocal lattice vector of the form

$$\mathbf{g} = h\mathbf{a}^* + k\mathbf{b}^* + l\mathbf{c}^*,$$

where h, k, l are always integers, identical to the Miller indices of the reflecting planes, and are the coordinates of the reciprocal lattice points (Fig. 2.2).

Formally the reciprocal lattice of unit cell vectors **a***, **b***, **c*** defined in terms of the real space lattice **a**, **b**, **c** by $\mathbf{a}^* = \mathbf{b} \wedge \mathbf{c}/\mathbf{a} \cdot \mathbf{b} \wedge \mathbf{c}$ (and cyclic permutations), from which the useful relationships $\mathbf{a} \cdot \mathbf{a}^* = \mathbf{b} \cdot \mathbf{b}^* = \mathbf{c} \cdot \mathbf{c}^* = 1$, $\mathbf{a}^* \cdot \mathbf{b} = \mathbf{a}^* \cdot \mathbf{c} = 0$, and so on, immediately follow. Similarly it may be shown that the reciprocal lattice vector $\mathbf{g} = h\mathbf{a}^* + k\mathbf{b}^* + l\mathbf{c}^*$ is perpendicular to the real lattice plane (hkl) and is of magnitude $1/d_{hkl}$, where d_{hkl} is the separation of the (hkl)

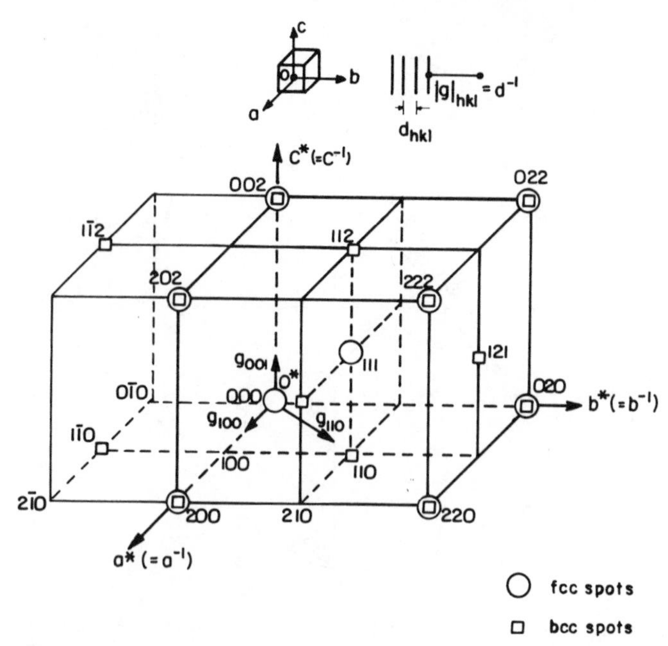

Fig. 2.2 The reciprocal lattice: for cubic crystals $a = b = c$, and the corresponding reciprocal lattice points for diffraction for fcc and bcc crystals are shown.

planes. These results hold for all crystal symmetries, not just the simple case of cubic crystals cited here. For hexagonal systems, however, it is often convenient to use a four-index system. A special notation has been developed for this situation which is particularly convenient for diffraction and microscopy. It is described fully in Appendix B.

1.2 The Structure Factor

In the diffraction of radiation by crystals, each reciprocal lattice point is associated with a diffracted beam whose intensity is related to the shape, volume, and perfection of the crystal and the structure factor F_g for the unit cell, defined by

$$F_g = \sum_j f_j\left(\frac{\sin\theta}{\lambda}\right) \exp\,(2\pi i \mathbf{g} \cdot \mathbf{r}_j), \tag{2.3}$$

where $f_j(\sin\theta/\lambda)$ is the atomic scattering factor of the jth atom for the angle concerned, this atom being at position $\mathbf{r}_j$. The summation is over the contents of the unit cell. The structure factor F is defined by eq. 2.3 whatever radiation

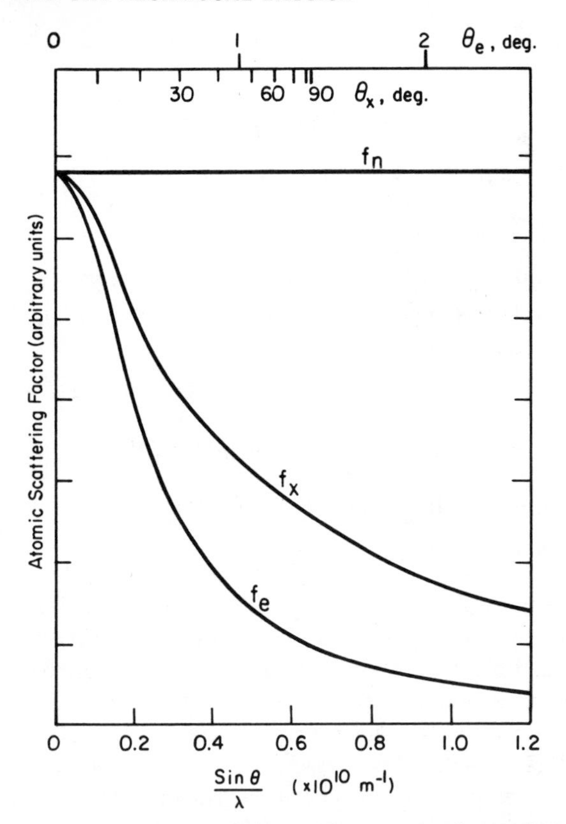

Fig. 2.3 Schematic angular variation of the atomic scattering factors for electron, X-ray, and neutron diffraction. Note the differences in angular scale when plotted for typical wavelengths λ[(Cu K_α) for X-rays and neutrons, 0.037 Å for electrons (100 keV)]. The three scattering factors have been arbitrarily scaled to coincide at zero scattering angle.

is being considered, the differences between radiations being represented by the atomic scattering factor f. The values of f for electrons and X-rays, both being associated with charge distributions within the atom, are closely related, while the value for neutron scattering may be widely different. The angular variations of each are shown schematically in Fig. 2.3, where it may be seen once again that for the typical electron case Bragg angles must be very small if appreciable diffraction is to take place. Neutron and X-ray diffraction are not restricted in this way.

The vector $\mathbf{r}_j$ is given by

$$\mathbf{r}_j = x\mathbf{a} + y\mathbf{b} + z\mathbf{c},$$

where x, y, z are the coordinates of the atom in the unit cell of basis vectors **a**, **b**, **c**. Thus

$$\begin{aligned}\mathbf{g}\cdot\mathbf{r}_j &= (h\mathbf{a}^* + k\mathbf{b}^* + l\mathbf{c}^*)\cdot(x\mathbf{a} + y\mathbf{b} + z\mathbf{c})\\ &= hx + ky + lz \qquad (\text{since } \mathbf{a}\cdot\mathbf{a}^* = \mathbf{b}\cdot\mathbf{b}^* = \mathbf{c}\cdot\mathbf{c}^* = 1)\\ &= \text{integral.}\end{aligned}$$

Thus "allowed" reflecting planes are those for which $F_g \neq 0$. For example, in the simplest fcc lattice (e.g., Al, Cu, γ-Fe) there are four atoms of the same kind per cell at coordinates 000, $\frac{1}{2}\frac{1}{2}0$, $\frac{1}{2}0\frac{1}{2}$, $0\frac{1}{2}\frac{1}{2}$, and thus

$$F = f[1 + e^{\pi i(h/2+k/2)} + e^{\pi i(h/2+1/2)} + e^{\pi i(k/2+1/2)}].$$

Since, it will be recalled, $e^{\pi i} = e^{3\pi i} = -1$ and $e^{2\pi i} = e^{4\pi i} = +1$, then

$$\begin{aligned} F &= 0 \quad \text{if } h, k, l \text{ are mixed (even and odd) integers,}\\ F &= 4f \quad \text{if } h, k, l \text{ are unmixed.}\end{aligned}$$

Thus the reciprocal lattice corresponding to diffraction includes only the points for which $F \neq 0$ (Fig. 2.2). Table 2.1 summarizes F values for some simple lattices. For cubic crystals these results are summarized in Table 2.2. The last column is useful when diffraction patterns are to be indexed. Since $2d \sin\theta = n\lambda$, $\sin\theta = n\lambda/2d$ and hence the sequence of order of reflections is $(h^2 + k^2 + l^2)$ (i.e., the smallest Bragg angle for reflection in bcc crystals corresponds to that for the 110 reflection), as listed in Table 2.2. Thus 220 is the second-order reflection of 110 in the bcc system ($n = 2$).

For hexagonal crystals the Miller-Bravais four-index notation is recommended, and it is especially helpful if all crystallographic data are expressed in direction indices. Details are given in Appendix B.

Table 2.1 Geometrical Structure Factor Rules for Basic Cells Containing Only One Kind of Atomic Species

Type of Crystal	Values	$F =$
Primitive	All h, k, l	f (1 atom per cell)
Body centered	$(h + k + l)$ even	$2f$ (2 atoms per cell)
Face centered	h, k, l unmixed	$4f$ (4 atoms per cell)
Base centered (e.g., ab face)	h, k, l unmixed	$2f$ (2 atoms per cell)
Hexagonal close-packed	$h + 2k = 3n$, l odd	0, e.g., 0001
	$h + 2k = 3n$, l even	$2f$, e.g., 0002
	$h + 2k = 3n \pm 1$, l odd	$\sqrt{3}\,f$, e.g., $01\bar{1}1$
	$h + 2k = 3n \pm 1$, l even	f, e.g., $01\bar{1}0$

Table 2.2 Allowed $\{hkl\}$ Values for Cubic Crystals[a]

Bcc		Fcc		Dc		
$(h^2+k^2+l^2)$	hkl	$(h^2+k^2+l^2)$	hkl	$(h^2+k^2+l^2)$	hkl	$\sqrt{h^2+k^2+l^2}$
2	110					1.414
		3	111	3	111	1.732
4	200	4	200			2.000
6	211					2.449
8	220	8	220	8	220	2.828
10	310					3.162
		11	311	11	311	3.317
12	222	12	222			3.464
14	321					3.742
16	400	16	400	16	400	4.000
18	411					4.243
	330	19	331	19	331	4.359
20	420	20	420			4.472
22	332					4.690
24	422	24	422	24	422	4.899
26	431					5.099
	510	27	511, 333	27	511, 333	5.196
30	521					5.477
32	440	32	440	32	440	5.659

[a]All values of $(h^2+k^2+l^2)$ are possible except $4^p(8n+7)$, where p and n are integers including zero; thus, for example, 7, 15, and 23 are not possible.

2 Indexing Diffraction Patterns

2.1 Assignment of Indices

If a sphere of radius $\chi = \lambda^{-1}$ is drawn from the specimen so as to pass through the origin of the reciprocal lattice, it can be seen from Fig. 2.4 that the reciprocal lattice vector $\mathbf{g} = O^*B$ will lie on this sphere if $\sin\theta = (|\mathbf{g}|/2) \div (1/\lambda)$, that is, $2d\sin\theta = \lambda$, since $|\mathbf{g}| = d^{-1}$. Thus an alternative way of describing Bragg's law is to state that diffraction can occur only when the reflecting sphere passes through (or very close to) a reciprocal lattice point for which $F \neq 0$.

As shown in Table 1.2, the reflecting sphere for high energy electrons has a very large radius, and a number of reciprocal lattice points from the same reciprocal lattice section can be quite close to the reflecting sphere and be observed on an electron diffraction pattern. The distances measured on the diffraction pattern are actually magnified reciprocal lattice vectors; the magnification factor for the microscope is λL, where L is the effective camera length resulting

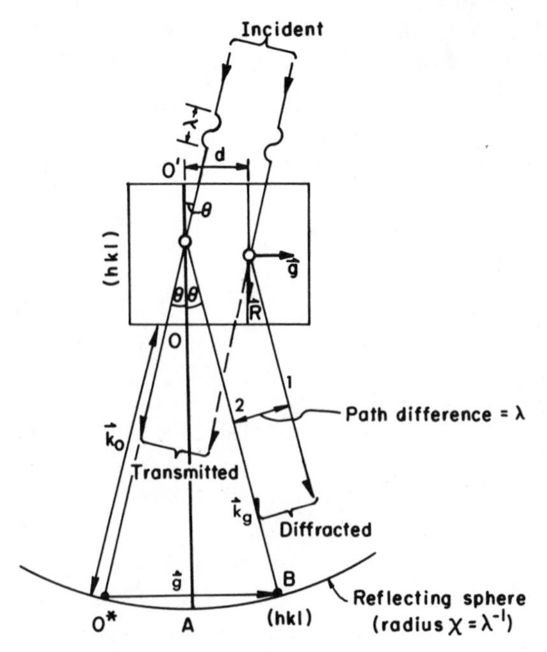

Fig. 2.4 Bragg's law and the reflecting sphere for fast electrons.

from the magnifications of the imaging lenses of the microscope column. In other words, r, the distance on the diffraction pattern from the origin to a diffracted spot (see Fig. 2.5), is given by $r = L \tan 2\theta \simeq L \sin 2\theta = L(\lambda/d) = g(\lambda L)$, using the fact that the angles involved are small. The last equality,

$$r = g(\lambda L), \tag{2.4}$$

makes electron diffraction patterns very easy to analyze; all that is needed is a ruler, a protractor, and a slide rule or pocket calculator.

The ratio between two reciprocal lattice vectors is

$$\frac{g_1}{g_2} = \frac{1/d_1}{1/d_2} = \frac{d_2}{d_1}. \tag{2.5}$$

If attention is restricted to the cubic system,

$$\frac{g_1}{g_2} = \frac{d_2}{d_1} = \frac{\sqrt{h_1^2 + k_1^2 + l_1^2}}{\sqrt{h_2^2 + k_2^2 + l_2^2}}, \tag{2.6}$$

where $h_1 k_1 l_1$ and $h_2 k_2 l_2$ are the Miller indices of the two diffracting planes giving rise to the spots on the diffraction pattern, g_1 and g_2. For example, in the fcc system the [112] diffraction pattern for electrons will be a rectangular net-

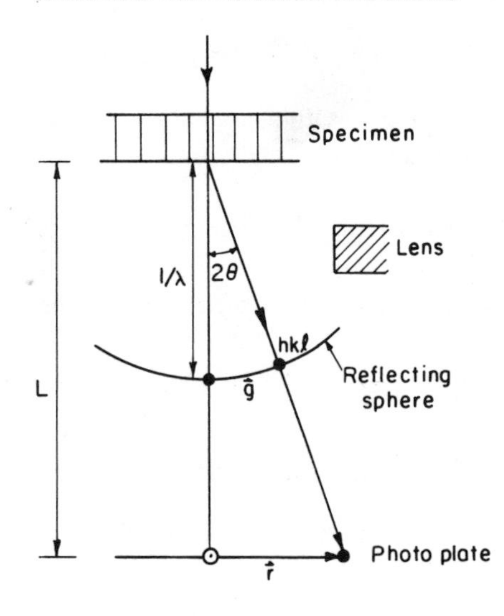

Fig. 2.5 Schematic diagram of the formation of a diffraction pattern in the TEM, with effective camera length L.

work of points lying on the reciprocal lattice section normal to [112] and containing the origin of reciprocal space. This can be seen by drawing the 112 plane in Fig. 2.2. The two shortest reciprocal lattice vectors will be in the ratios

$$\frac{g_1}{g_2} = \frac{\sqrt{3}}{\sqrt{8}} = \frac{\sqrt{1^2 + 1^2 + 1^2}}{\sqrt{2^2 + 2^2 + 0^2}}.$$

From Table 2.2 it can be seen that these arise from planes of the forms {111} and {220}. The signs on the indices and the positions of the indices in the brackets must now be chosen so that (*a*) the indices are so arranged and signed that both reciprocal lattice vectors are normal to [112], and (*b*) the angle between the two reflecting planes matches the measured angle between the two reciprocal lattice vectors, that is,

$$\cos\theta = \frac{\mathbf{g}_1 \cdot \mathbf{g}_2}{|\mathbf{g}_1|\,|\mathbf{g}_2|} = \cos(\mathbf{g}_1 \cdot \mathbf{g}_2) = 90^\circ.$$

A set of reflecting planes matching these conditions is $(11\bar{1})$ and $(\bar{2}20)$. A final check is made by taking the cross product $\mathbf{g}_1 \times \mathbf{g}_2 = [11\bar{1}] \times [\bar{2}20] = [224] \parallel [112]$. Therefore the indices $\mathbf{g}_1 = 11\bar{1}$, $\mathbf{g}_2 = \bar{2}20$ may be assigned. All reciprocal lattice points on this section can be indexed using proper multiples of these two, for example, $\mathbf{g}_3 = \mathbf{g}_1 + \mathbf{g}_2 = \bar{1}3\bar{1}$, $\mathbf{g}_4 = 2\mathbf{g}_1 + \mathbf{g}_2 = 04\bar{2}$, and $\mathbf{g}_5 = \mathbf{g}_1 - \mathbf{g}_2 = 3\bar{1}\bar{1}$. These are all normal to [112], and the resulting pattern is then completely indexed.

In similar procedure the ratios of the angles subtended by different sets of rel-vectors are compared. Examples of calculated symmetrical spot patterns are given in several of the texts listed in the Bibliography at the end of Chapter 1, and also in the exercises at the end of this chapter.

Usually, in solving the diffraction pattern, it is not known what the normal to the pattern is, and the steps for solution are as follows:

1. Measure a number of nonlinear rel-vectors, find their ratios, and assign tentative *hkl* values.
2. Measure the angle between two rel-vectors, and assign signs and position indices such that the angle is correct.
3. Perform the cross-product of the resulting rel-vectors, and determine the zone axis of the reflecting planes.
4. Assign consistent indices to the rest of the spots on the pattern.

To assist in the first of these procedures, tables of relative reciprocal lattice lengths for cubic materials are presented in Appendix C.

2.2 Curvature of the Reflecting Sphere

That the reflecting sphere for high energy electrons is not completely flat can be illustrated by using the microscope with a very small value of camera length L (Fig. 2.5). For 100 kV electrons and ordinary metal specimens, the reflecting sphere intersects rel-points on the reciprocal layer containing the origin out to about 2 or 3°, so that many reflections appear within the zero layer. At about 4 or 5° it intersects points on the next higher rel-layer (or Laue zone; eq. 2.1 satisfied with nonzero, integral, right-hand side), and at about 7°, the third rel-layer. Thus the reflecting sphere is not at all flat; but since the pattern is such a highly magnified view of a small section of reciprocal space, the approximation of a flat reflecting sphere is reasonable. As can be seen from the λ^{-1} values of Table 1.2, the reflecting sphere becomes flatter the higher the accelerating voltage of the electrons.

3 The Shape of Reciprocal Lattice Points and the Occurrence of Diffraction Spots

3.1 The Crystal Shape Factor

Since the diffraction pattern is, as will be seen below, related to the Fourier transform* of the object, the intensities of the diffracted beams depend on the shape and volume of the diffracting crystal, that is, there is a shape-factor effect.

*See, for example, ref. 1.

A spherical crystal will produce a spherical distribution of electrons about the rel-point. A disk-shaped specimen, which is roughly the shape of the irradiated part of the specimen in the electron microscope, will produce electrons distributed along a rod passing through the rel-point. These are called rel-rods and have their long axes normal to the thinnest direction of the crystal, that is, usually normal to the plane of the diffraction vectors. This phenomenon is simply a reflection of the fact that Bragg's law is rigid only for a crystal of infinite dimensions. Relaxation of the law occurs for thin crystals, and crystal planes can diffract waves at angles slightly different from the Bragg angle, although not as efficiently (see Fig. 2.7). For specimens containing planar or volume defects (e.g., small particles), two factors must be considered:

1. The shape factor due to the thinness of foils in the direction of the beam.
2. The shape factor due to the defects in the foil.

A useful physical picture of these effects is obtained by considering the diffraction pattern to be a composite formed by superposition of the diffraction patterns of the dispersed phases and the pattern of the matrix. If the structures are different, then, of course, the structure factors are also different and mixed reciprocal lattices must be considered.

The shape factor can be derived by means of the kinematical theory of diffraction, in which thin crystals and single scattering are assumed. The approximations mean that strong Bragg diffraction does not occur, that is, that the reciprocal lattice point does not coincide with the reflecting sphere.

Suppose that the rel-point is at a distance s from the sphere (Fig. 2.6*b*). The amplitude of the scattered wave is proportional to the atomic scattering amplitude times the phase factor summed over all atoms:

$$\phi_g = \phi_0 \times \sum_{\text{atoms}} f \exp(2\pi i \mathbf{k} \cdot \mathbf{r}). \tag{2.7}$$

Consider a parallelepiped crystal made up of unit cells, each of dimensions **a**, **b**, **c**, numbering N_1, N_2, N_3 along the principal axes **a**, **b**, **c** (Fig. 2.6*a*).

Let $\mathbf{r}$ be the position of atoms with respect to the origin of the crystal, $\mathbf{r}_n$ be the position of the nth unit cell with respect to the origin of the crystal, $\mathbf{r}_j$ be the position of an atom with respect to the origin of its unit cell, and $\mathbf{k} = \mathbf{g} + \mathbf{s}$ (Fig. 2.6*b*). Thus,

$$\phi_g = \phi_0 \times \sum_{\substack{\text{all} \\ \text{atoms}}} f_j \exp [2\pi i(\mathbf{g} + \mathbf{s}) \cdot (\mathbf{r}_j + \mathbf{r}_n)] \tag{2.8}$$

$$= \phi_0 \times \sum_{\substack{\text{all} \\ \text{unit} \\ \text{cells}}} \left(\sum_{\substack{\text{all} \\ \text{atoms} \\ \text{per cell}}} f_j \{\exp [2\pi i(\mathbf{g} + \mathbf{s}) \cdot \mathbf{r}_j]\} \cdot \exp [2\pi i(\mathbf{g} + \mathbf{s}) \cdot \mathbf{r}_n] \right). \tag{2.9}$$

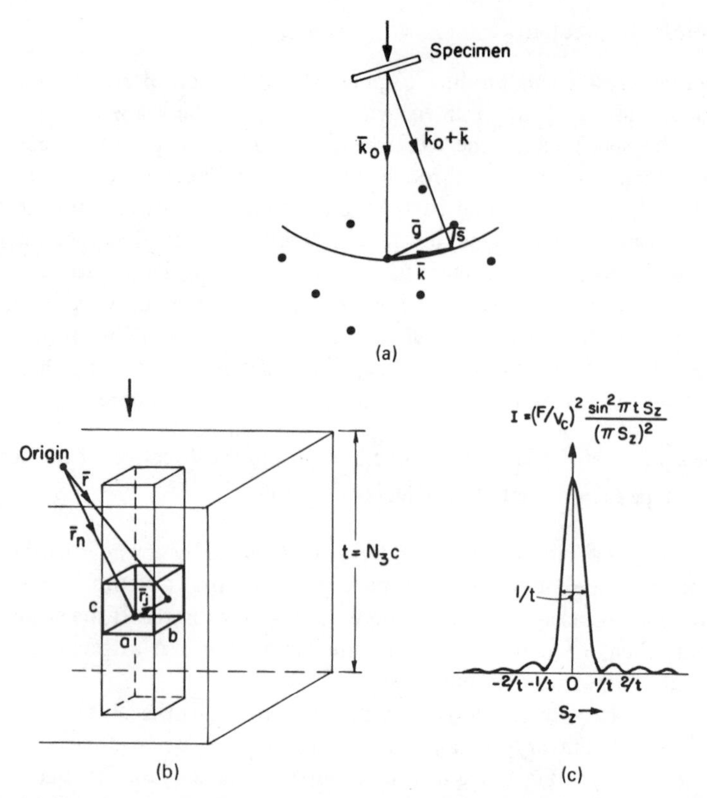

Fig. 2.6 (*a*) Showing the relation between the reciprocal lattice and reflecting sphere when Bragg's law is not exactly satisfied. (*b*) Sketch of a column of unit cells in a parallelepiped crystal for calculating the interference function. (*c*) The interference function along the s_z direction, showing the kinematical intensity distribution for thin foils.

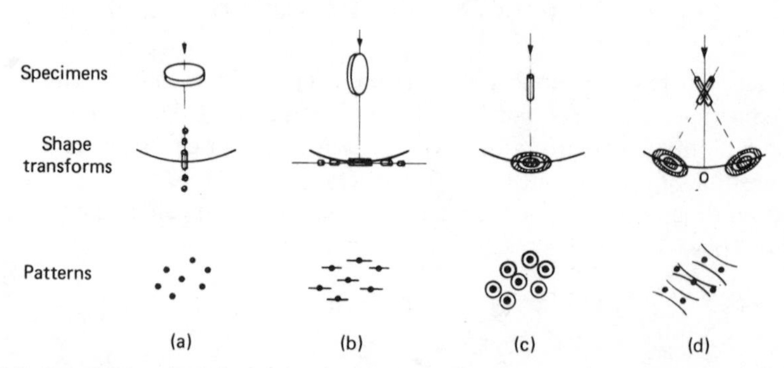

Fig. 2.7 Interference functions for crystals of small sizes, showing the effect of the shape factor on the diffraction pattern. Thin plates (*a*) normal to the beam and (*b*) parallel to the beam; needles (*c*) parallel to the beam and (*d*) inclined to the beam. Notice curved streaks in (*d*) (compare with Fig. 2.8).

Now the term

$$\sum_{\substack{\text{all}\\\text{atoms}\\\text{per cell}}} f_j\{\exp[2\pi i(\mathbf{g}+\mathbf{s})\cdot\mathbf{r}_j]\}$$

does not depend on the shape of the crystal, and at $s = 0$ this becomes the structure factor F_g (eq. 2.3). Since $|s| \ll |g|$, the dependence on s is not very strong; hence

$$\sum_{\text{cell}} f_j\{\exp[2\pi i(\mathbf{g}+\mathbf{s})\cdot\mathbf{r}_j]\} \simeq F_g.$$

Thus

$$\phi_g = \phi_0 \times \sum_{\substack{\text{all}\\\text{unit}\\\text{cells}}} F_g\{\exp[2\pi i(\mathbf{g}\cdot\mathbf{r}_n)]\exp[2\pi i(\mathbf{s}\cdot\mathbf{r}_n)]\}. \qquad (2.10)$$

Since $\mathbf{g}\cdot\mathbf{r}_n$ = integer, $\exp[2\pi i(\mathbf{g}\cdot\mathbf{r}_n)] = 1$.

Also, the quantity $\mathbf{s}\cdot\mathbf{r}_n$ does not change appreciably from cell to cell. Thus approximating the sum by an integral gives

$$\phi_g = \phi_0 F_g \int_{\text{crystal}} \frac{1}{V_c}\exp[2\pi i(\mathbf{s}\cdot\mathbf{r}_n)]\cdot dV, \qquad (2.11)$$

where V_c is the volume of the unit cell.

By definition $\mathbf{r}_n = u\mathbf{a} + v\mathbf{b} + w\mathbf{c}$ and

$$\mathbf{s} = s_x\frac{\mathbf{a}^*}{a^*} + s_y\frac{\mathbf{b}^*}{b^*} + s_z\frac{\mathbf{c}^*}{c^*}.$$

Hence

$$\frac{\phi_g}{\phi_0} = \frac{F_g}{V_c}\int_0^{N_1 a}\int_0^{N_2 b}\int_0^{N_3 c}\exp\left[2\pi i\left(s_x\frac{u}{a^*} + s_y\frac{v}{b^*} + s_z\frac{w}{c^*}\right)\right]dx\,dy\,dz, \qquad (2.12)$$

which is a Fourier transform. Equation 2.12 integrates to

$$\frac{\phi_g}{\phi_0} = \frac{F_g}{V_c}\frac{\sin\pi s_x N_1 a}{\pi s_x}\frac{\sin\pi s_y N_2 b}{\pi s_y}\frac{\sin\pi s_z N_3 c}{\pi s_z},$$

which is known as the interference function.

For a crystal in the form of a thin plate $N_3 \ll N_2, N_1$, and for $s_x = s_y = 0$

$$\frac{\phi_g}{\phi_0} = \frac{F_g}{V_c}\frac{\sin\pi s_z N_3 c}{\pi s_z} \qquad (2.13)$$

Now $N_3 c = t$, the thickness of the plate, and thus the intensity is

$$\frac{I_g}{I_0} = \left(\frac{F_g}{V_c}\right)^2 \frac{\sin^2 \pi t s_z}{(\pi s_z)^2}. \tag{2.14}$$

This well-known function is shown in Fig. 2.6*c*. It means that for the thin foils used in electron microscopy the intensity distribution about the reciprocal lattice points is in the form of a rod of length $(1/t)$. Likewise the shape transform (intensity distribution) for needles is in the form of disks (Fig. 2.7). These effects are very important in studies of precipitation; two well-known examples are the formation of small platelets of Guinier-Preston (GP) zones and θ'' on {001} in Al-Cu and needles in ⟨001⟩ in Al-Mg-Si. Examples are shown in Fig. 2.8.

The reciprocal lattice for the matrix as shown in Fig. 2.2 should thus be modified for thin foils so that the "points" are rel-rods with the axis of the rods parallel to the incident beam, as shown in Figs. 2.9*a* and *b*. If the crystal is very

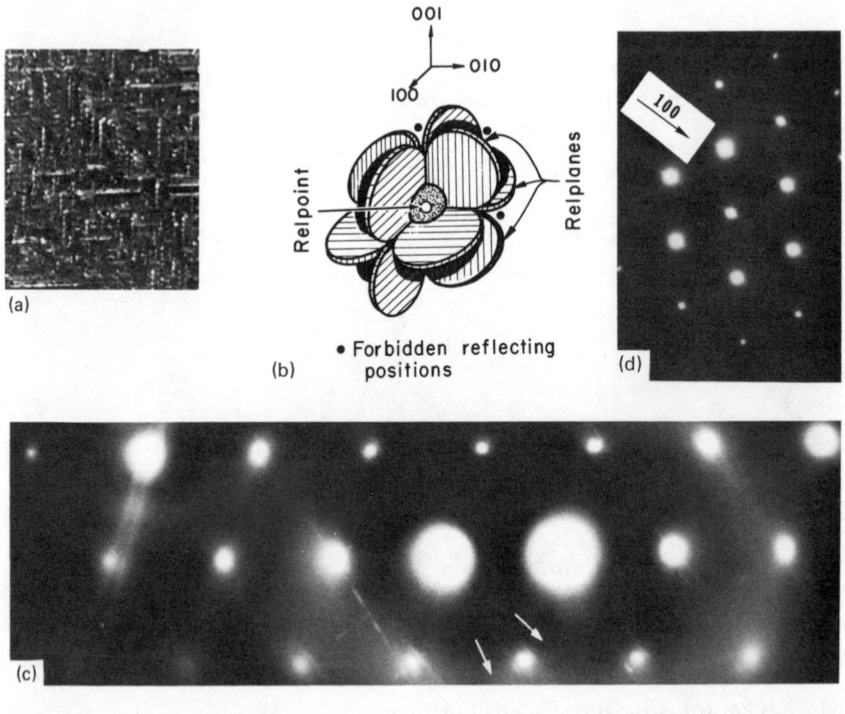

Fig. 2.8 (*a*) Dark field image of needles in ⟨001⟩ in an aged Al-Mg_2-Si alloy. (*b*) Interference function for three ⟨001⟩ needles. (*c*) [011] Diffraction pattern showing curved streaks for needles along [001] and [010] at 45° to beam (cf. Fig. 2.7*d*). (*d*) [011] Diffraction pattern for Al-Cu containing thin plates (of GP zones) on [001]; notice straight streaks in [001] from (100) plates (cf. Fig. 2.7*b*). Courtesy North-Holland.[35]

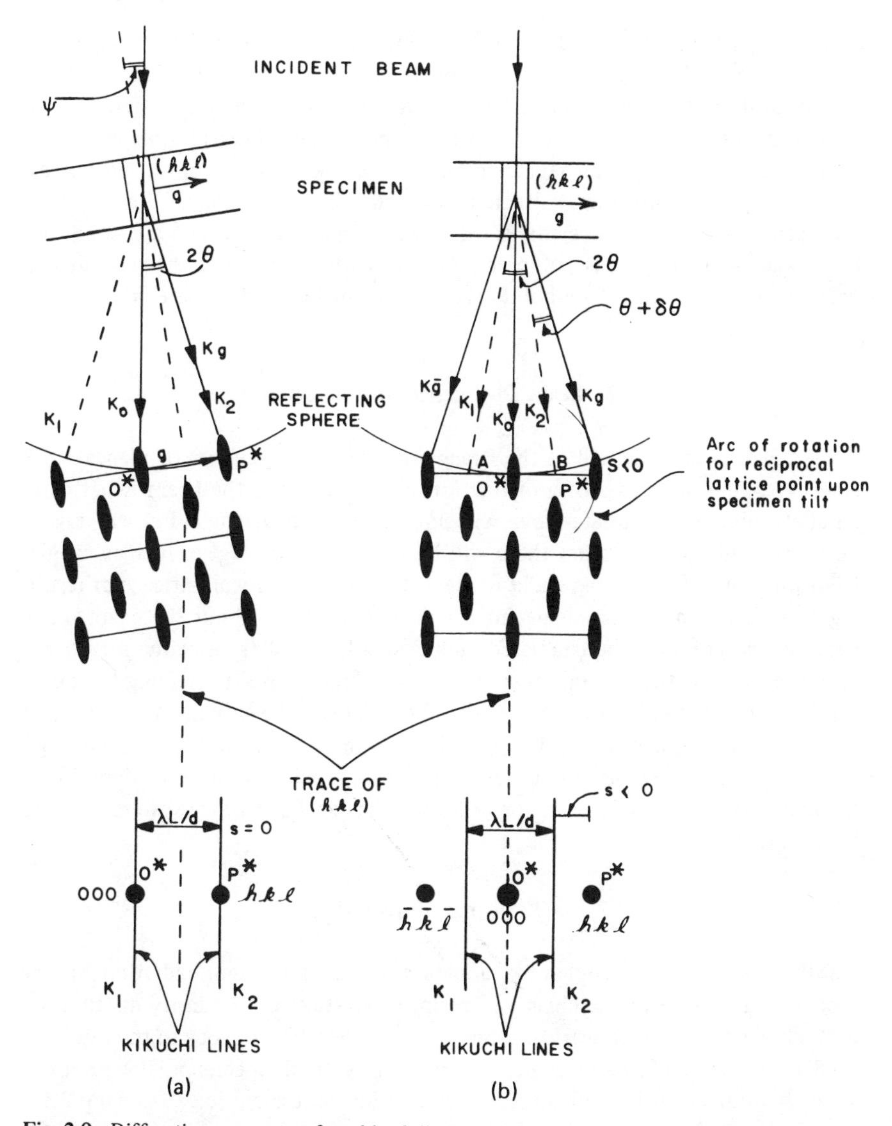

Fig. 2.9 Diffraction geometry for thin foils (*a*) oriented for two-beam excitation and (*b*) symmetric orientation. For thick foils, Kikuchi lines will also appear as shown. Courtesy *Physica Status Solidi.*[3]

thin, it is possible for rel-rods from the upper and lower levels of the reciprocal lattice to extend sufficiently to cut the reflecting sphere, giving spots at positions that do not correspond to allowed reflections. To check for this effect, one merely moves into thicker parts of the foil, where such spots will disappear.

Thus sufficiently thin planar defects (twins, stacking faults, or second phases) will give rise to rel-rods in the reciprocal lattice. The direction of the rel-rods is normal to the platelet since this is the direction along which the Laue condition is relaxed, and in the case of precipitation in the *early* stages the rel-rods will pass through the matrix rel-points (i.e., the form factor dominates the diffraction pattern).

3.2 Beam Divergence or Convergence

A crystal may be oriented at the Bragg angle (for a certain set of planes) with respect to the optic axis of a diffraction unit, but unless the beam is perfectly parallel, the entire irradiated area will not be at the Bragg angle. For a divergent or convergent incident beam there will be a continuous range of incident angles from one side of the Bragg angle to the other. The reciprocal lattice spot resulting from such an irradiated specimen will be spread out about the exact reciprocal lattice point. If the irradiated area is small, subsidiary maxima may occur on either side of the central maximum (exact Bragg condition) along the direction of the reciprocal lattice vector (Fig. 2.6*c*). This is the situation of convergent beam diffraction, which was discussed in Chapter 1, Section 4.9; compare Figs. 2.6*c* and 1.29. If the irradiated area is too large, so that it is of nonuniform thickness, or contains defects, for example, the idealized fringe systems tend to wash out.

3.3 Primary Extinction

Another possibility is to have a constant angle of incident radiation on the specimen and observe the change in reciprocal lattice spot intensity distribution with changing foil thickness. In a manner completely analogous to the intensity oscillations with changing angle in a uniformly thick specimen, the intensity varies periodically with thickness at constant incident angle [e.g., from eq. 2.14, the kinematical intensity formula, whenever $s = n/t$ for constant t, or $t = n/s$ for constant s, there are intensity minima (Fig. 2.6*c*)]. The kinematical theory predicts a central maximum at the Bragg angle for any foil thickness; actually, because of dynamical effects, there can be an intensity minimum surrounded by two maxima for certain thicknesses of foil. Thus the shape of the reciprocal lattice spot and the intensity distribution about the exact rel-point depend critically upon the thickness of the irradiated specimen. Although the rel-rods, due to small thickness, are still present, the distribution of intensity along them can be vastly altered by changing the thickness by a relatively small amount. This phenomenon, called primary extinction, gives rise to fringe contrast at inclined defects, holes, wedges, and so on.

4 Information from Diffraction Spot Patterns

The diffraction patterns provide basic crystallographic information such as the orientation of the specimens, orientation relationships between crystals, and qualitative phase identification; and if Kikuchi patterns are utilized, geometrical structure factor analysis is possible, as well as the obtaining of other information (see Section 5). The diffraction pattern is the starting point for all electron microscopy; and, in general, one of the two orientation conditions shown in Fig. 2.9 is utilized. The Kikuchi lines denoted in Fig. 2.9 are discussed in Section 5.

The accuracy of analysis of a diffraction pattern depends upon the accuracy of measurement and is discussed in many of the books cited in Chapter 1 and in several other publications (e.g., refs. 2–5). Some factors of importance are (*a*) the shape factor as described above, which determines the shape of the intensity distribution about the rel-points and the orientation of the rel-rods in relation to the reflecting sphere (Figs. 2.9*a* and *b*); (*b*) instrumental alignment and beam divergence (the condenser lens should be defocused to obtain, as closely as possible, parallel illumination; (*c*) specimen perfection (elastic and plastic strains); (*d*) curvature of the reflecting sphere and relative orientation of the foil; and (*e*) double diffraction, giving rise to reflections of zero structure factor.

Two other points should be made with regard to the interpretation of spot patterns:

1. The assignment of directions is arbitrary; for example, a fourfold symmetrical pattern can be indexed in six possible ways: 100, 010, 001 (and negatives). Thus the diffraction pattern of any crystal lying within the zero Laue zone (basal plane of the reciprocal lattice) will appear the same even if the crystal is rotated 180° about the incident beam so that the orientation has changed. Thus all spot patterns contain a 180° ambiguity. This can be removed only by trace analysis or by using Kikuchi patterns (see Section 5.5.1).
2. With high index patterns some patterns of different zone axes are identical (e.g., $[\bar{5}\bar{4}7]$ and $[815]$ fcc, $u^2 + v^2 + w^2 = 90$). Although one rarely works in high index orientations, the effect should be recognized.

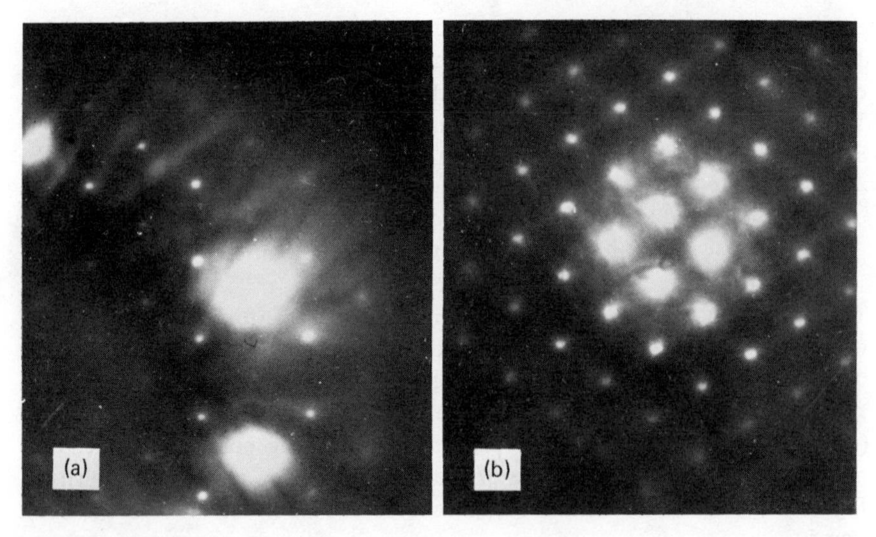

Fig. 2.10 Spot patterns from silicon corresponding to Figs. 2.9*a* and *b*; notice the appearance of the forbidden 200 spots in the [011] pattern in (*b*) due to double diffraction, which is absent in (*a*).

4.1 Double Diffraction

The structure factor determines that certain reciprocal lattice points, on an otherwise regular reciprocal lattice, have systematic absences, that is, zero intensity. It is necessary to determine these missing reflections to characterize the structure. However, each diffracted beam in the crystal behaves to some extent as an incident beam and can diffract electrons to a point on the diffraction pattern forbidden by the structure factor rules, especially in simultaneous orientations where several different reflections are excited. For example, the 002 reflection in the diamond cubic structure is not allowed, but if a [110] foil is viewed the reflection $1\bar{1}1$ acts as a primary beam, diffracting electrons from the $(\bar{1}11)$ planes; this gives $\mathbf{g}_1 + \mathbf{g}_2 = 1\bar{1}1 + \bar{1}11 = 002$, as shown in Fig. 2.10*b*. If the foil has [100] orientation, the two smallest reciprocal lattice vectors allowed are 022 and $02\bar{2}$, so that the 002 reflection cannot appear in this pattern. Similarly, the 0001 reflection in hcp crystals, which is not allowed, arises through the combination of two reciprocal lattice vectors, as in the $[11\bar{2}0]$ orientation $\mathbf{g}_1 + \mathbf{g}_2 = 0\bar{1}10 + 01\bar{1}1 = 0001$ and in $[\bar{1}2\bar{1}3]$, $\bar{1}101 + \bar{1}010 = \bar{2}111$. This phenomenon of rediffracting a diffracted beam is very common and is called double diffraction. Although it is sometimes difficult to avoid these effects, tilting the foil so as to remove one of the diffracted beams, which is required for double diffraction, will remove the forbidden reflection (e.g., Fig. 2.10*a*).

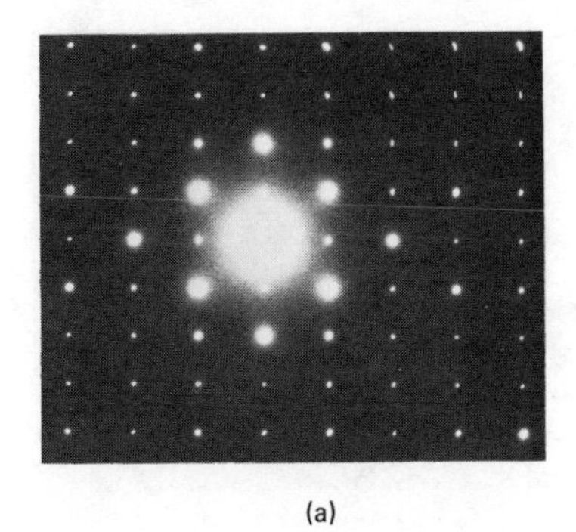

(a)

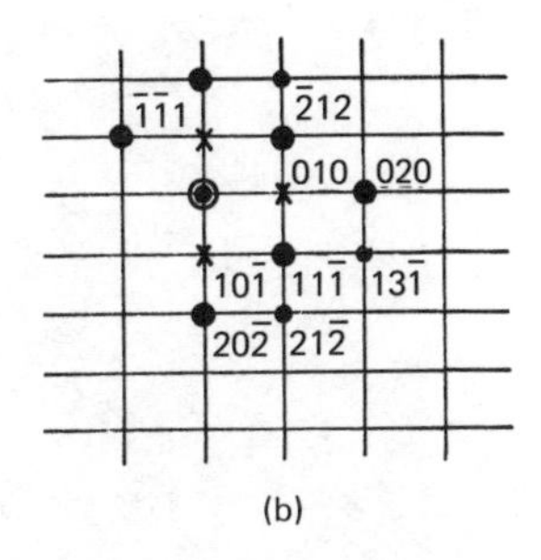

(b)

Fig. 2.11 (*a*) Experimental diffraction pattern from anthracene (monoclinic $P_{21/a}$) for [101] orientation. Courtesy G. M. Parkinson. (*b*) Reciprocal lattice section relevant to (*a*) to illustrate "allowed" (·) and "forbidden" (x) diffraction spots. The size of the "allowed" spot is an indication of the strength of the reflection.

Further examples of the occurrence of double diffraction are illustrated in Fig. 2.11. Fig. 2.11*a* shows the [101] pattern observed in anthracene, which is monoclinic. The allowed spots (space group $P_{21/a}$) are sketched in Fig. 2.11*b*, where it can be seen that the forbidden spots (e.g., 010, $10\bar{1}$) may be produced by a number of allowed double events (010 by $13\bar{1} + \bar{1}\bar{1}1$ or $20\bar{2} + \bar{2}12$, etc.). Figure 2.11*a* also illustrates the highly variable structure factors for the allowed reflections found in complicated, low symmetry structures; the spot intensities in a sufficiently thin, symmetrically bent crystal are roughly proportional to F^2.

Double diffraction can be especially problematic when working with specimens containing more than one phase. Interaction between diffracted beams from matrix and precipitates can give rise to extra spots which, if indexed as due to second phases, can lead to erroneous results such as fictitious *d*-spacings.[6] Figure 2.12 shows an example from a maraging steel in which dark field analysis has been used to interpret the pattern. The presence of doubly diffracted beams is seen from these analyses.

It should be noted that, unlike the monoclinic system of Figs. 2.11*a* and *b*, in fcc or bcc crystals double diffraction does not introduce extra spots since the combination of any two diffracted beams generates only allowed reflections. For example, for bcc $110 \pm 200 \rightarrow 310$ or $\bar{1}10$. Although no extra spots will be observed, double diffraction will obviously increase the intensity of spots where

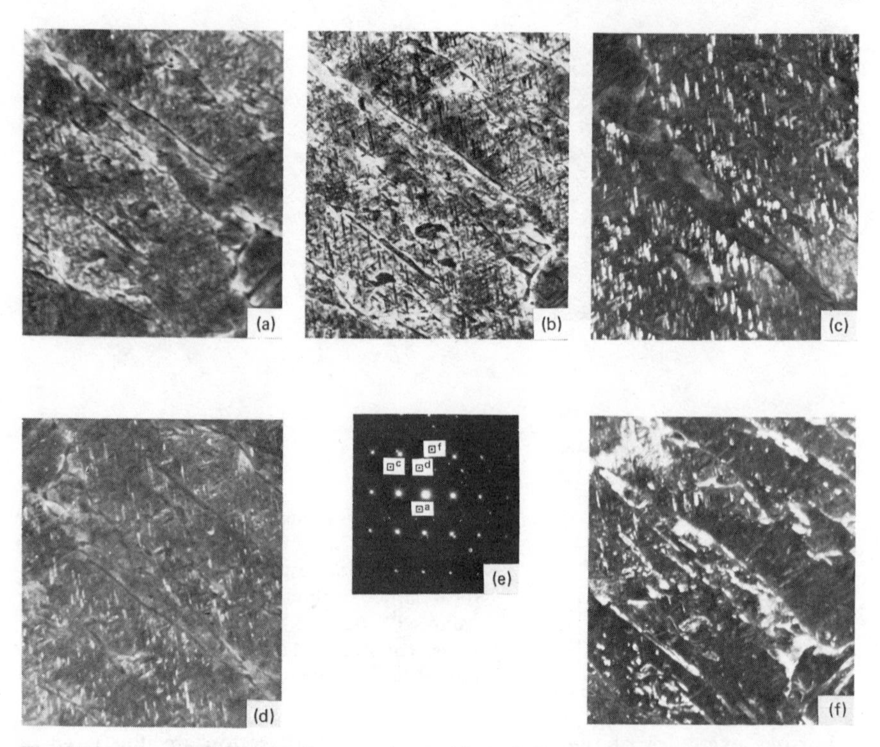

Fig. 2.12 Maraged Fe-Ni-Co alloy specimen. (*a*) Dark field image from double diffraction spot *a*. (*b*) Bright field image. (*c*) Dark field image from spot *c* (fcc 220 type) shows only one of the austenite sets on (110). (*d*) dark field image from double diffraction spot *d*. (*e*) Symmetrically oriented $[0\bar{1}1]$ bcc diffraction pattern. Spots not at bcc or fcc positions are caused by multiple diffraction. (*f*) Dark field image from spot *f* (fcc 220 type). Copyright American Society for Metals, 1969.[6]

superposition occurs. However, double diffraction in twinned fcc or bcc crystals can introduce extra spots.

4.2 Moiré Patterns

A special case of double diffraction or, rather, of two single diffraction events, one from each layer, occurs from overlapping crystals, as in composite films or in two- (or more) phase systems. Two general cases occur.

First, parallel moirés are formed from parallel reflecting planes of different spacings corresponding to reciprocal lattice vectors $\mathbf{g}_1$ and $\mathbf{g}_2$, differing only in magnitude. In practice it may be difficult to resolve two separate rel-points if $\mathbf{g}_1$ and $\mathbf{g}_2$ are almost equal. The effective reciprocal lattice vector for the composite is thus $\Delta\mathbf{g} = \mathbf{g}_1 - \mathbf{g}_2$ (for $\mathbf{g}_1 > \mathbf{g}_2$) corresponding to a moiré image spacing

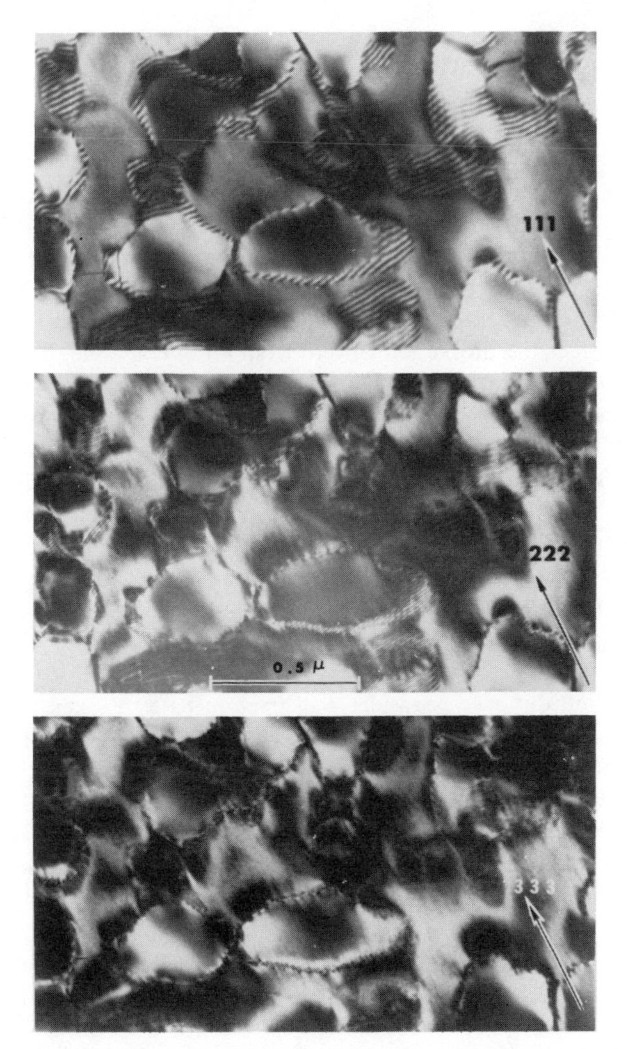

Fig. 2.13 Parallel moiré patterns in aged Cu-Mn-Al alloy. Notice how the fringe spacing varies inversely as Δg for the 111, 222, and 333 reflections. Courtesy M. Bouchard.

$D_p = |\mathbf{g}_1 - \mathbf{g}_2|^{-1}$ and lying normal to $\mathbf{g}_1$ and $\mathbf{g}_2$. An example for a Cu-Mn-Al alloy containing coherent phases utilizing 111, 222, and 333 reflections is shown in Fig. 2.13. It can be seen that D_p varies inversely as $|\mathbf{\Delta g}|$.

Second, a rotational moiré is formed when planes having equal spacing d, but mutually rotated through an angle α, diffract together. In this case the effective reciprocal lattice vector is $g \sin \alpha$, and the moiré image spacing D_r is $(g \sin \alpha)^{-1}$, that is, d/α. Mixed moirés can occur because of two overlapping gratings of different spacings which are relatively twisted by a small angle.

4.3 Faulted Crystals

Since in electron diffraction it is possible to consider diffraction from a narrow column of crystal (e.g., in the case of cubic close-packed crystals), then even for a single intrinsic stacking fault, where four layers are in hcp stacking: $ABCAB \mid CACA \mid BCA$, there may be a sufficient volume of hcp material present so that the layers can be regarded as a thin platelet of hcp structure. One can then consider the diffraction pattern from a foil containing a fault in terms of two reciprocal lattices: one "normal" pattern corresponding to the matrix, and the other consisting of the streaked hcp reciprocal lattice corresponding to the thin fault, superimposed.[3] Since the fault plane is one of the four $\{111\}$ in fcc and because unique crystallographic relations exist between fcc and hcp lattices, the rel-rods will lie along $\langle 0001 \rangle$ hcp parallel to and coincident with $\langle 111 \rangle$ fcc continuously throughout reciprocal space. These streaks, which can be considered to originate from hcp reciprocal lattice points, will thus pass through all matrix reflections contained in the particular $[111]$ and parallel zones. Examples for $[\bar{1}01]$ and $[112]$ orientations are sketched in Figs. 2.14*a* and *b*.

In support of this view, consider a diffraction pattern taken from a single fault lying parallel to the incident beam (Fig. 2.14*a*). Figure 2.15 illustrates such a case for growth faults in silicon. The foil is oriented with $[\bar{1}01]$ parallel to the incident beam, and the "edge-on" faults on (111) and $(\bar{1}1\bar{1})$ are joined by a fault on $(11\bar{1})$ or $(\bar{1}11)$, inclined at 35° to the beam. By placing a small field-limiting aperture over the edge-on fault at A, the diffraction pattern in Fig. 2.15*b* was obtained. It can be seen that streaks are visible along $[111]$ and pass through 000 and the other rel-points. These streaks can be attributed to the stacking fault acting as a thin (hcp) platelet; the structure is indeterminate in the direction of streaking, as no maxima are visible along $[111]$. If a larger fraction of edge-on faults were to contribute to the pattern, or if there were regular faulting on alternate (111) planes over a sufficient volume of crystal, the $[\bar{1}01]$ pattern would contain resolvable hcp maxima as sketched in Fig. 2.14*a*.

Figure 2.15*c* is a diffraction pattern taken across the inclined fault B shown in Fig. 2.15*a* after the foil was tilted into a strong 202 beam case (i.e., the reflecting sphere passed exactly through the 202 rel-point). Doublets (arrowed) can be resolved in the pattern. These doublets are due to the effects of both matrix and fault rel-rods. The orientation of Fig. 2.15*c* is then almost exactly $[\bar{3}\bar{1}3]$, and this may also be taken as the normal to the foil surface. There are two possible streaking directions, depending on whether the fault is on $(\bar{1}11)$ or $(11\bar{1})$. By

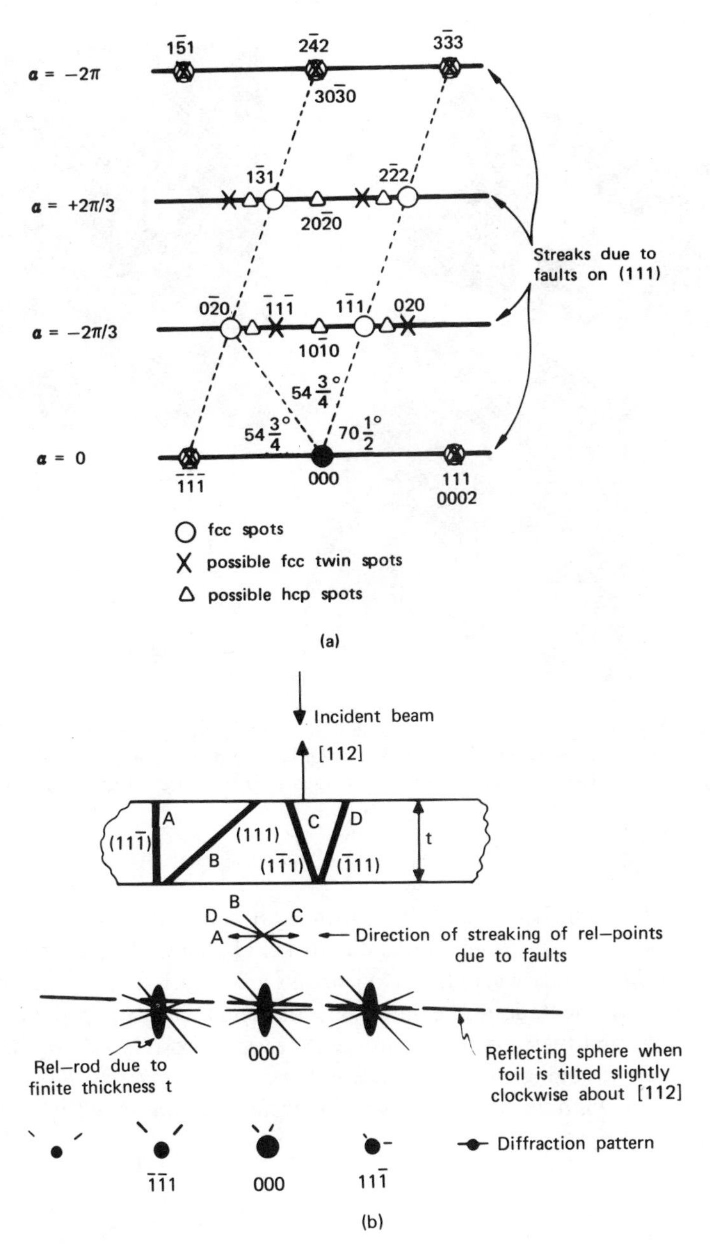

Fig. 2.14 (*a*) Calculated [$\bar{1}$01] pattern for a fcc crystal containing thin faults or hcp precipitates (with ideal *c*/*a* ratio) on (111). Twin spots are also shown. Courtesy *Physica Status Solidi.*[3] (*b*) As (*a*) but for a [112] foil containing faults or precipitates on all four {111}, showing the effect of rel-rods on the diffraction pattern.

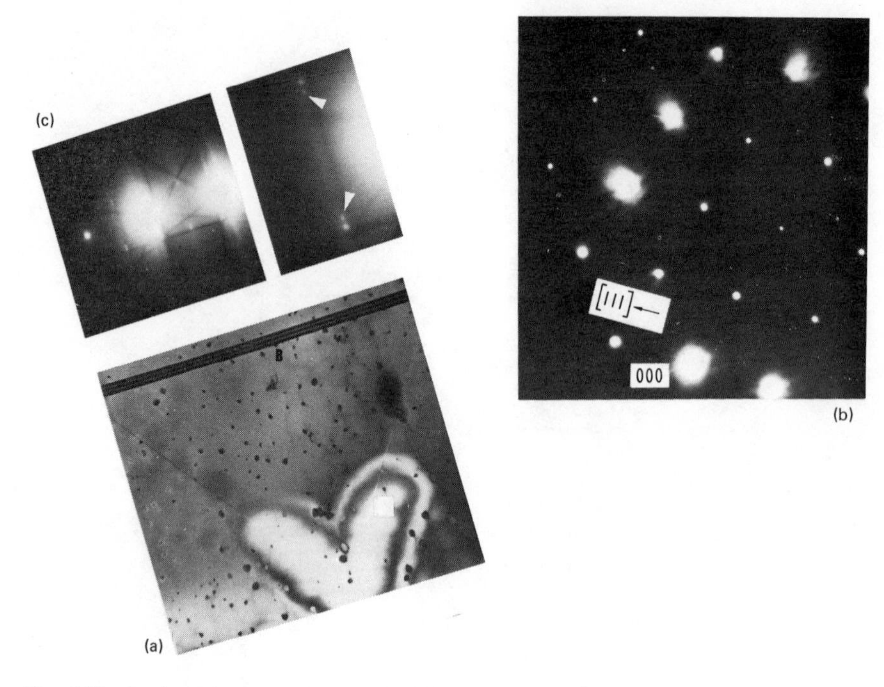

Fig. 2.15 Stacking faults in silicon: (*a*) bright field image, foil near [101]; (*b*) diffraction pattern from region *A* (compare with Figure 2.14*a*); (*c*) diffraction pattern (after tilting) corresponding to region of inclined fault *B* (compare with Figure 2.14*b*). Courtesy *Physica Status Solidi.* [3]

suitable projection, it can easily be shown that the spots to the inside of the doublets can arise only from streaks due to a $(11\bar{1})$ fault. The positions of these inner spots are exactly where the $[11\bar{1}]$ streaks are expected to cut the reflecting sphere for this orientation. The outer spots of the doublets arise from foil thickness rel-rods, and their distance from the inner spots corresponds to that calculated for rel-rods along $[\bar{3}\bar{1}3]$. *This example serves to emphasize the importance of precise orientation determinations for explaining fine detail in diffraction patterns.*

4.4 Twin Patterns

Twinning is a phenomenon commonly observed as a result of plastic deformation, and recrystallisation as well as occurring during martensitic and bainitic phase transformation (e.g., in ferrous alloys and steels), and is of considerable importance in controlling mechanical properties. The analysis of twinned diffraction patterns is facilitated by stereographic projection or by calculation using matrix algebra, together with dark field microscopy.[7-9]

A twin can be obtained by shear such that all atomic sites on one side of the twin boundary are in mirror image relationship to those on the other. For fcc and bcc crystals the twin planes are {111} and {112}, respectively. An alternative description of twinning is 180° rotation about the twin plane normal (twin axis). These relationships enable the matrix of the transformation to be readily derived. Since the indices of reciprocal lattice points are the Miller indices of the diffracting planes, the reciprocal lattice for the twinned crystal will be related to that of the original crystal by the twinning matrix. The indices of a reciprocal lattice point *PQR* for the twinned crystal will be related to the point *pqr* in the reciprocal lattice of the original crystal after twinning on the (*hkl*) plane [rotation of 180° about the normal to (*hkl*)] by the following general expressions:

$$P = \frac{p(h^2 - k^2 - l^2) + q(2hk) + r(2hl)}{(h^2 + k^2 + l^2)}, \tag{2.15}$$

$$Q = \frac{p(2hk) + q(-h^2 + k^2 - l^2) + r(2kl)}{(h^2 + k^2 + l^2)}, \tag{2.16}$$

$$R = \frac{p(2hl) + q(2kl) + r(-h^2 - k^2 + l^2)}{(h^2 + k^2 + l^2)}. \tag{2.17}$$

In matrix form this can be written as $(PQR) = T_{hkl} \cdot (pqr)$. Therefore the general twinning matrix for the cubic system is

$$T_{(hkl)} = \frac{1}{(h^2 + k^2 + l^2)} \begin{pmatrix} h^2 - k^2 - l^2 & 2hk & 2hl \\ 2hk & -h^2 + k^2 - l^2 & 2kl \\ 2hl & 2kl & -h^2 - k^2 + l^2 \end{pmatrix}. \tag{2.18}$$

The indices for reflection in the twin plane are obtained by multiplication of this matrix by

$$\begin{vmatrix} \bar{1}00 \\ 0\bar{1}0 \\ 00\bar{1} \end{vmatrix},$$

that is, the indices are *PQR* after rotation, but $\overline{P}\overline{Q}\overline{R}$ after reflection. (The reader can prove this by using Fig. 2.1.)

In the fcc systems twinning occurs on {111} planes, and so, in general,

$$T_{\{111\}} = \frac{1}{3}\begin{pmatrix} -1 & 2hk & 2hl \\ 2hk & -1 & 2kl \\ 2hl & 2kl & -1 \end{pmatrix},$$

For example,

$$T_{(111)} = \frac{1}{3}\begin{pmatrix} -1 & 2 & 2 \\ 2 & -1 & 2 \\ 2 & 2 & -1 \end{pmatrix}.$$

In bcc materials twinning occurs on {112} planes; hence

$$T_{(112)} = \frac{1}{3}\begin{pmatrix} \dfrac{h^2 - k^2 - l^2}{2} & hk & hl \\ hk & \dfrac{-h^2 + k^2 - l^2}{2} & kl \\ hl & kl & \dfrac{-h^2 - k^2 + l^2}{2} \end{pmatrix}.$$

The appropriate plane indices can be substituted for each case (e.g., Table 2.3). For either fcc or bcc twinning the elements inside the matrices shown are

Table 2.3 Twin Indices *PQR* of Matrix Reflections *pqr* for the First Two Sets of Allowed Reflections in Fcc and Bcc Crystals[a]

Fcc		Bcc	
pqr	*PQR* [(*hkl*) = (111)]	*pqr*	*PQR* [(*hkl*) = (112)]
333	$\bar{3}\bar{3}\bar{3}$	330	$\bar{1}\bar{1}4$
$33\bar{3}$	$\bar{1}\bar{1}5$	303	033
$3\bar{3}3$	$\bar{1}5\bar{1}$	033	303
$\bar{3}33$	$5\bar{1}\bar{1}$	$\bar{3}30$	$3\bar{3}0$
600	$\bar{2}44$	$30\bar{3}$	$4\bar{1}1$
060	$4\bar{2}4$	$03\bar{3}$	$\bar{1}\bar{4}1$
006	$44\bar{2}$	600	$\bar{4}24$
		060	$2\bar{4}4$
		006	442

[a]The indices are obtained by the rotation matrix. For reflection (as in stereograms) the *PQR* indices are the negatives of those shown.

integers. Then, because of the factor $\frac{1}{3}$ outside the matrix, all third-order reciprocal lattice spots for the matrix will coincide with allowed reciprocal lattice spots for the twin. Table 2.3 illustrates, for the third order of the two shortest reciprocal lattice vectors in each system, the twin reciprocal lattice points obtained by transformation of the original reciprocal lattice points, that is, in a diffraction pattern from both crystals these diffracted spots will coincide (as plotted in Fig. 2.14*a*). It should be noted that, if the twinned region is very narrow, the twin reciprocal lattice will be streaked along $\langle 111 \rangle$.

Reflections of different indices coincide when the sum of the squares of their indices are the same since then the Bragg angles are also the same. Thus {333} and {511} coincide in fcc patterns ($h^2 + k^2 + l^2 = 27$), and {411} and {330} in bcc patterns ($h^2 + k^2 + l^2 = 18$). Notice that, if a 511 fcc matrix spot twins to 333, then 222 twin and 111 twin spots also appear in the pattern. Such a case exists with $\langle 105 \rangle$ orientations, and an example for double twinning in shock-loaded copper is given in Fig. 2.16. This example also illustrates how dark field imaging enables twin spots to be identified. The interpretation by stereographic projection is shown in Fig. 2.17. Figure 2.18 shows another example of (112) twinning in Fe-Ni martensite (bcc), in which extra spots from double diffraction also occur.[10] The spots are streaked because of the fineness of the twin lamellae (shape-factor effect).

In fcc crystals it can be seen that twin spots either coincide with matrix spots or are positioned one third along $\langle 111 \rangle$ directions. For example, in fcc with twin plane (hkl)

$$(\bar{1}11) \longrightarrow \tfrac{1}{3}(5\bar{1}\bar{1})_T = 200 - \tfrac{1}{3}(111),$$
$$200 \longrightarrow \tfrac{1}{3}(\bar{2}44)_T = \bar{1}11 + \tfrac{1}{3}(111),$$
$$\bar{2}44 \longrightarrow (600)_T; \quad \text{see also Table 2.3.}$$

The twin points that do not coincide with matrix points are thus displaced from matrix points by vectors of $\pm\frac{1}{3}\langle 111 \rangle$. However, not all of the one-third points are occupied, as can be shown from the following. The twin points $u_1v_1w_1$ of matrix points uvw are $u_1 = (u \pm \frac{1}{3}h)$, $v_1 = (v \pm \frac{1}{3}k)$, $w_1 = (w \pm \frac{1}{3}l)$. In view of the structure factor rule for fcc crystals ($h_1k_1l_1$ or $u_1v_1w_1$ must be all odd or all even), and the fact that $\mathbf{g}(uvw)$ must equal $\mathbf{g}_1(u_1v_1w_1)$, then

$$(u_1^2 + v_1^2 + w_1^2) - (u^2 + v^2 + w^2) = 4N, \tag{2.19}$$

where N = integer for $u_1v_1w_1$ and uvw, both all even or both all odd.

Also,
$$(u_1^2 + v_1^2 + w_1^2) - (u^2 + v^2 + w^2) = 2N + 1 \tag{2.20}$$

for $u_1v_1w_1$ all even, uvw all odd, and vice versa. Hence, if uvw are all even, the twin point $u_1v_1w_1$ must have all odd indices, and vice versa. It therefore follows

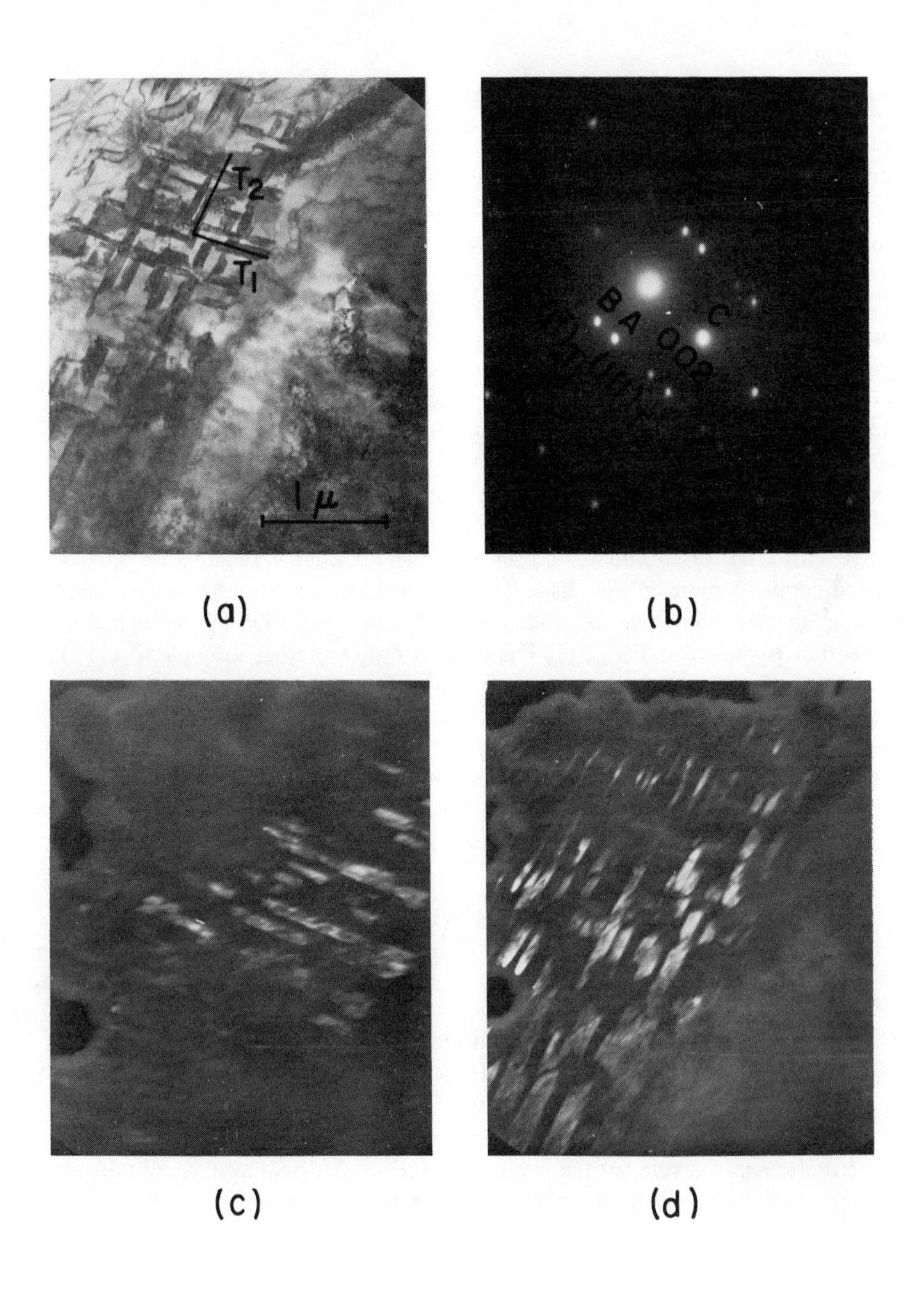

Fig. 2.16 Twins in copper as a result of explosive deformation: (*a*) bright field image, (*b*) diffraction pattern indexed, (*c*, *d*) dark field images using the different twin spots T_1 and T_2, showing contrast reversal of twins; Orientation is $[5\bar{1}0]$. Courtesy O. Johari.

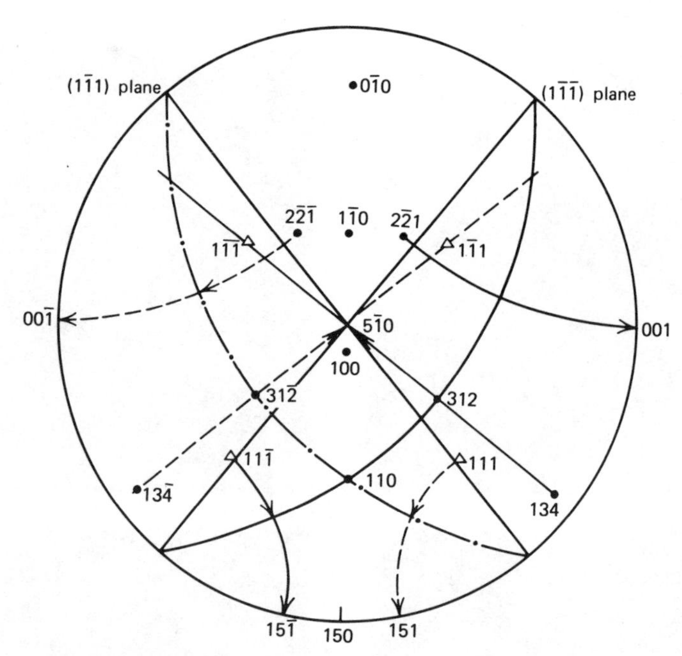

Fig. 2.17 Stereographic projection corresponding to Fig. 2.16, showing that the twin planes must be $(1\bar{1}\bar{1})$ and $(1\bar{1}1)$. Courtesy O. Johari.

that the allowed twin point $(u \pm \frac{1}{3}h)$, $(v \pm \frac{1}{3}k)$, $(w \pm \frac{1}{3}l)$ for the $2N + 1$ condition is given by the following selection rule:

$$hu + kv + lw = 3N + 1. \tag{2.21}$$

This rule determines whether a point $(u \pm \frac{1}{3}h)$, $(v \pm \frac{1}{3}k)$, $(w \pm \frac{1}{3}l)$ is occupied by a twin spot. Similar arguments apply to bcc and hcp lattices, and the topic is discussed in detail in refs. 7 and 8.

The stereographic projection is also convenient for analyzing twinning.[9] The twin plane great circle is drawn, and the reflections that will appear in the diffraction pattern because of twinning will be those that, after reflection, fall on the basic circle. This is illustrated in Fig. 2.1 for (111) twinning in the [001] fcc orientation, and in Fig. 2.17 for double twinning on $(1\bar{1}\bar{1})$ and $(1\bar{1}1)$ in the $[5\bar{1}0]$ fcc orientation. The required poles are those that lie at an angle to the twin plane equal to the angle of the twin plane from the basic circle. Thus these poles also lie on a great circle. This is shown in Fig. 2.1 as the dotted circle for (111) twinning, and demonstrates how the $\bar{2}40$ matrix reflection coincides with the 204 twin spot.

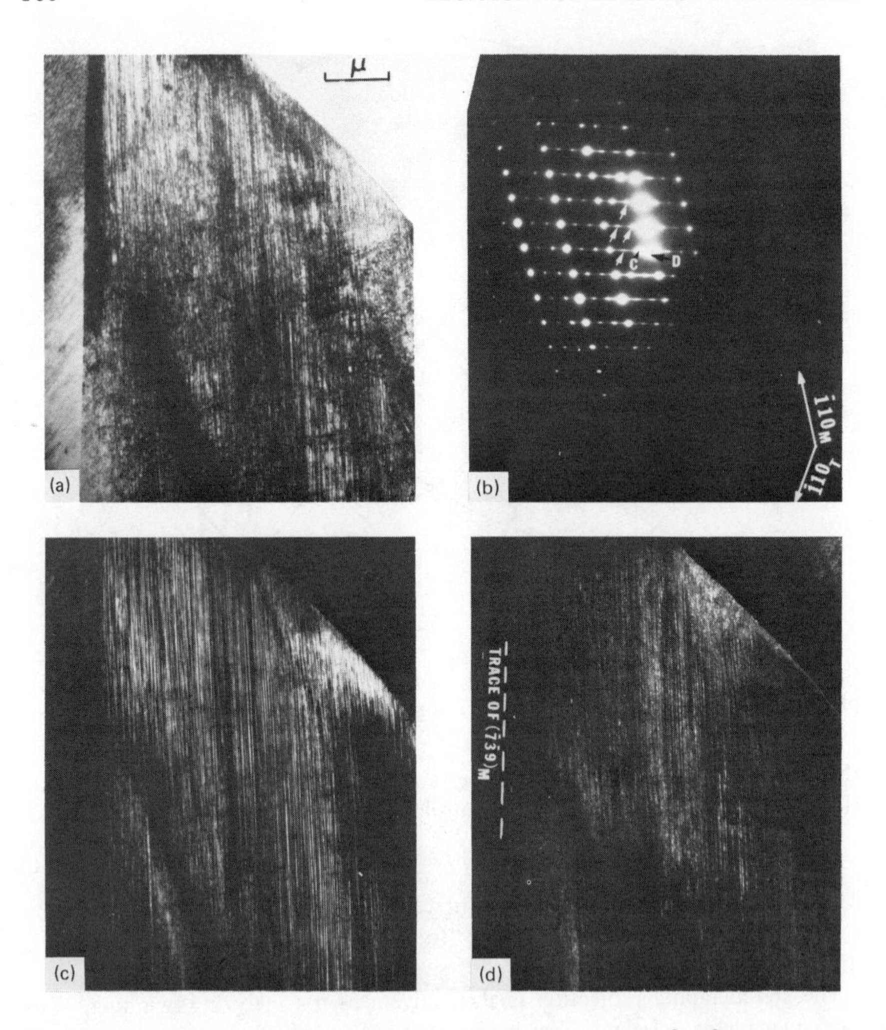

Fig. 2.18 (*a*) Bright field image. (*b*) Diffraction pattern showing {112} transformation twinning in Fe-32% Ni martensite. The $\bar{1}\bar{1}3$ primary twin spot pattern superimposes on the [113] matrix; double diffraction spots are arrowed. Primary twin planes in $(21\bar{1})_\alpha$ (*c*, *d*) are dark field images of $(110)_T$ and $(1\bar{1}0)_{\text{matrix}}$, respectively. Courtesy The Metals Society.[10]

4.5 Sidebands–Modulated Structures

Another important effect arises when periodic modulations, wavelength λ_p, in composition occur, as in spinodal alloys. Such modulations produce modulations in lattice parameter. The period of modulation appears in the diffraction pattern as sidebands[11,12] whose spacing is inversely proportional to λ_p. Examples are shown in Fig. 2.19.

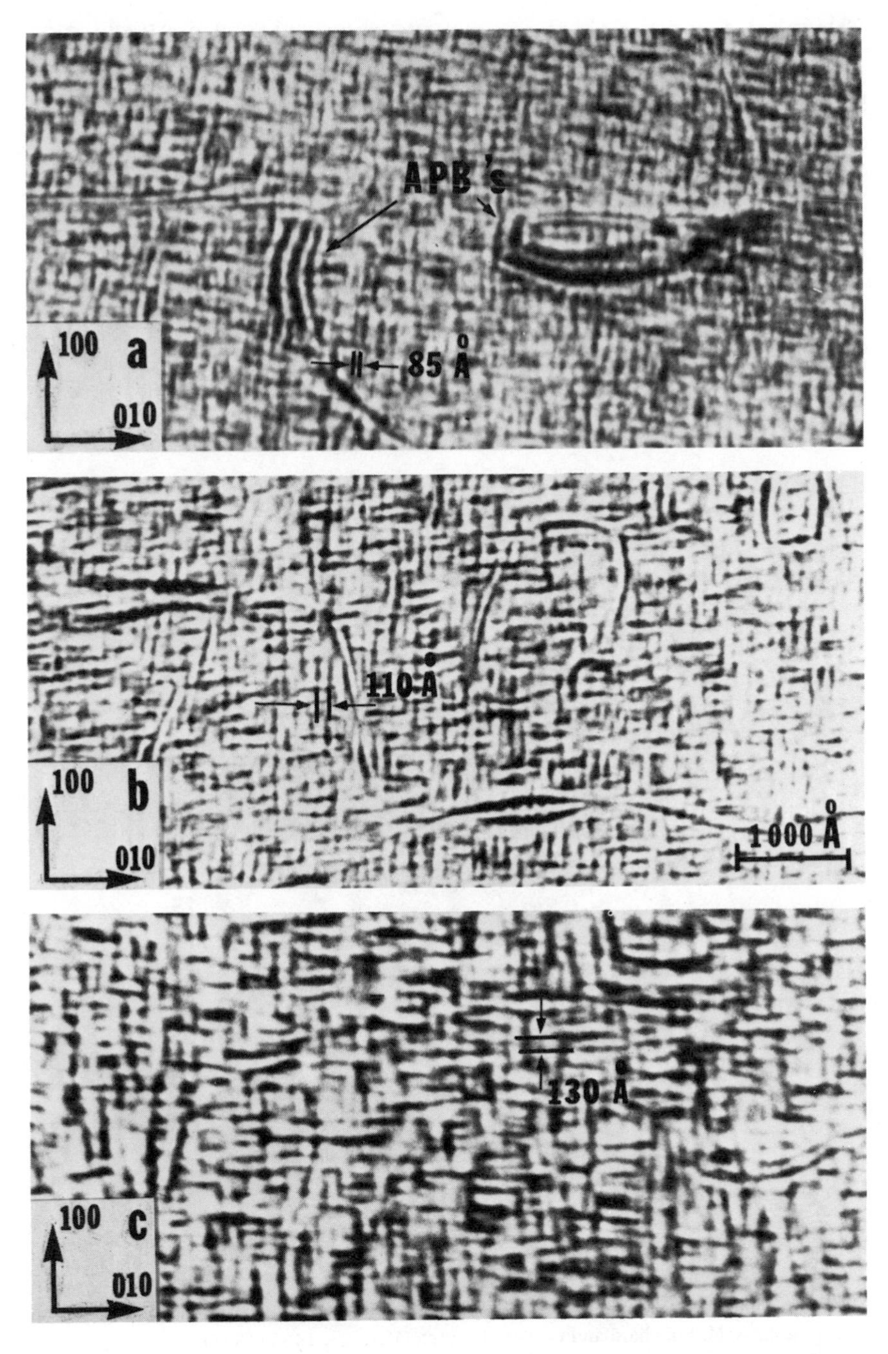

Fig. 2.19 (*a–c*) Bright field images of spinodal microstructure of $Cu_{2.2}Mn_{0.8}Al$, showing coarsening of the wavelength with aging at 350°C for (*a*) 30 sec, (*b*) 1 min, (*c*) 2 min, **g** = 220.

For cubic crystals the periodicity λ_p of the $\langle h00 \rangle$ interplanar spacing produced by fluctuations, assumed for simplicity to be along the x-axis, is given by

$$\lambda_p = \frac{ah}{(h^2 + k^2 + l^2)} \frac{\tan \theta}{\Delta \theta}, \tag{2.22}$$

where $\Delta\theta$ is the angular separation between a main spot and its sideband spot. In the diffraction pattern the corresponding sideband spacing is Δ_p. Since θ is

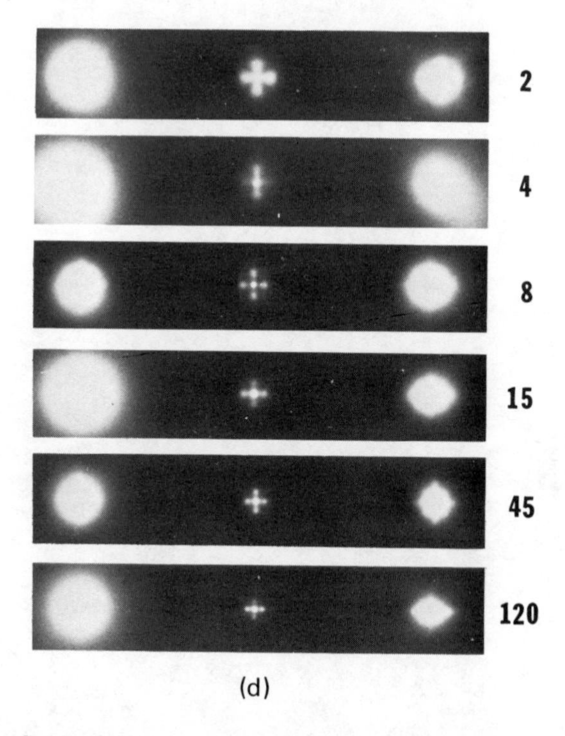

Fig. 2.19 (*Continued*) (*d*) Diffraction patterns of 200 superlattice and 400 fundamental spots, showing $\langle 100 \rangle$ sidebands corresponding to modulations visible in (*a*), (*b*), (*c*) aged at 200°C for times shown in minutes. Structure corresponds to $L21$ superlattice. Courtesy Pergamon Press; M. Bouchard and G. Thomas, *Acta Met*., **23,** 1485 (1975).

small for $h00$ reflections, it follows[12] that

$$\lambda_p = \frac{ha}{h^2 + k^2 + l^2} \cdot \frac{g}{\Delta_p}, \tag{2.23}$$

where g/Δ_p is obtained directly from the pattern. Examination of eq. 2.23 shows that Δ_p is independent of **g**, that is, the order of reflection. For accurate measurements, symmetrical orientations such that the diffraction pattern contains the direction(s) of modulation should be chosen; otherwise Δ_p is projected.

The direction of Δ_p is parallel to the direction of the modulation (generally ⟨001⟩ in cubic systems, e.g., [100] and [010] in Fig. 2.19), whereas Δ_g, the difference in lattice parameter between the two phases, is parallel to **g**. Thus the spacing of the doubling of spots due to the two different lattice spacings increases with increasing magnitude of **g** for all spots, radially outward from the origin. As shown in Fig. 2.13, this effect produces moiré fringes of smaller spacing as Δ_g increases with **g**. Consequently, it is easy to distinguish sidebands (due to periodic fluctuations in composition) from doubling of reflections (due to difference in lattice parameter) just by observation of the diffraction pattern out to several orders.

An interesting geometrical feature of diffraction in this case is the distinction between low angle satellites and those about the Bragg diffracted beams. Modulations in lattice parameter affect only the Bragg diffracted beams (unless double diffraction occurs), so that careful analysis of the small angle regions (as, e.g., by increasing the camera length) can reveal information on composition via modulations in structure factor. This points to the possibility of obtaining kinetic data from measurements of the small angle scattered electron intensities.

The example of sidebands in the diffraction patterns shown in Fig. 2.19*d* is from a modulated ordered alloy of $Cu_{2.5}Mn_{0.5}Al$. They are well resolved in the central superlattice spots. The modulations are shown in the images of Figs. 2.19*a–c*, and faint antiphase domain boundaries are also visible (Chapter 3, Section 8.3).

4.6 Mixed Patterns–Identification of Phases

If n orientations or phases in a specimen contribute to a diffraction pattern, n spot patterns will appear and each pattern can be individually indexed. Although in certain cases of matching, such as twinning, superposition of patterns can occur, dark field imaging will generally allow this to be recognized.

If a foil contains a second phase that is large enough to produce an identifiable pattern, this second phase can be identified by calibration of the pattern using the matrix spots. To avoid difficulties from rel-rod projections and to obtain the

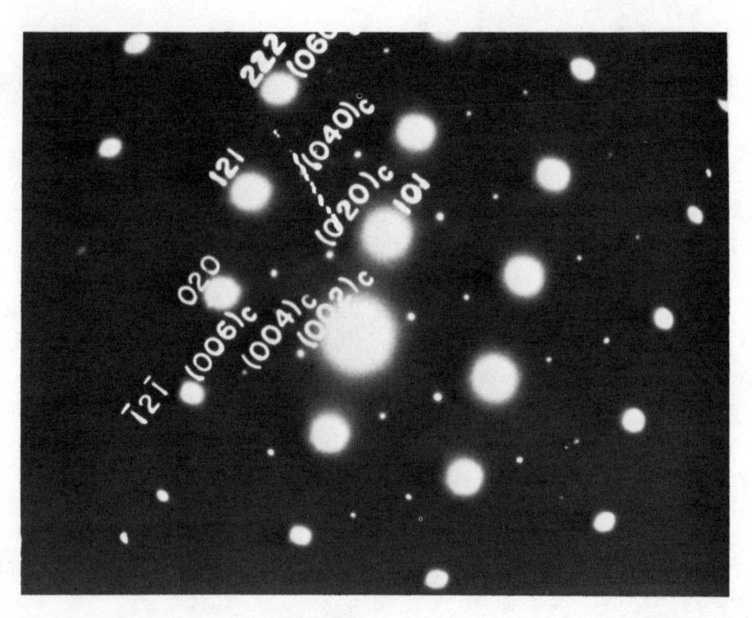

Fig. 2.20 Diffraction pattern of a tempered 0.3% carbon steel, showing [100] orthorhombic (Fe_3C) pattern superimposed on the $[\bar{1}01]$ bcc matrix pattern.

most accurate measurements, the foil should be oriented to give a symmetrical diffraction pattern (Fig. 2.9*b*).

As an illustration consider Fig. 2.20, which shows a pattern from a quenched and tempered Fe-Ni-C alloy, containing carbide particles in addition to the bcc α-matrix. The matrix spots are indexed as shown. They are in $[\bar{1}01]$ orientation. From the measured r values on the original negative (Table 2.4) and with a_0 taken as 2.861 Å for bcc iron, an average value of the camera constant $\lambda L = 2$ Å-cm is obtained by plotting r versus $\{h^2 + k^2 + l^2\}^{1/2}$ since the slope of this plot is $\lambda L/a$.

Table 2.4 Some *d*-Spacings from Fig. 2.20 as Measured on the Original Negative

Distance of Spots from Origin r (cm)	$d = \lambda L/r$	Reflection (Compared to ASTM Card 6-0688 for Fe_3C Reflections)
0.59	3.39	002
1.18	1.69	004
0.83	2.5	020
1.66	1.25	040

The spots of lower intensity correspond to carbide reflections. Their d-spacings are $\lambda L/r_c$, where r_c represents the distances of carbide spots from the origin. The d values so obtained are listed in Table 2.4. By comparison of these d values with those listed in the ASTM card index file, it can be seen that the carbide is cementite, which is orthorhombic. These spots may now be indexed as shown in Fig. 2.20. The cementite spots give a single crystal pattern in [100]. It can be seen that the cementite pattern is crystallographically oriented with respect to the α-bcc pattern. It follows that the orientation relationships are $[\bar{1}01]_\alpha \| [100]_c$, $[\bar{1}2\bar{1}]_\alpha \| [001]_c$, $[111]_\alpha \| [010]_c$.

In general, identification of second phases and orientation relationships with the matrix requires examination of different reciprocal lattice sections in order to eliminate double diffraction and Laue layer effects (intersections of reflections extending from adjacent layers of reciprocal space, e.g., streaks). It should also be noted that second phases that occur on foil surfaces can exist in different orientation variants within one orientation relationship as a consequence of the additional degree of freedom. Likewise, phase transitions in thin foils can be different from those of bulk because of the removal of constraints in the third dimension. An example is martensitic transformations.

4.7 Patterns from Ordered Crystals

If an alloy contains different kinds of atoms, which have attractive interactions, their equilibrium configuration will be an ordered structure that maximizes the number of unlike neighbors. This results in relaxation of the structure factor rules for allowed reflections, and in many cases the symmetry is lowered to primitive. For a perfectly ordered material there will be superlattice reflections at positions that are forbidden for the disordered structure. Their intensity is related to the difference between the atomic scattering factors of the atoms involved, as opposed to the intensity of fundamental reflections, which is related to their sum. A comprehensive review has been given by Marcinkowski.[13] A simple example is the $B2$ superlattice based on the bcc structure of the binary CsCl containing one kind of atom (A) at 000 and the other (B) at $\frac{1}{2}\frac{1}{2}\frac{1}{2}$. Upon complete ordering the structure factor becomes

$$F = f_A + f_B \quad \text{for } h+k+l \text{ even: fundamental reflections}$$

or

$$F = f_A - f_B \quad \text{for } h+k+l \text{ odd: superlattice reflections.}$$

Hence the reflections appearing in a diffraction pattern satisfy primitive symmetry (see Table 2.2), with fundamental reflections of higher intensity separated by weaker superlattice reflections halfway between. In the disordered state the probable occupancy of any site is 50% A or B; hence when $h + k + l =$ odd, $F = 0$. The reciprocal lattice (Fig. 2.2) is again useful in interpreting the geometry of diffraction patterns, although the spot intensities will depend on the degree of order and the particular type and distribution of ordered regions (domain size, composition, and orientation).

In general, the positions of superlattice reflections depend on the type of ordering present in the crystal. The periodicity of a perfect superlattice can only be an integral multiple of the periodicity of the fundamental lattice. Thus superlattice spots will in general appear at fractional positions between fundamental reflections. For each crystal structure (fcc, bcc, hcp, etc.) there exist only a limited number of possible short-period superlattices.[14] Figure 2.21 shows a [011] diffraction pattern of Fe_3Al ordered in the DO_3 type structure, and a more complex [112] diffraction pattern of Ni_3Mo with two types of superstructures, Ni_2Mo and Ni_4Mo, present simultaneously.[15] Superlattices can be identified from their diffraction patterns either by comparison with structure factor calculations for a number of possible superstructures, or by an analytical method which can be viewed as the reversal of the structure factor calculation.[14] The latter method is based on the representation of an ordered lattice by a superposition of concentration waves. It allows the unique determination of the real space lattice from knowledge of the experimentally determined superlattice diffraction vectors.

This reversible relationship between the reciprocal and its real space lattice can be used to identify superlattices in substitutional[15] and interstitial solutions[16] alike. However, it should be noticed that superstructures cannot always be uniquely identified by diffraction patterns alone. If an ordered structure has noncubic symmetry, as in most interstitial alloys, a number of orientational variants may exist. As an example, Fig. 2.22 shows a ⟨100⟩ pattern of an interstitially ordered tantalum phase.[16] The superlattice spots arise from very small ordered domains. The selected area contributing to the diffraction pattern is large compared to the domain size. In this case, therefore, it is impossible to distinguish between diffraction from ordered domains of composition $Ta_{64}X$, where X is the interstital (probably oxygen in this example), and that from a mixture of different ordered domains of other compositions. The use of lattice imaging in a high resolution microscope may help to solve the problem.

At nonstoichiometric concentrations or above the critical ordering temperature, the order in an alloy may be imperfect or short range. Superlattice reflections will then be weaker or more diffuse, and in some cases they will not coincide with the positions expected from the long-range order at this composition. The diffuse reflections at $[1\frac{1}{2}0]$ positions in some alloys (e.g., Ni-Mo) may be explained by the presence of ordered domains having a structure that is not stable in the long-range ordered state at low temperature. Electron diffraction

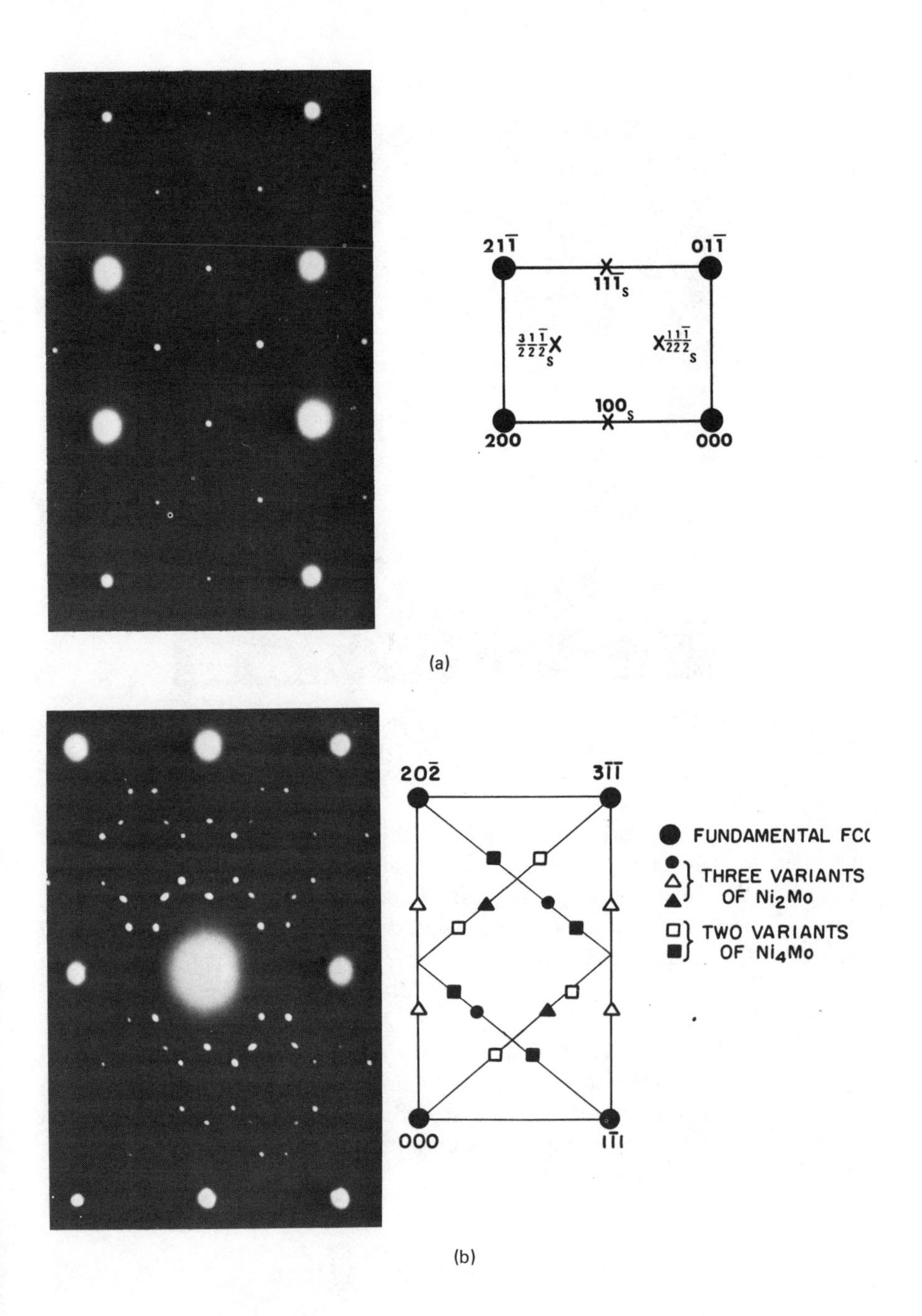

Fig. 2.21 (*a*) [011] Pattern of ordered Fe_3Al (DO_3 structure). (*b*) [112] Pattern of Ni_3Mo showing reflections due to the ordered phases Ni_2Mo and Ni_4Mo. Courtesy U. Dahmen.

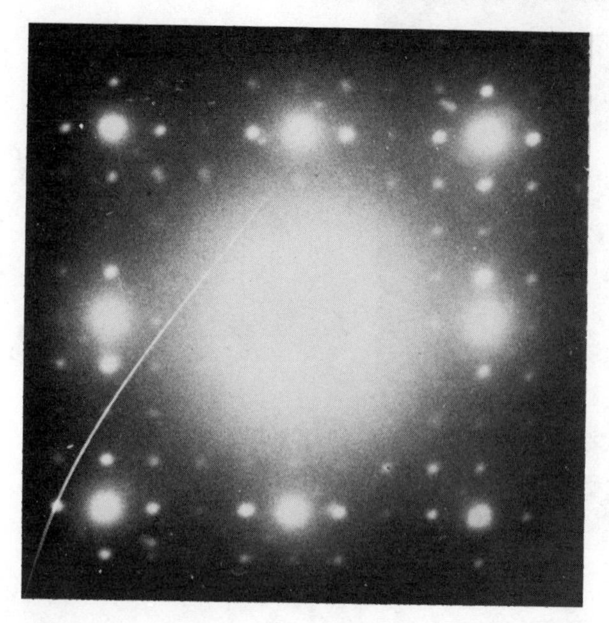

Fig. 2.22 [100] Pattern of interstitially ordered phase in Ta-1.53% C. Courtesy Pergamon Press; B. V. N. Rao and G. Thomas, *Acta. Met.*, **23**, 314 (1975).

and high resolution dark field imaging in this system and a number of others have led to the "multiple domain" model of short-range order.[15] This model describes the structure of short-range order as very small, highly ordered, coherent domains of one or several superstructures existing in a disordered matrix.[17] The model has been confirmed for several systems using lattice imaging methods,[18] but may not be valid for Ni_4Mo, for which more high resolution studies will be needed.

Unexpected superlattice spots at nonintegral positions can also be obtained from long-period superlattices containing domains of regular arrays of antiphase boundaries equally, or nearly equally, spaced apart. Such structures may lead to "splitting" of the diffraction spots, the magnitude of the splitting being inversely proportional to the domain size.

4.8 Surface Contamination Effects

Because of their high affinity for oxygen, most metals readily absorb it on their surfaces. This may lead to the spontaneous formation of oxide nuclei on thin

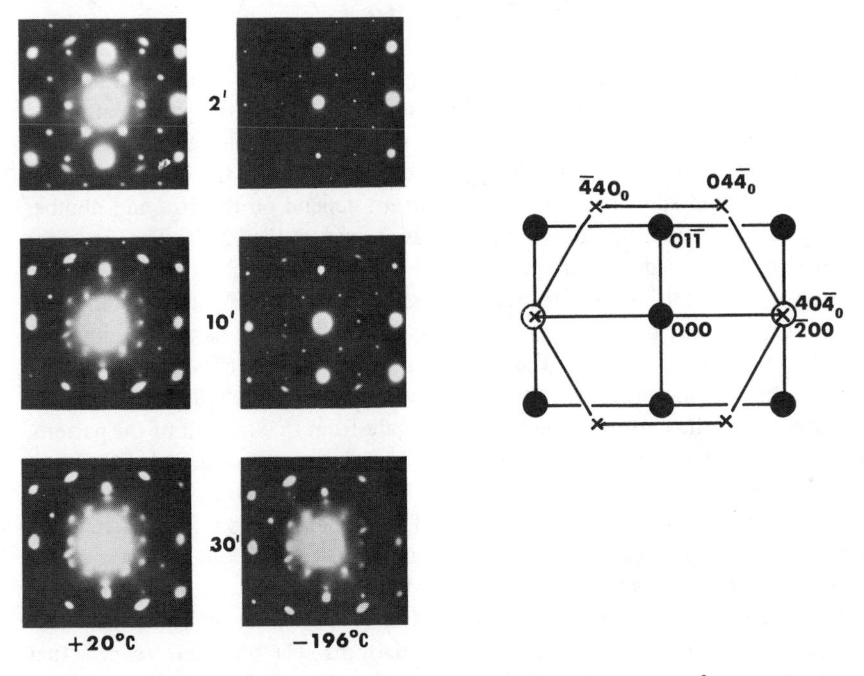

Fig. 2.23 Foil of Fe_3Al on exposure to the microscope atmosphere at 20°C (left-hand side) and −196°C (right-hand side). Note growing intensity of (111) epitaxial oxide film reflections. Courtesy U. Dahmen.

foils prepared for electron microscopy. In many cases these oxides grow epitaxially, giving rise to diffuse reflections in the diffraction patterns instead of the rings observed for amorphous or randomly distributed oxide nuclei. The growth of these oxides may be enhanced by the electron beam even in a vacuum better than 10^{-5} torr. Fig. 2.23 shows an example of the changes in a [011] diffraction pattern of Fe_3Al during beam irradiation in a conventional 100 kV microscope. The intensity of the oxide reflections increases rapidly even at liquid nitrogen temperature. The growth and morphology of surface oxides depend critically on the surface orientation and roughness and the quality of the microscope vacuum. Although electron beam-stimulated processes at solid surfaces are well known,[19] their implications for electron microscopy are often neglected. In short-range ordered Fe-Al, for example, oxide and superlattice reflections are very close together or overlap (Fig. 2.23), so that care is needed in interpreting high resolution dark field images obtained with a superlattice reflection in this system. Thus it is important to distinguish diffuse surface oxide reflections from bulk diffraction effects due to ordering, clustering, or elastic instabilities.

4.9 Ring Patterns

For a randomly oriented set of crystalline grains, diffraction produces a series of cones, each cone of angle 4θ, where θ is the Bragg angle for the particular reflection. These cones intersect the reflecting sphere on a circle; and, as in a spot pattern, the radius of the circle r is given by $\lambda L/d$. This is true for all crystals. The breadth and spottiness of the ring pattern depend on the size and number of crystals contributing to the pattern, as shown in Fig. 2.24. The finer the grain size, the broader is the pattern; and there is a lower limit beyond which the physical definition of "crystalline" grains becomes difficult. Thus in some cases very diffuse, broad ring patterns may be referred to as originating from "amorphous" solids, whereas the material may actually be crystalline, but only a few unit cells in grain size. Thus great care must be exercised in drawing conclusions about the nature of the material merely from the breadth of the pattern, even if dark field images of the individual crystals are also obtained (Fig. 2.24). In such cases lattice imaging and optical diffraction techniques can be utilized,[20] but even then great care is necessary. These considerations are particularly important in regard to thin films, for example, in applications of the so-called amorphous solids, which have useful properties.

Ring patterns can be useful for qualitative identification of materials; their analysis is similar to that of X-ray powder patterns. The main difference is that in electron diffraction the small angle approximation $\sin 2\theta \simeq 2\theta$ is applicable. If the camera constant λL is known, the d values can be calculated from the relation $\lambda L = rd$. One way of fixing λL is to evaporate a known material, such as gold, onto part of the specimen being investigated. This pattern can be superimposed on the unknown by doubly exposing the plate or film. Then λL can be evaluated from the gold pattern, for example, by plotting the ring radii, which will appear in the sequence $\{111\}$, $\{200\}$, $\{220\}$, $\{311\}$, and so on, (Table 2.2) against $\sqrt{h^2 + k^2 + l^2}$ and finding the slope. Since a_0 for gold is known, λL is found. The unknown d values are then found by dividing λL by the measured radii of rings of the unknown substance, and using the ASTM card index file.

Considerable care must be exercised in such work. For example, if the grain structure is not perfectly random, not all reflections will appear. Suppose, for example, a fcc polycrystal had a [001] preferred fiber orientation; then reflections for which $[001] \cdot [hkl] \neq 0$ would not appear. Hence 111, 311 rings would be absent. Also, the structure of thin films may not necessarily be the same as that of the bulk specimen, especially if produced under nonequilibrium conditions (such as very fast rates of evaporation or "splat" cooling). For example, the tin patterns of Fig. 2.24 do not conform to the normal tetragonal sequence of reflections. Double diffraction may also cause the appearance of "forbidden" reflections.

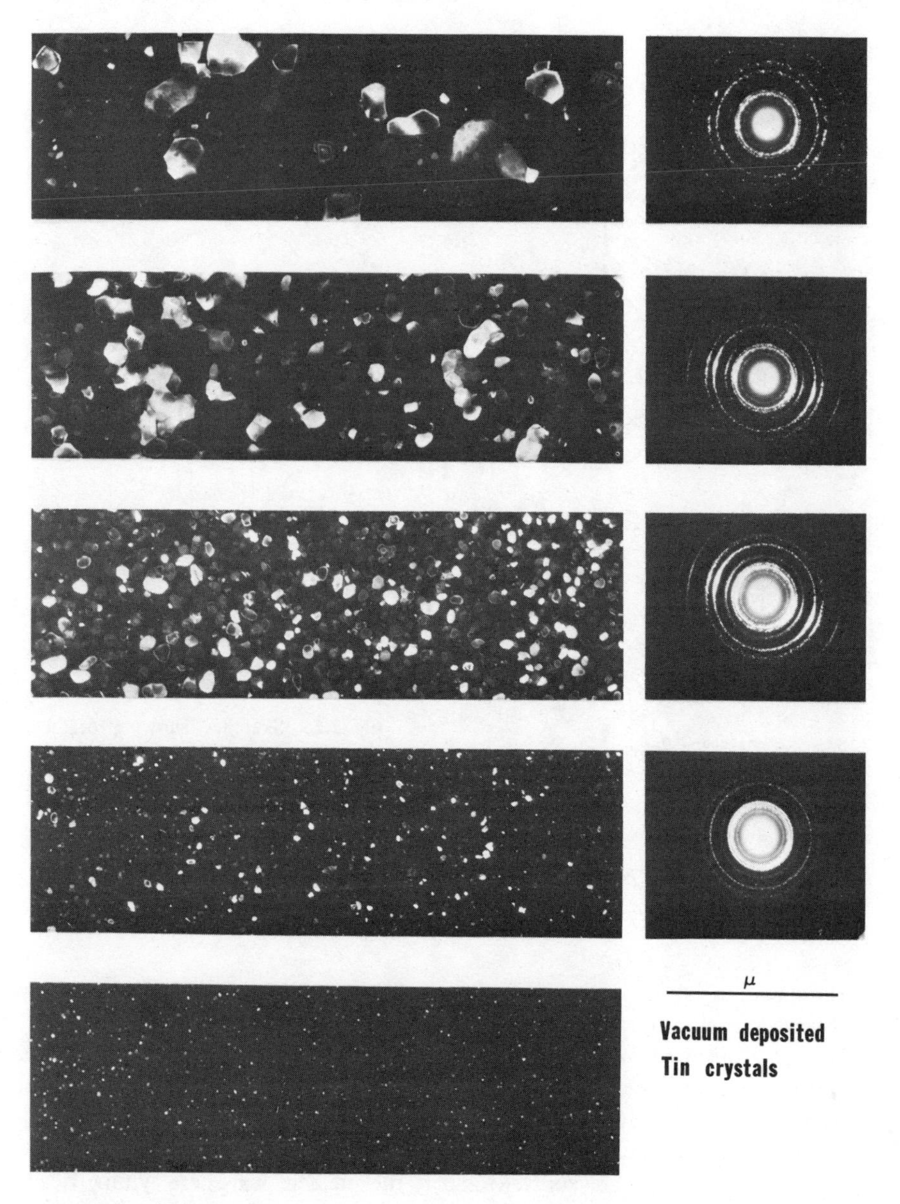

Fig. 2.24 Dark field images and ring patterns from vacuum-evaporated tin, showing change in grain size. Arcing of rings indicates preferred orientation. Courtesy W. L. Bell.

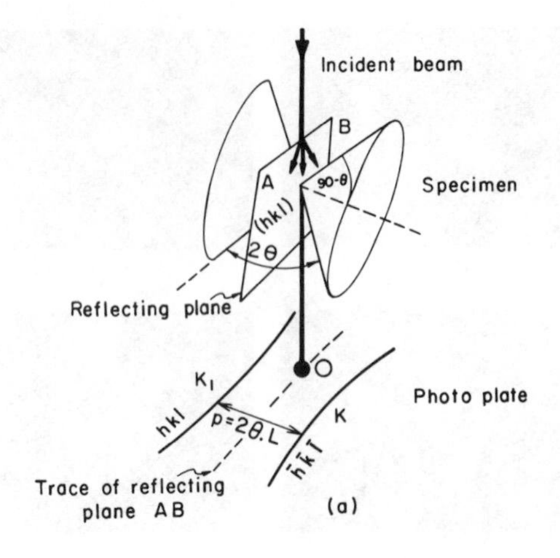

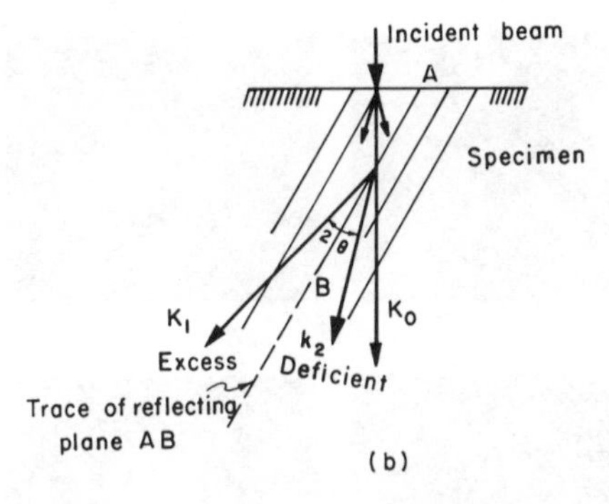

Fig. 2.25 Sketches showing origin of Kikuchi lines: (*a*) symmetrical case, (*b*) exact Bragg case (compare with Figs. 2.9*a* and *b*). Courtesy North-Holland.[35]

5 Information from Kikuchi Patterns*

5.1 Geometry of Formation

The name "Kikuchi electron diffraction" is given to the patterns of lines that are observed from fairly thick crystals, after their discovery in 1928 by Kikuchi.[21] The mechanism of formation is as follows. The electron beam, on entering a specimen, suffers inelastic and incoherent scattering by interaction with the atoms. These electrons can be subsequently rescattered coherently when Bragg's law is satisfied at a suitable set of reflecting planes.

Cones of radiation are emitted; and if the incident waves are symmetrically impinging on the plane *AB*, cones of equal intensity are scattered, with semi-vertex angles of (90 - θ) (Fig. 2.25) to each side, bisecting the reflecting plane

*Sections 5.1 to 5.5 appeared previously, in slightly different form, as the chapter "Kikuchi Electron Diffraction and Applications" by G. Thomas in *Modern Diffraction and Imaging Techniques in Material Science*, edited by S. Amelinckx et al. (North Holland, 1970).

AB. If, however, the waves impinge on an inclined reference plane *AB* (Fig. 225b), most of the electrons are initially scattered in the direction K_1 and relatively few in the forward direction K_0. Under normal conditions and on a positive print, one then observes a bright line corresponding to K_1 near the Bragg spot and a dark line corresponding to K_0 near the origin. Since for most applications a knowledge of the geometry is sufficient, the dynamical behavior[22,23] and intensities of the Kikuchi lines[24] will not be discussed here.

The intersection of the cones of Kikuchi radiation with the reflecting sphere produces slightly hyperbolic lines because of the small angles θ and large radius of the sphere for fast electrons. These lines are actually straight on the photographic plate for the usual angles recorded in a pattern ($\approx 9^\circ$ at 100 kV for $\lambda L = 2$ Å · cm). Each reflection thus gives rise to a pair of Kikuchi lines *hkl* and

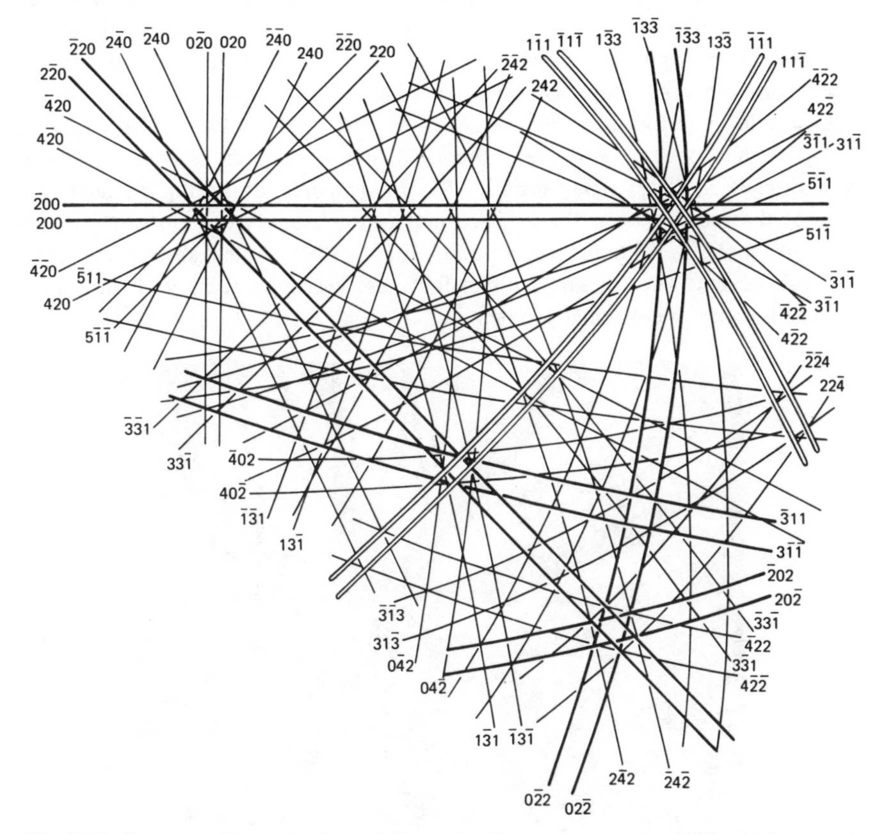

Fig. 2.26 Stereographic projection of first-order Bragg contours (and Kikuchi line pairs), drawn to scale for 100 kV electrons in aluminum up to $(h^2 + k^2 + l^2) = 27$. Courtesy K. Ashbee and J. Heavens, *UCRL Report* 17614.

$\bar{h}\bar{k}\bar{l}$, whose respective intensities depend principally upon the orientation, perfection, and thickness of the crystal. Figures 2.9*a* and *b* show the geometry of crystals oriented symmetrically and oriented at the exact Bragg condition. These are really the only two orientations that are needed in electron microscopy applications; case *b* should be used when crystallographic data are needed, and case *a* is the two-beam or systematic situation necessary for contrast analysis.

It should be noted that in the rescattering of inelastic beams the specimen acts as a monochromator, that is, the planes (*hkl*) reflect electrons which are satisfied by Bragg's law for the wavelengths involved: $2d \sin \theta' = n\lambda'$. Since typically the characteristic energy losses are of the order of tens of volts, $\lambda' \approx \lambda$ (incident), and the same reflecting sphere-reciprocal lattice construction describing the spot pattern can be used for the Kikuchi pattern (Figs. 2.9*a* and *b*).

Kikuchi patterns are always produced, even in thin crystals, but the specimen must be thick enough so that a sufficiently intense Kikuchi cone to be observed on the photographic plate is produced. Furthermore, the specimen should be relatively free from long-range internal strains (e.g., elastic buckling, a high dislocation density); otherwise the Kikuchi cones will be incoherently scattered and may become too diffuse to be observed. The absence of observable Kikuchi lines or the appearance of very broad diffuse lines from heavily dislocated structures (e.g., ferrous martensite) is due to incoherent scattering. As the thickness of the foil increases, the diffraction pattern changes from spots, to Kikuchi lines and spots, to Kikuchi lines or bands, until finally complete absorption within the foil occurs. The thickness limits for these events increase with increasing voltage because of enhanced penetration.

Fig. 2.27 Bragg extinction contours in copper foil symmetrically deformed about [011]. Compare with [011] region of Fig. 2.26. Courtesy W. L. Bell.

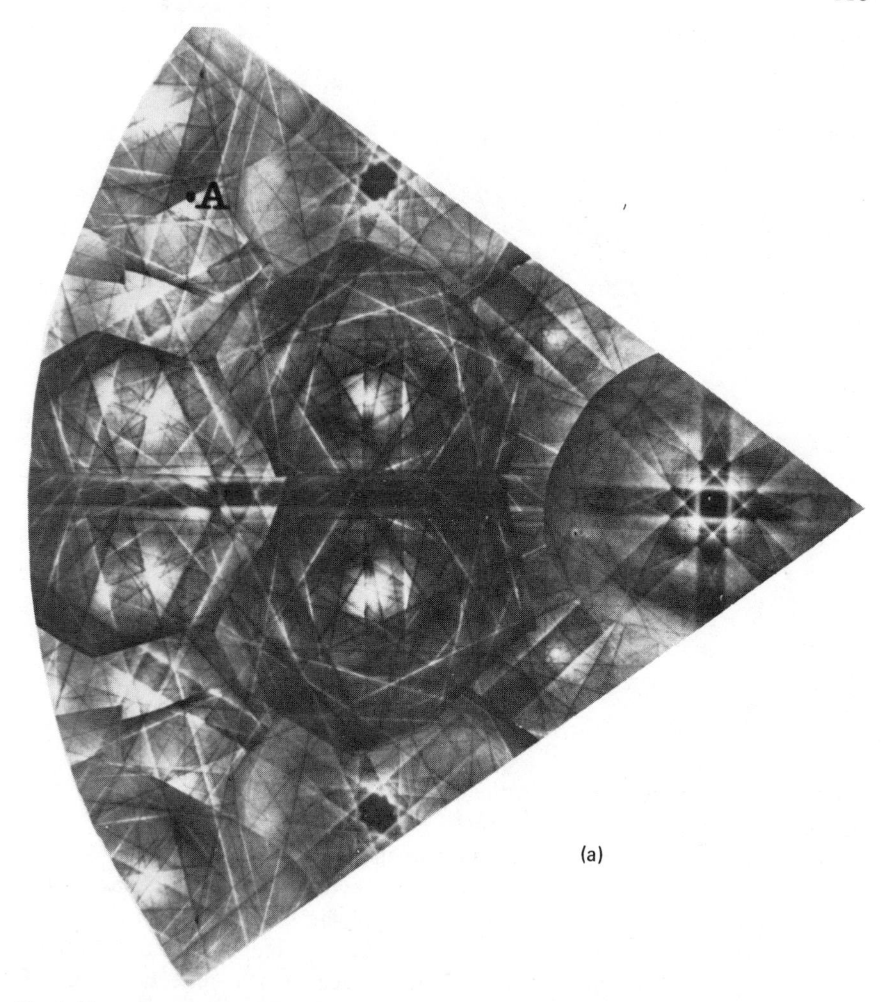

Fig. 2.28 (*a*) Part of the Kikuchi map for silicon corresponding to the area near [001] of reciprocal space. (*b*) Indexing of (*a*) together with scale factor. Courtesy *Journal of Applied Physics*.[26]

It can be seen from Figs. 2.9*a* and *b* that, on tilting the specimen in one sense, the Kikuchi lines sweep across the pattern in the same sense. In Fig. 2.9*a* the crystal has been tilted by θ_{hkl} anticlockwise so as to excite the first-order (hkl) reflection. The Kikuchi origin is fixed in the crystal so that, as the crystal is tilted, the cones sweep across the pattern as if rigidly "fixed" to the specimen. Thus the Kikuchi pattern is extremely useful in determining the precise orientation, as well as for calibrating tilt angles, and so on. Since each Kikuchi line in a

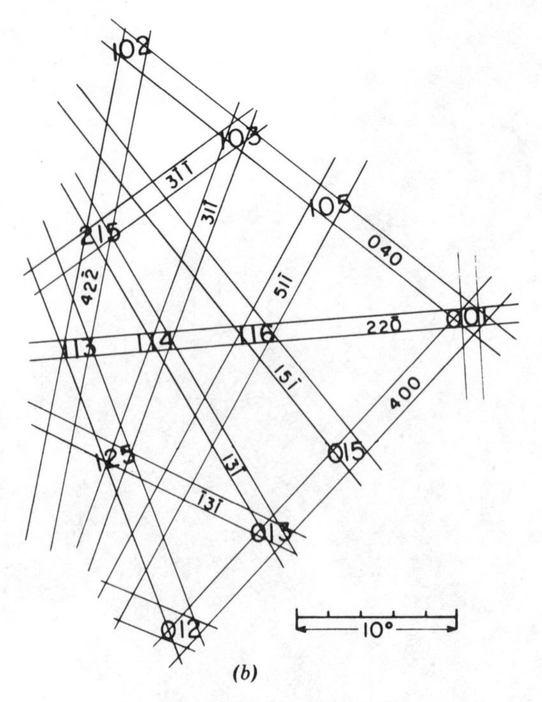

(b)

Fig. 2.28 (*Continued*)

pair bisects the reflecting plane, the angle subtended by each pair is always 2θ independently of crystal orientation.* Furthermore, the Kikuchi pattern represents the traces of all reflecting planes in the crystal and can thus be directly compared to the appropriate stereographic projection. The Kikuchi lines are also parallel to the Bragg extinction contours. Applying Bragg's law and the appropriate structure factor rules enables one to plot the complete Kikuchi pattern. Figure 2.26 is derived to scale for the first-order Kikuchi reflections for aluminum at 100 kV. This can be compared to the actual pattern shown in Fig. 2.28, and the Bragg contour pattern of a bent foil in Fig. 2.27.

Tilting the crystal tilts the reciprocal lattice in the same sense and magnitude. The spot pattern thus translates only slightly on tilting since each spot then rotates about an arc of radius $|\mathbf{g}|$ centered at the origin (Fig. 2.9*b*). The Kikuchi pattern, however, shifts in an easily observable manner ($\approx$1 cm per 1° tilt for $\lambda L \approx 2$ Å · cm).

The Kikuchi lines associated with a particular reflection *hkl* always lie per-

*Special situations arise when this rule is not strictly correct, especially for large angle reflections (see ref. 23).

pendicular to $\mathbf{g}(hkl)$, that is, on a line through the origin and normal to the Kikuchi pair.

The centers of symmetry of the spot pattern and Kikuchi pattern thus coincide only in symmetrically oriented foils. For accurate determinations of orientation relationships the foil can be tilted until the symmetrical situation appears on the screen.

5.2 Relationship to Spot Patterns and Determination of Deviation Parameters

The Bragg deviation parameter s is important in contrast analysis (see Chapters 3, 4, and 5). Both the sign and the magnitude of s can be readily obtained by noting the orientation of the Kikuchi pattern to that of the spot pattern.[25]

Since by definition $s > 0$ occurs when the reciprocal lattice point lies inside the reflecting sphere, the Kikuchi line lies outside its corresponding diffraction spot (Fig. 2.9). Thus, in a symmetrical orientation, $s < 0$ (Fig. 2.9*b*). The sense of tilt of a foil is thus immediately apparent from the relation between the Kikuchi and spot patterns.

Figure 2.29 shows the effect of tilting the foil to produce an $s > 0$ orientation (such as is required for maximum bright field contrast in absorbing crystals). As the foil is tilted from the exact Bragg condition by an angle ϵ, the Kikuchi lines move outward a distance x along $\mathbf{g}_{hkl}$, whereas the spot moves inside the sphere on the arc O^*PP_1. In the pattern the Kikuchi line now lies outside the corresponding spot, as shown in Fig. 2.29*b*.

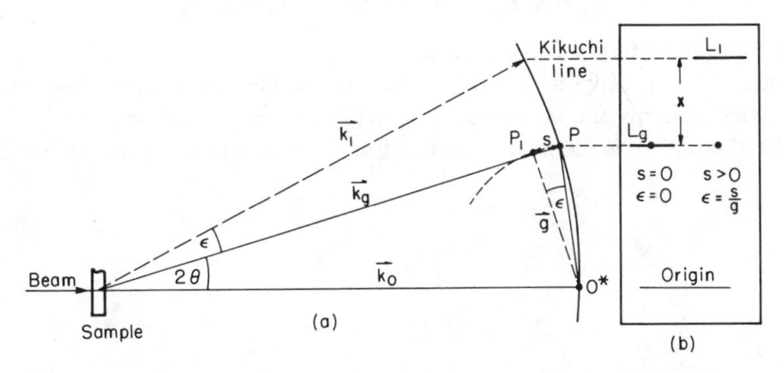

Fig. 2.29 Sketch showing shift in Kikuchi lines produced by a tilt ϵ. Position of Kikuchi lines at L_g and L_1 correspond to the cases $s = 0$ and $s > 0$, respectively. The tilt ϵ moves the Kikuchi line through x, and the spot through distance s about the radius $O{*}P$. Courtesy *Journal of Applied Physics.*[25]

Since the Bragg angles are small,

$$\phi = \frac{x}{L} = \frac{x\lambda}{rd},$$

where L is the camera length.

Since x can be measured on the plate, and since r and d are known from the indexed pattern, ϕ is calculated. Also,

$$\phi = \frac{s}{g}.$$

Hence

$$s = \frac{xg}{L} = \frac{x}{Ld}$$

for a given Kikuchi band of separation p_1, $2\theta_1 \cong p_1/L = \lambda/d_1$; hence

$$s = \frac{x}{Ld} = \frac{x\lambda}{d(p_1 d_1)}. \tag{2.24}$$

5.3 The Precise Determination of Orientations

The general method of solving a Kikuchi pattern and determining the precise foil orientation is similar to the method for solving spot patterns. Since the spacing of each Kikuchi pair is proportional to 2θ (Fig. 2.24) then, for different sets of Kikuchi pairs of spacings, p_1, p_2, and so on,

$$p_1 = K2\theta_1, \quad p_2 = K2\theta_2, \quad p_n = K2\theta_n,$$

where K is the effective camera length L.

Thus, if the reflections $h_1k_1l_1 \ldots h_nk_nl_n$ are identified, the pattern can be calibrated in terms of distances on the plate and corresponding angles.

The Kikuchi reflections are identified as follows. Suppose that in Fig. 2.30

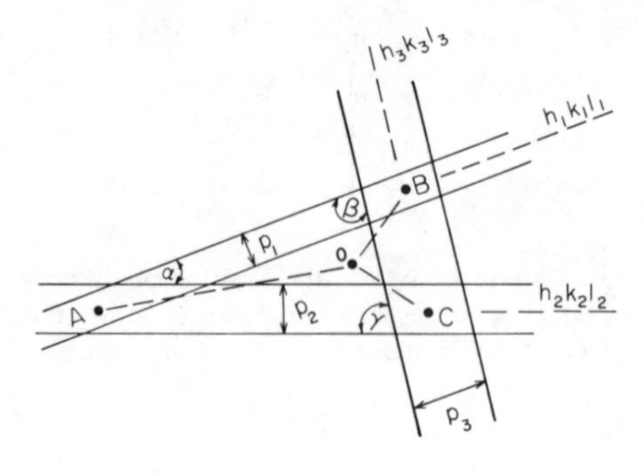

Fig. 2.30 Sketch to illustrate indexing of any Kikuchi pattern. If the poles A, B, C do not appear on the plate, use tracing paper to extend the Kikuchi lines through points of intersection. Courtesy North-Holland.[35]

there are three sets of intersecting Kikuchi lines at angles α, β, γ, and the points of intersection A, B, C are zone axes (Kikuchi poles). If the crystal is cubic, then, since $p_1 \propto 1/d_{h_1k_1l_1}$, and so on, $p_1d_1 = \lambda L$, $p_2d_2 = \lambda L, \ldots, p_nd_n = \lambda L$, or

$$\frac{p_1}{p_2} = \frac{\sqrt{h_1^2 + k_1^2 + l_1^2}}{\sqrt{h_2^2 + k_2^2 + l_2^2}} \quad \text{and} \quad \frac{p_1}{p_3} = \frac{\sqrt{h_1^2 + k_1^2 + l_1^2}}{\sqrt{h_3^2 + k_3^2 + l_3^2}},$$

and so on.

Measure the spacings p_1, p_2, p_3, take their ratios, and then, by using either tables of d-spacing ratios or a slide rule, assign the tentative indices $h_1k_1l_1$, and so on. Then check the correctness of the assignment by measuring the angles α, β, γ, and comparing the results to the calculated values based on $h_1k_1l_1$, $h_2k_2l_2$, and so on:

$$\cos\alpha = \frac{h_1h_3 + k_1k_3 + l_1l_3}{\sqrt{h_1^2 + k_1^2 + l_1^2} \cdot \sqrt{h_3^2 + k_3^2 + l_3^2}}.$$

This process can be time consuming, as it is often a question of trial and error in order to obtain the correct solution. The results can be checked by measuring other Kikuchi lines in the pattern.

Once the lines are indexed, the poles A, B, C are obtained by taking the respective cross products, for example, $\mathbf{A} = [h_1k_1l_1] \times [h_2k_2l_2]$. Let these poles be $p_1q_1r_1$, $p_2q_2r_2$, and $p_3q_3r_3$. The indices of the direction of the beam through the crystal (i.e., where the transmitted beam intersects the pattern at O) can be found either by calculation or by stereographic analysis. Both require measurement of the angles $\widehat{OA}$, $\widehat{OB}$, $\widehat{OC}$, (Fig. 2.30). Measure the distances OA and OB and convert to angles either by using the calibration $p_1 = K2\theta$, or by measuring the distances AB, BC, CA and convert into angles since the angles $\widehat{AB}$, $\widehat{AC}$, $\widehat{BC}$ can be calculated once A, B, C are indexed. Let $[uvw]$ be the axis O; then, if $\theta_1, \theta_2, \theta_3$ are the angles $\widehat{OA}$, $\widehat{OB}$, $\widehat{OC}$,

$$\left.\begin{aligned}
\cos\theta_1 &= \frac{up_1 + vq_1 + wr_1}{\sqrt{u^2 + v^2 + w^2} \cdot \sqrt{p_1^2 + q_1^2 + r_1^2}}, \\
\cos\theta_2 &= \frac{up_2 + vq_2 + wr_2}{\sqrt{u^2 + v^2 + w^2} \cdot \sqrt{p_2^2 + q_2^2 + r_2^2}}, \\
\cos\theta_3 &= \frac{up_3 + vq_3 + wr_3}{\sqrt{u^2 + v^2 + w^2} \cdot \sqrt{p_3^2 + q_3^2 + r_3^2}},
\end{aligned}\right\} \tag{2.25}$$

and uvw is determined by solving these equations. The solution can also be found by the use of stereographic projection as described in the book by Johari and Thomas (see Chapter 1, Bibliography).

For patterns containing one or no Kikuchi pole the solution can be obtained only by reference to the appropriate Kikuchi map.

5.4 Kikuchi Maps

When no Kikuchi poles are present, the solution to a Kikuchi pattern can be obtained by comparing an unknown pattern with standard Kikuchi projections, called Kikuchi maps.[26,27] The case of no Kikuchi poles occurs most commonly when one works in two-beam orientations. However, in general the Kikuchi map is suitable and convenient for solving any unknown pattern since the maps eliminate the three-pole (or two-pole) solutions described above, as well as the usual trial and error procedures involved in indexing.

If one is working with a particular crystal system, all one needs is a Kikuchi map of that system. The maps are obtained so as to cover completely the standard triangle of the appropriate stereographic projection. In order to do this either single crystals or large, randomly oriented polycrystalline specimens should be used. The specimen is then tilted into a symmetrical low index orientation; and by tilting outward successively along principal Kikuchi lines, successive photographs are obtained so that some overlap of the pattern occurs from one plate to the next. After all the plates are printed, they can be glued onto a board so that successive photographs are matched. It is necessary to completely cover the area of reciprocal space in order to complete the map. If a complete [001]-[011]-[111] map is plotted, the curvature of the Kikuchi lines causes some distortion of the map. The distortion is not noticeable if one works with regions of about 20° around a principal pole. Further details can be found in refs. 26 and 27.

Maps can also be obtained by computation or from stereographic projection to obtain traces of reflecting planes, as shown in Fig. 2.25 to scale for aluminum at 100 kV.

Figures 2.28, 2.31, and 2.32 show composite maps for diamond cubic bcc and hcp crystals. Figure 2.28 can also be used for fcc crystals (compare with Fig. 2.25) by allowing for the differences in structure factor (e.g., 200 is normally missing in diamond cubic). However, both the dc and hcp maps include reflections of zero structure factor which appear due to double diffraction.[27] In general, for noncubic crystals separate maps are needed for each material, so that axial ratios and angles are identical.

Many difficulties can be avoided in the case of hexagonal crystals if the four-index notation is used as described in detail in Appendix B. For example, any pole [*uvtw*] is determined from the intersection of two Kikuchi bands

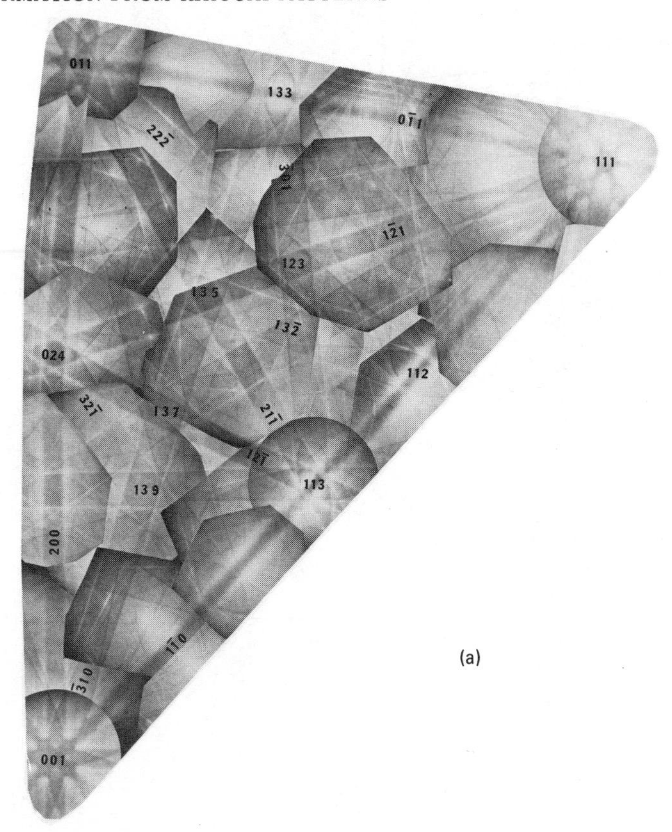

Fig. 2.31 (*a*) Composite Kukuchi map for bcc crystal. (*b*) Indexing and scale factor. Courtesy *Journal of Applied Physics.*[27]

$(h_1k_1i_1l_1)$ and $(h_2k_2i_2l_2)$ from the equation

$$uvtw = \left[\begin{vmatrix} l_1 k_1 i_1 \\ l_2 k_2 i_2 \\ 0\ 1\ 1 \end{vmatrix},\quad \begin{vmatrix} h_1 l_1 i_1 \\ h_2 l_2 i_2 \\ 1\ 0\ 1 \end{vmatrix},\quad \begin{vmatrix} h_1 k_1 l_1 \\ h_2 k_2 l_2 \\ 1\ 1\ 0 \end{vmatrix},\quad \begin{vmatrix} h_1 k_1 i_1 \\ h_2 k_2 i_2 \\ 1\ 1\ 1 \end{vmatrix} \right]. \tag{2.26}$$

As a rule, the more useful working orientations coincide with one of the prominent poles of the Kikuchi map. However, when the foil orientation coincides with one of the less prominent poles, such as A in Fig. 2.32, its indices $[u_A v_A t_A w_A]$ can be obtained immediately. Since A is the intersection of several Kikuchi bands,

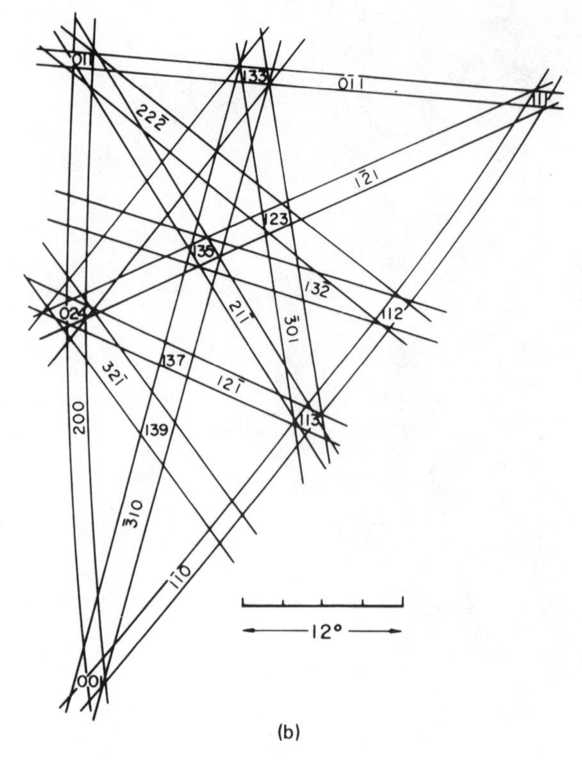

(b)

Fig. 2.31 (*Continued*)

any two, such as the $[02\bar{2}1]$ and $[\bar{3}121]$, may be used to obtain $[u_A\, v_A\, t_A\, w_A] = [5, \bar{7}, 2, 18]$ from eq. 2.26.

In the worst possible situation, corresponding to a completely arbitrary foil orientation, a simple solution is still possible provided that the center of the unknown diffraction pattern lies at the intersection of two lines passing through pairs of poles that are or can be indexed. In view of the large numbers of such poles, the probability of finding two suitable lines is large. For example, suppose that the unknown diffraction pattern is compared with the map and its center found to lie at B_p in Fig. 2.32. It is found to be at the intersection of lines 1 and 2 where line 1 passes through poles A and $[\bar{1}013]$, and line 2 through poles $[10\bar{1}4]$ and $[\bar{2}\bar{2}49]$. But these lines are themselves parallel to Kikuchi bands; and if they can be indexed, the orientation of pole B_p can be determined as before.

Since $\mathbf{g} \cdot \mathbf{r} = 0$ for any Kikuchi band passing through the zone axis of $\mathbf{r}$, any two such zones, $\mathbf{r}_1 = [u_1\, v_1\, t_1\, w_1]$ and $\mathbf{r}_2 = [u_2\, v_2\, t_2\, w_2]$, will suffice for the

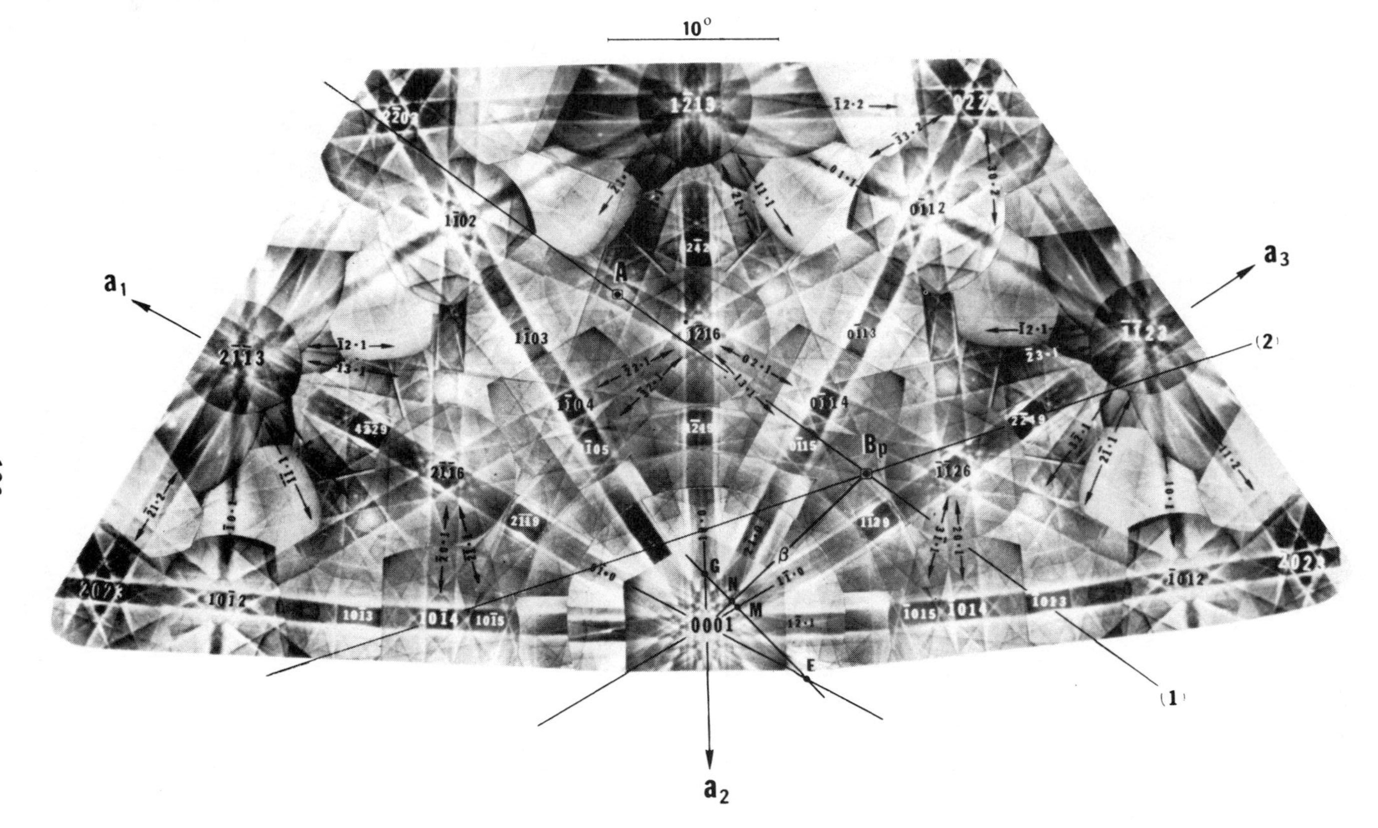

Fig. 2.32 Composite Kukuchi map for hcp crystals centered about [0001] pole. All poles are given in directional indices (see Appendix B). The scale factor shown corresponds to a c/a ratio of 1.588 (e.g., Ag_2Al, Ti). Courtesy *Physica Status Solidi*.[2]

determination of its indices. They are the solution to the system of equations

$$\left.\begin{aligned} \mathbf{g}\cdot\mathbf{r}_1 &= hu_1 + kv_1 + it_1 + lw_1 = 0, \\ \mathbf{g}\cdot\mathbf{r}_2 &= hu_2 + kv_2 + it_2 + lw_2 = 0, \\ h + k + i &= 0. \end{aligned}\right\} \tag{2.27}$$

Therefore the indices of the unknown Kikuchi band can be written as

$$(hkil) = \left(\begin{vmatrix} \overline{w}_1 v_1 t_1 \\ \overline{w}_2 v_2 t_2 \\ 0\ 1\ 1 \end{vmatrix},\quad \begin{vmatrix} u_1 \overline{w}_1 t_1 \\ u_2 \overline{w}_2 t_2 \\ 1\ 0\ 1 \end{vmatrix},\quad \begin{vmatrix} u_1 v_1 \overline{w}_1 \\ u_2 v_2 \overline{w}_2 \\ 1\ 1\ 0 \end{vmatrix},\quad \begin{vmatrix} u_1 v_1 t_1 \\ u_2 v_2 t_2 \\ 1\ 1\ 1 \end{vmatrix} \right). \tag{2.28}$$

For the present problem, using the orientation of pole A calculated earlier, the indices of lines 1 and 2 are $(h_1k_1i_1l_1) = (3, \overline{15}, 18, \overline{7})$ and $(h_2k_2i_2k_2) = (11, \overline{14}, 3, \overline{2})$, respectively. When these indices are substituted into eq. 2.26, the orientation of pole B_p is found to be $[u_Bv_Bt_Bw_B] = [\overline{1}, \overline{1.85}, 2.85, 11.7]$. It should be pointed out, however, that in view of the large angular range covered by the Kikuchi map. Kikuchi bands are actually curved lines. Therefore, to justify the use of straight lines, the two pairs of poles defining lines 1 and 2 should be chosen as close together as possible.

5.5 Some Applications of Kikuchi Patterns

In the foregoing it has been shown how Kikuchi patterns facilitate obtaining the precise orientation of the foil. If the Kikuchi pattern is sharp, the orientation may be obtained to within 0.01° and without the 180° ambiguity inherent in spot patterns. Thus the Kikuchi pattern is obviously very useful for obtaining crystallographic data, such as orientations, orientation relationships, trace analysis and habit planes, lattice parameters and axial ratios, identification of phases, and detection of deviations from random solid solutions. Kikuchi patterns also greatly simplify contrast analysis by facilitating the selection of particular reflections and control of orientation and the obtaining of stereoscopic images quickly and accurately. In addition, the Kikuchi pattern provides the most accurate method for calibrations, such as image rotation, electron wavelength, and foil thickness, and can provide information on nonisotropic elastic strains, as well as angular misorientation across subgrains and dislocation walls.

Since space is limited, a comprehensive survey will not be attempted, but included in the following subsections are representative examples of some of these applications. The details can be found in the references listed.

5.5.1 *Structure Analysis.* Since the Kikuchi patterns map out reciprocal space very accurately, they provide a useful means of structure analysis by electron diffraction. Examples have been described previously.[28] The maps can be constructed as described in the preceding section. The main advantage of the Kikuchi pattern over the spot pattern is that the symmetry of the Kikuchi pattern is precisely that of the crystal giving rise to the pattern. The same is not true of spot patterns, at least if they are in symmetrical orientation or are confined to a single Laue zone. This is illustrated by Fig. 2.33, which compares the symmetrical dc and hcp patterns from silicon and magnesium. Although both spot patterns show sixfold symmetry, the Kikuchi pattern of [111] silicon is three-

Fig. 2.33 (*a*) Symmetrical [111] silicon diffraction pattern. (b) Symmetrical [0001] magnesium pattern. Notice the sixfold symmetry of both spot patterns, but in (*a*) the Kikuchi symmetry is threefold. Courtesy North-Holland.[35]

Fig. 2.33 (*Continued*)

fold, whereas the [0001] pattern of magnesium is sixfold. This is particularly clear upon examinating the high order lines near the center of the patterns.

5.5.2 *Contrast Work.* The Kikuchi map can be utilized similarly to a road map when one is operating the microscope. Examination of a low index pattern on the screen and comparison with the appropriate map (which can be placed in front of the operator) immediately locates the orientation. Once this is established, suitable diffraction vectors via the shortest tilting paths can be quickly chosen to examine the contrast behavior of defects, for example, determination of Burger's vectors.[26,27] The map is particularly useful for studies of dislocations in noncubic crystals, where spot patterns are often difficult to analyze by inspection under the microscope. Examples of the use of hexagonal maps have been given for the system Ag-Al.[29] Contrast theory predicts that under two-beam conditions and $s \approx 0$, a defect characterized by a displacement vector **R** is invisible (or weakest) when $\mathbf{g} \cdot \mathbf{R} = 0$. Thus, any Kikuchi pair (**g**) which converges at a pole **R** satisfies $\mathbf{g} \cdot \mathbf{R} = 0$, and the Kikuchi map indicates the sense and amount of tilt needed to obtain that two-beam orientation (Fig. 2.9*a*). For example, Fig. 2.31 is the appropriate map for bcc crystals; dislocations whose Burger's vector are along [111] will thus be invisible for all reflections (Kikuchi lines) which pass through the [111] pole, (e.g., $0\bar{1}1$, $1\bar{2}1$, $1\bar{1}0$). Similar arguments apply to strain contrast images in general.

Such applications enable one to carry out the somewhat tedious process of contrast analysis at the microscope in the minimum of time.

5.5.3 *Stereomicroscopy.* The electron microscope image is a two-dimensional projection of the volume of the specimen being examined. The true three-dimensional image of the specimen can be obtained, as outlined in Chapter 1, Section 4.10, by taking stereo pairs (two pictures) of the same area, without changing **g** (or s), by simply tilting along the Kikuchi band corresponding to **g** by about 10°. The Kikuchi map greatly simplifies and facilitates this process. Suppose, for example, that one wanted a stereo image near the 0001 zone of a hcp crystal, using $\mathbf{g} = [0\bar{1}10]$. Reference to Fig. 2.32 shows that a tilt from [0001] to $[2\bar{1}\bar{1}9]$ left along the $0\bar{1}10$ Kikuchi band shifts the viewing angle by about 13°, which is ample for stereo images. Although the scale factor for hcp (and noncubic in general) maps will depend on c/a, the tilt angle is not critical merely for obtaining three-dimensional information. Sometimes, however, the tilt angle does need to be known accurately (e.g., for accurate thickness measurements or depth determinations, as in the analysis of small defects after irradiation). The tilt angle is accurately determined by taking two diffraction patterns, one before and one after tilting, and after locating these on the map, measuring the tilt angle directly from the map, using the appropriate scale factor.

5.5.4 *Measurements from Intensity Distributions.* Normally, diffraction patterns are obtained using a defocused condenser lens system in which the illumination is nearly parallel. When a fully focused condenser lens system is employed, or special alignment procedures are used as mentioned in Chapter1, Section 4.9, a convergent beam pattern is obtained. The diffraction spot then becomes a disk, being the image of the condenser aperture, exhibiting intensity fluctuations (the "rocking curve") across its diameter. The control of orientation and the unique indexing of the diffraction spot used for intensity measurements are facilitated by utilizing the Kikuchi pattern. For example, Figs. 2.34*a–c* show a convergent beam pattern for the 440, 220, and 000 reflections in silicon. The microphotometer traces corresponding to these are shown in Fig. 2.34*d*.

Amelinckx's relation between foil thickness and intensity distribution is [30]

$$t = 2^{1/2}(s_1^2 - 2s_2^2 + s_3^2)^{-1/2}, \tag{2.29}$$

where s_1, s_2, s_3 are the deviation parameters of three successive dark field intensity minima in the dynamical intensity distributions.

For small angles this equation can be rewritten[31] as

$$t = \frac{2^{1/2}d^2}{\lambda}\left[\left(\frac{\Delta\theta_1}{2\theta_B}\right)^2 - 2\left(\frac{\Delta\theta_2}{2\theta_B}\right)^2 + \left(\frac{\Delta\theta_3}{2\theta_B}\right)^2\right]^{-1/2}, \tag{2.30}$$

where θ_B is the Bragg angle and $2\theta_B$ is the angle between diffraction "spots." Thus $\Delta\theta_1, \Delta\theta_2, \Delta\theta_3$, and $2\theta_B$ can be measured in any convenient units and are independent of magnification.

Further applications of this method include the determination of extinction distances[23] and magnification calibrations.

5.5.5 *Calibration of Electron Wavelength (and Accelerating Voltage).* The wavelength of electrons is related to the magnification of the diffraction pattern through the camera constant equation

$$\lambda L = pd,$$

where p is the measured width of the Kikuchi band in the pattern (Fig. 2.30). To find λ it is necessary to know the value of L. This can be determined by using the Kikuchi pattern as first indicated by Uyeda et al.[32] Actually, by utilizing a known pattern which has been accurately calibrated as to the angle-distance scale factor (Figs. 2.28, 2.31, 2.32), a simple method of analysis can be followed.

Suppose that on a given Kikuchi pattern two poles, A and B, have been uniquely indexed (e.g., by comparing to the appropriate map). Since the angle ϕ between these poles can easily be calculated or obtained from tables of angles, and the distance y between poles measured on the actual pattern, the scale factor is ϕ/y (rad cm^{-1}). Now take any known low index (hkl) Kikuchi band of spacing p; then since, by Bragg's law, for small angles

$$\lambda = 2\theta d,$$

$$2\theta = \frac{\lambda}{d} = \left(\frac{\phi}{y}\right)p$$

or

$$\lambda = \frac{d\phi p}{y},$$

which for cubic crystals is written as

$$\lambda = \frac{\phi a p}{y\sqrt{h^2 + k^2 + l^2}}. \qquad (2.31)$$

Hence λ can be found without knowing the plate-specimen distance L.

It should be noted that a Kikuchi map, for example, Figs. 2.28, 2.31, and 2.32, for a particular known voltage may be used as a standard to calibrate other voltages (e.g., for high voltage microscopes). If λ_1 is the true wavelength determined for a given map for a single crystal pattern identical in orientation to a part of a map, the unknown

$$\lambda_2 = \lambda_1 \left(\frac{p_2}{p_1} \cdot \frac{y_1}{y_2}\right),$$

where p_1, y_1 and p_2, y_2 are measured on the map and pattern, respectively.

5.5.6 *Phase Transformations.* The Kikuchi pattern is helpful in identifying changes in microstructure produced by phase transformations, for example, in detecting modulations in composition (and/or structure) due to ordering, clustering, and spinodal reactions. Two types of approach can be utilized: (*a*) analysis of symmetrical patterns, and (*b*) analysis of overlapping patterns in which line pairing from reflections of identical indices but different *d*-spacings are examined.[33] Examples have been given in ref. 28. Also, since Kikuchi patterns enable one to obtain orientations very accurately, they should always be used where possible for determining orientation relationships between phases and for measurements of orientation changes across boundaries.

5.5.7 *Grain Boundaries.* As noted above, orientation changes across boundaries are easily and accurately measured. Determining the relationship between the orientations of the grains on both sides of a grain boundary (as well as the grain boundary plane and displacements) is of fundamental importance in studying the properties of grain boundaries. A number of techniques based on the analysis of Section 5.3 have been used to determine this relationship; see, for example, refs. 28 and 34-37.

5.5.8 *Critical Voltage Measurements.* One other application worth noting is the measurement of critical voltage—the voltage at which second (or higher) orders of reflection go through a minimum in intensity (see, e.g., ref. 38 or 39). The effect is a consequence of many-beam dynamical interactions (see Chapter 5, Section 6), and values of the critical voltage can be easily measured (if a high voltage electron microscope is available) or calculated. For example, Table 2.5 lists values of critical voltage using systematic interactions (see, e.g., refs. 38-45).

Table 2.5 Critical Voltages for Several Elements

Structure	Element	B	Reflection	Calculated Three Beam V_c (kV)[a]	Experimental V_c (kV)
Fcc	Al	0.85[b]	222	–, 465	425,[c] 430[b]
			400	–, 1010	–
			440	–, 3700	–
	Cu	0.54[b]	222	130, 405	310,[c] 325[b]
			400	380, 840	600[c]
			440	1530, 1900	–
	Ni	0.40[b]	222	145, 380	295[b]
			400	410, 670	610[d]
			440	1680, 1870	–
	Ag	0.58[e]	222	25, 75	–
			400	215, 230	225[c]
			440	1075, 910	–
			622	1770, 1500	–
	Au	0.62[e]	222	-110, –	–
			400	35, –	–
			440	630, –	715[f]
			622	1100, –	–
Bcc	Fe	0.36[b]	220	240, 390	305[b]
			400	1240, 1350	–
	W	0.20[e]	220	-80, -40	–
			400	520, 590	550[c]
			422	1175, 1280	–
Dc	Si	0.24[e]	440	–, 1440	1400[g]
	Ge	0.87[e]	440	680, 960	–

[a] Electron scattering data from International Tables (Ibers and Vainshtein). Values on the left from the Thomas-Fermi-Dirac statistical model data. Values on the right from self-consistent field data (W. L. Bell). See eq. 5.27 for formula.
[b] Reference 38.
[c] Reference 39.
[d] Experimental observation, W. L. Bell.
[e] International Tables: Table 3.3.5, 1A (1962).
[f] Reference 43.
[g] Reference 48.

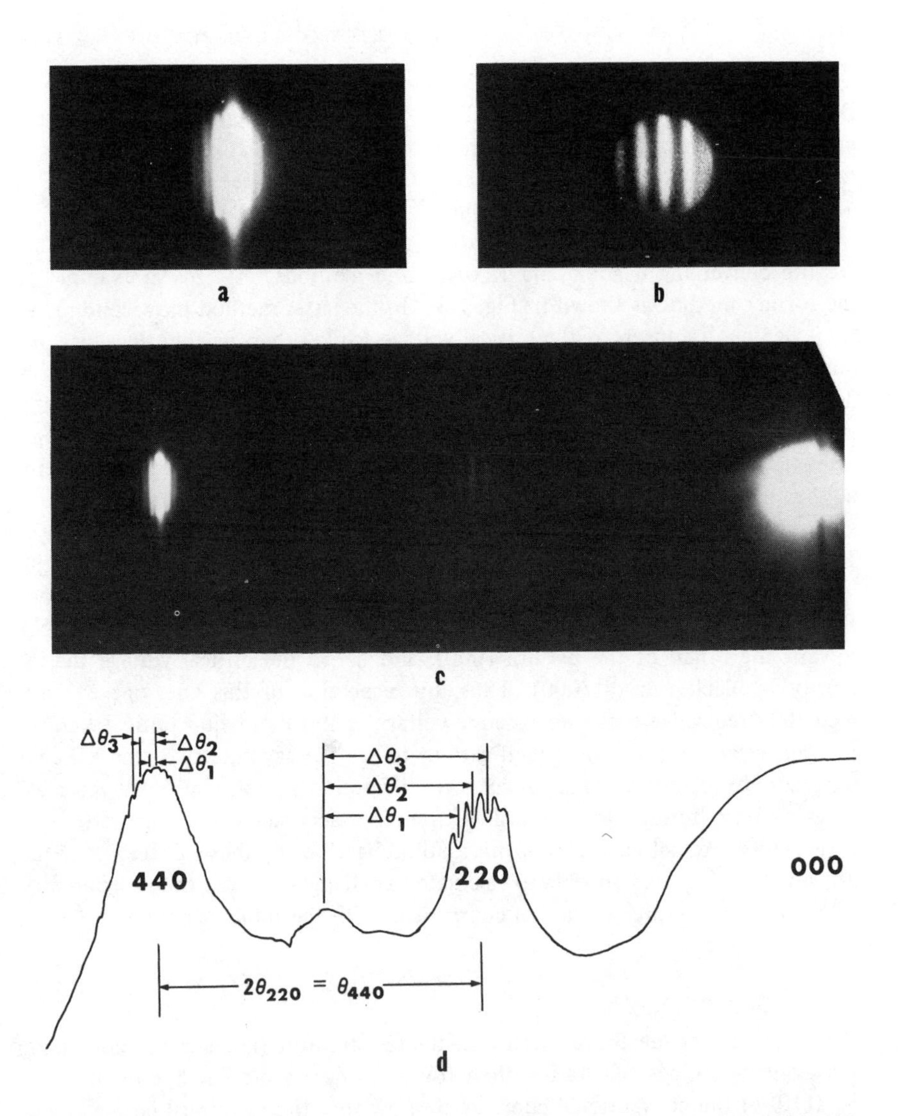

Fig. 2.34 (*a*–c) convergent beam photograph of the diffraction pattern of the 000, 220, and 440 maxima in silicon; (*a*) and (*b*) are enlargements of the 220 and 440 reflections in (*c*). (*d*) Microphotometer trace of (*c*). Courtesy North-Holland.[35]

The critical voltage depends sensitively on the *d*-spacings of reflecting planes, Debye temperature, and structure factors. Thus measurements of critical voltage provide data on local chemical composition, scattering factors, Debye temperature, improved contrast at lattice defects, and so on (see, e.g., refs. 38, 39, 43-47) and appear to be very promising in studies of phase transformations.[44]

Experimentally, the two most accurate methods of measuring critical voltage are the converging beam[41] and Kikuchi line methods.[38,42,47] An example of the former method is shown in Fig. 2.34. In the latter method the second-order Kikuchi line disappears at the critical voltage while other, neighboring lines are unaffected. Calibration of the operating voltage is essential and is done as described in Section 5.5.5. Measurements are made using a small field-limiting aperture (e.g., 20-50 μm diameter) with a fully focused second condenser and a slightly off-focus diffraction lens (this makes it possible to observe a converging beam, as in Fig. 2.34). The specimen is tilted so that the Kikuchi pattern is at s slightly positive of 2g (so the Kikuchi intensity is more easily observed). There is an asymmetry in the intensity of the line occurring at the center of the second-order Kikuchi band, which will pass close to the first-order reflection. Below the critical voltage *on the plate or film negative* the intensity is light toward the origin of the pattern (000), and above the critical voltage the intensity is black toward (000). Thus, by inspection of this line, one can immediately recognize if the microscope voltage is above or below critical voltage. It is necessary to use high magnifications to see the asymmetry in the Kikuchi line; and by carefully taking successive diffraction patterns at 5 kV intervals so as to pass through the reversal in intensity, accuracies of 1% or better can be achieved. Actual examples of this method have been published (see, e.g., refs. 38, 44, 46). It is also possible to calculate the changes in Kikuchi line symmetry and display the results for ease of comparison with the actual experiments.[42]

Exercises

2.1 Noting that the Bravais lattice of the fcc structure (i.e., a lattice containing only one point per cell, rather than four) has base vectors **a**, **b**, **c** of the form $a/2\langle 110\rangle$ in the conventional cubic axes, show that the reciprocal lattice is bcc. What is the length of the side of the conventional bcc cell of this reciprocal lattice?

2.2 Do a similar exercise, starting from the bcc structure, to show that its reciprocal lattice is fcc.

2.3 Using the results of Exercises 2.1 and 2.2, confirm that the allowed reflections for fcc and bcc in Table 2.2 and Fig. 2.2 are correct.

2.4 Index the idealized symmetrical diffraction patterns of Fig. 2.35, which are for (*a*) fcc, (*b*) bcc, and (*c*) hcp (hexagonal with $c/a = \sqrt{8/3}$), respectively.

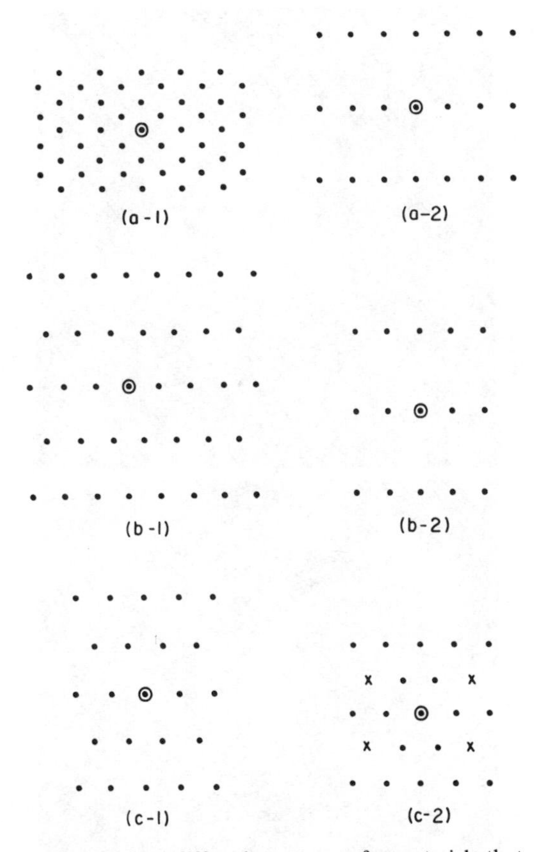

Fig. 2.35 Idealized symmetrical diffraction patterns for materials that are (*a*) fcc, (*b*) bcc, and (*c*) ideal hcp (see Exercise 2.4).

What is the approximate beam direction in each case? Explain the spots marked x in Fig. 2.35*c* (ii).

2.5 The diffraction patterns of Fig. 2.36 are from a specimen of a polycrystalline, cubic material with lattice parameter $a = 4.078 \times 10^{-10}$ m, taken with fixed camera length L, using electrons of energy (*a*) 60 keV, (*b*) 80 keV, (*c*) 100 keV. Determine:

(*a*) The lattice type of the specimen.

(*b*) The value of $\lambda L/a$ for each pattern.

(*c*) The camera length (assuming for simplicity that 60 keV electrons are nonrelativistic).

(*d*) The relativistic correction to the wavelength for 100 keV electrons.

Compare your result for (*d*) with eq. 1.5.

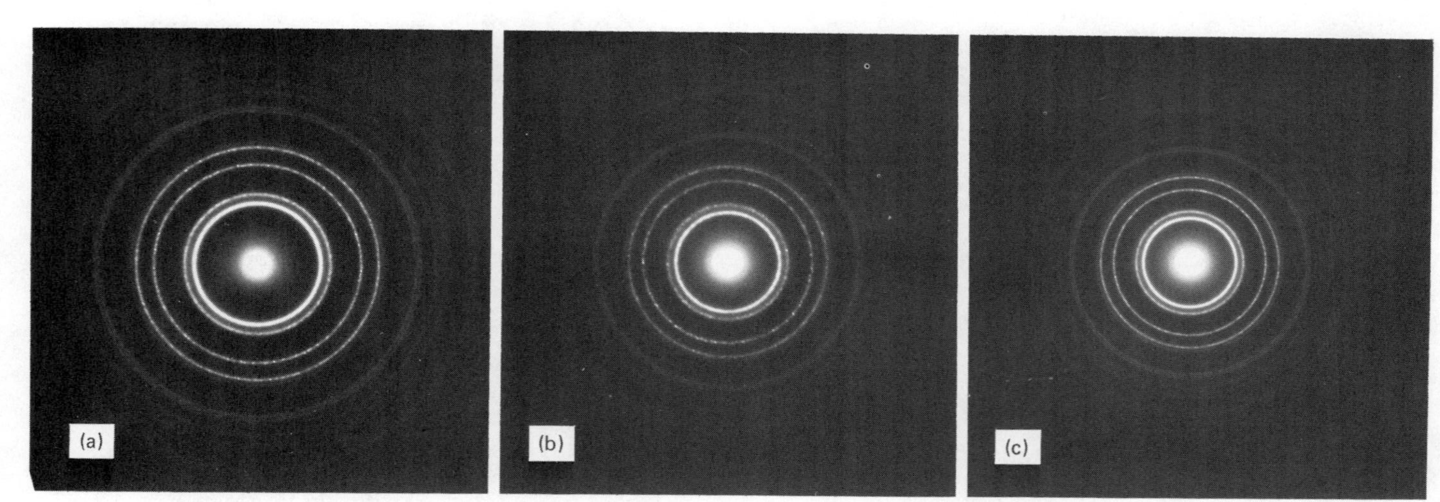

Fig. 2.36 Ring diffraction patterns from a polycrystalline specimen at a) 60 keV, b) 80 keV and 3) 100 keV (see Exercise 2.5).

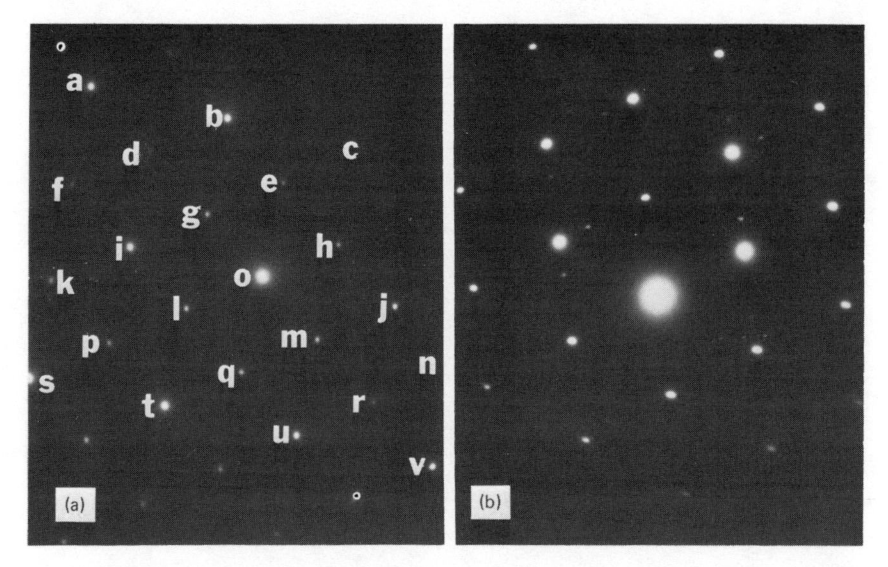

Fig. 2.37 Diffraction patterns from twinned fcc specimens (see Exercise 2.7).

2.6 Figure 2.1 is the [001] stereographic projection of a fcc matrix with, superimposed, the effect of twinning on (111). Confirm that, if the electron beam direction with respect to the matrix were [001], it would be $[\bar{2}\bar{2}1]$ in the twin, and the diffraction pattern would contain no *extra* spots.

2.7 The diffraction patterns of Fig. 2.37 were produced from fcc crystals containing twins. Explain in the case of Fig. 2.37*a* why some of the reflections (e.g., *a*, *b*, *i*, *j*, *s*, *t*, *u*, *v*) have such large intensity, and in the case of Fig. 2.37*b* how the *extra* spots may have been produced.

2.8 The lines shown in Fig. 2.38 represent the three narrowest Kikuchi bands intersecting at a Kikuchi pole of an fcc crystal. Check that they have been correctly indexed.

2.9 The diffraction pattern of Fig. 2.39 is of silicon, which has the diamond cubic structure with lattice parameter 5.428×10^{-10} m. Index a number of prominent Kikuchi lines and hence determine:

(*a*) The incident beam direction.

(*b*) The wavelength of the electrons (and thus the operating voltage of the microscope.

2.10 Estimate the thickness of the specimen of silicon that produced the convergent beam diffraction pattern of Fig. 2.34 (electron energy-100 keV).

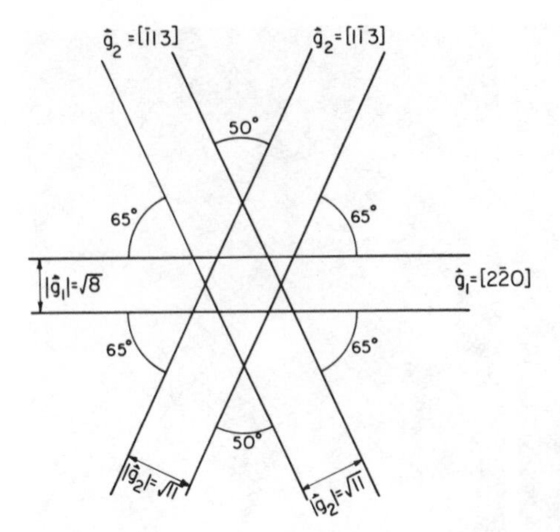

Fig. 2.38 Schematic Kikuchi line pattern to illustrate indexing (see Exercise 2.8).

Fig. 2.39 Diffraction pattern from a silicon single-crystal specimen (see Exercise 2.9).

References

1. Champeney, D. C., *Fourier Transforms and Their Physical Applications*, Academic, London, 1973.
2. Okamoto, P. R. and Thomas, G., *Phys. Status Solidi.*, **25,** 81 (1968).
3. Thomas, G., Bell, W. L., and Otte, H. M., *Phys. Status Solidi.*, **12,** 353 (1965).
4. Ryder, P. L. and Pitsch, M., *Phil. Mag.*, **18,** 807 (1968).
5. Samudra, A. V., Johari, O., and Heimendahl, M. V., *Pract. Metallogr.*, **9,** 516 (1972).
6. Thomas, G., Cheng, I-Lin, and Mihalisin, J. R., *Trans. ASM*, **62,** 852 (1969).
7. Pashley, D. W. and Stowell, M. J., *Phil. Mag.*, **8,** 1605 (1963).
8. Bullough, R. and Wayman, M., *Trans. AIME*, **236,** 1704 (1966).
9. Johari, O. and Thomas, G., *Stereographic Projection and Applications*, Wiley-Interscience, New York, 1969.
10. Thomas, G. and Das, S. K., *J. Iron Steel Inst.*, **209,** 801 (1971).
11. Daniel, W. and Lipson, H., *Proc. R. Soc.*, **A182,** 378 (1943).
12. Butler, E. P. and Thomas, G., *Acta Met.*, **18,** 347 (1970).
13. Marcinkowski, M. J., in *Electron Microscopy and Strength of Crystals* (Ed. G. Thomas), University of California Press, 1971, p. 333.
14. Khachaturyan, A. G., *Order-Disorder Transformations in Alloys* (Ed. H. Warlimont), Springer, New York, 1974, p. 114.
15. Das, S. K., Okamoto, P. R. Fisher, P. M. J., and Thomas, G., *Acta Met.*, **21,** 913 (1973).
16. Dahmen, U. and Thomas, G., 1979, Script. Met., (in press).
17. Das, S. K. and Thomas, G., *Order-Disorder Transformations in Alloys*, (Ed. H. Warlimont), Springer, New York, 1974, p. 332.
18. International Conference on Ordering, 1977. *J. Phys. Colloq.*, 7 **C7-165** (1977) No 12, Tome 38.
19. Madden, H. H., *J. Vac. Sci. Technol.*, **13,** 228 (1976).
20. Krivanek, O. L., Gaskell, P. H., and Howie, A., *Nature*, **202,** 454 (1976). See also Krivanek, O. L. and Howie, A., *J. Appl. Crystallogr.*, **8,** 213 (1975).
21. Kikuchi, S., *Jap. J. Phys.*, **5,** 83 (1928).
22. Thomas, G. and Bell, W. L., *Proceedings of the European Electron Microscopy Congress*, Rome, 1968, p. 285.
23. Tan, T. Y., Bell, W. L., and Thomas, G., *Phil. Mag.*, **24,** 417 (1971).
24. Kainuma, Y., *Acta Crystallogr.*, **8,** 247 (1955).
25. Von Heimendahl, M., Bell, W. L., and Thomas, G., *J. Appl. Phys.*, **35,** 361 (1964).
26. Levine, E., Bell, W. L., and Thomas, G., *J. Appl. Phys.*, **37,** 2141 (1966).
27. Okamoto, P. R., Levine, E., and Thomas, G., *J. Appl. Phys.*, **38,** 289 (1967).
28. Thomas, G., *Trans. AIME*, **233,** 1608 (1965).
29. Okamoto, P. R. and Thomas, G., *Acta Met.*, **15,** 1325 (1967).
30. Amelinckx, S., *Direct Observation of Dislocations*, Academic, New York, 1964, p. 193.
31. Bell, W. L. and Thomas, G., *Proceedings of the 27th Conference of the Electron Microscopy Society of America*, Claitors, Baton Rouge, LA, 1969, p. 158.

32. Uyeda, R., Nonoyama, M., and Kogiso, M., *J. Electron Microsc. (Jap.)*, **14**, 296 (1965).
33. Villagrana, R. E. and Thomas, G., *Phys. Status Solidi*., **9**, 499 (1965).
34. Pumphrey, P. H. and Bowkett, K. M., *Proceedings of the 7th International Congress on Electron Microscopy*, Grenoble, 1970, p. 189.
35. Thomas, G., *Modern Diffraction and Imaging Techniques in Materials Science* (Eds. S. Amelinckx et al.), North Holland, Amsterdam, 1970, p. 159. 2nd Ed. (rev.), 1978, p. 399.
36. Young, C. T., Steele, J. H., and Lytton, J. L., *Met. Trans*., **4**, 2081 (1973).
37. Smith, D. A., *J. Phys*., **36**, C4-1 (1975).
38. Watanabe, D., Uyeda, R. and Fukuhara, A., *Acta Crystallogr*., **A24**, 580 (1968); **A25**, 138 (1969).
39. Lally, J. S., Humphreys, C. J., Metherell, A. J. F., and Fisher, R. M., *Phil. Mag*., **25**, 321 (1972).
40. Bell, W. L., *Proceedings of the High Voltage Electron Microscopy Conference*, AERE Harwell, England, 1970, p. 35.
41. Bell, W. L., *Proceedings of the 1971 Conference of the Electron Microscope Society of America*, Claitors, Baton Rouge, LA, 1971, p. 184.
42. Thomas, L. E. and Humphreys, C. J., *Phys. Status Solidi*., **3**, 599 (1970).
43. Lally, J. S., Thomas, G. and Fisher, R. M., *Proceedings of the 1973 Conference of the Electron Microscope Society of America*, Claitors, Baton Rouge, LA, 1973, p.4.
44. Butler, E. P., *Phys. Status Solidi*., **18**, 71 (1973).
45. Bell, W. L. and Thomas, G., *Electron Microscopy and Structure of Materials* (Ed. G. Thomas), University of California Press, Berkeley, 1971, p. 23.
46. Fisher, R. M., *Electron Microscopy and Structure of Materials* (Ed. G. Thomas), University of California Press, 1971, p. 60.
47. Watanabe, D., Uyeda, R., and Kogiso, M., *Acta Crystallogr*., **A24**, 249 (1968).
48. Thomas, G. and Lacaze, J.-C., *J. Microsc*., **97**, 301 (1973).

THREE

INTRODUCTION TO CONTRAST ANALYSIS AND ITS APPLICATIONS

In this chapter some simple geometrical approaches are discussed that may help the beginning electron microscopist to better understand the nature of contrast and use the microscope effectively for characterizing the structure of materials. The Bibliography at the end of Chapter 1 also applies to this chapter. In recent years developments have been made in what is now often referred to as nonconventional techniques such as weak beam dark field imaging[1] (Chapter 1, Section 4.7), an example of which is shown in Fig. 3.1, and special effects from many-beam interactions that become important at high energies. These phenomena are discussed fully in Chapters 4 and 5 and in recent symposia (e.g., refs. 2-6), but some mention of the latter is made in this chapter. Some important and useful applications of the effects of anomalous absorption that affect the symmetry properties of the image are also described.

The contrast in the image depends on the electron distribution leaving the bottom surface of the specimen. Just as the diffraction pattern is the Fourier transform of the object, so the image is the Fourier transform of the diffraction pattern, so that all the points discussed in Chapter 2 apply directly to the understanding of image contrast phenomena. There are two methods of general applicability by which the electron distribution is modified on passing through the microscope lenses to produce an image. These are phase contrast imaging (e.g., lattice imaging), which depends on the recombination of two or more beams emitted from the specimen (see Fig. 1.2*a*), and amplitude contrast imaging, in which only one beam forms the image (Fig. 1.3). Although phase contrast imaging is a very powerful method, with resolutions better than 2 Å (Fig. 1.2*b*), the more generally applied technique is that of amplitude contrast. Thus in this

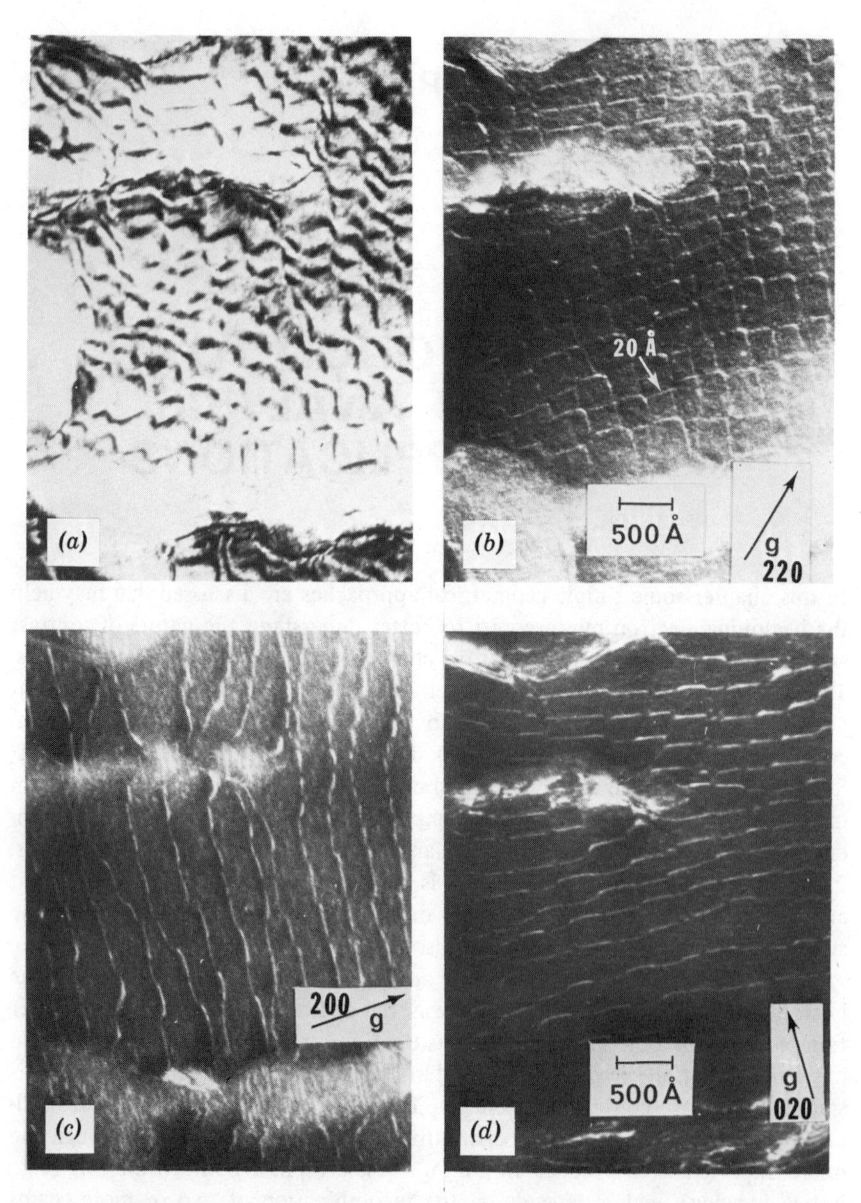

Fig. 3.1 (*a*) Bright field and (*b–d*) weak beam dark field images of interphase boundary dislocations in overaged Cu-Mn-Al alloy showing images ~20 Å wide. These are high resolution, kinematical images ($s >> 0$) formed in $-\mathbf{g}$ with $[s_{n\mathbf{g}} > 0]$ excited, where $n = 3$ or larger (see Fig. 3.2). D. J. H. Cockayne and G. Thomas (unpublished). Courtesy D. J. H. Cockayne.

chapter amplitude contrast will be described first, and the phase contrast imaging method will be discussed in Sections 12 and 13.

AMPLITUDE CONTRAST IMAGING

1 Introduction

Amplitude contrast is achieved in either of two ways: (*a*) formation of the bright field image by using only the transmitted beam, or (*b*) formation of a dark field image by using only one strong diffracted beam. These operations are carried out by means of the objective aperture, which is inserted at the back focal plane of the objective lens (Fig. 1.3). The dark field image is best obtained by gun tilting or by deflection so as to allow the diffracted beam to pass along the optic axis, thereby reducing the errors from chromatic and spherical aberrations that occur if the objective aperture is moved off the optic axis (Figs. 1.3 and 1.4). An important point to realize is that for the axial dark field as shown in Fig. 1.3 the gun translation or beam deflection is such that the direction of **g** is reversed (i.e., if **g** is excited for bright field, the gun tilt dark field obtained is $-\mathbf{g}$). This must be remembered when making the correct alignment between images and diffraction patterns.

2 Kinematical Approximation—Perfect Crystals

As explained earlier, the kinematical theory is applicable only to thin specimens and for conditions away from the exact Bragg position ($s \neq 0$). In Chapter 2 the kinematical intensities for a crystal of thickness t were derived, assuming that all the energy is conserved (no absorption), that is (eq. 2.14),

$$\text{Diffracted:} \qquad |\psi|_D^2 \cong \left(\frac{F}{t}\right)^2 \frac{\sin^2 \pi t s}{(\pi s)^2}, \tag{3.1}$$

$$\text{Transmitted:} \qquad |\psi|_T^2 \cong 1 - \left[\left(\frac{F}{t}\right)^2 \frac{\sin^2 \pi t s}{(\pi s)^2}\right]. \tag{3.2}$$

Since $|\psi|_T^2 + |\psi|_D^2 = 1$, the kinematical theory predicts that bright and dark field images are complementary. In practice, absorption occurs and this symmetry property is modified. However, the kinematical function predicts periodic variations in intensity with thickness for constant s (thickness fringes), or variations in intensity with s for constant thickness, which lead to fringes about the Bragg contours. The (*hkl*) Bragg contour in the image corresponds to the *hkl* spot or Kikuchi line in the diffraction pattern and is identified easily in dark field. Intensity minima occur in dark field whenever $s = n/t$ (constant t) or when $t = n/s$

for constant s. The periodic variation of $|\psi|^2$ with t leads to primary extinction $t_0 = 1/s$, as discussed in Chapter 2. In the dynamical theory (see Chapters 4 and 5) the extinction distance ξ_g is defined as $\pi V/\lambda F_g$, where V is the volume of the unit cell and F_g the structure factor for the particular reflection. The deviation from Bragg's condition is then often referred to by the dimensionless parameter $w = \xi_g s$, and in the kinematical case $w \gg 1$.

At every integral number of extinction distances all the electrons end up in the forward direction (transmitted), and at every odd half-multiple of extinction distances all electrons end up in the diffracted direction. The extinction distance is inversely proportional to the scattering factor, which decreases with increasing scattering angle. Thus electrons have short extinction distances (100 to 1000 Å), whereas X-rays have very long extinction distances.

The exchange of electron intensity between the transmitted and diffracted beams is exactly analogous to the motion of two coupled harmonic oscillators, which periodically exchange all the vibrational energy of the system. This forms the basis of the dynamical theory. The theory also shows that subsidiary maxima occur when $[s^2 + (\xi_g)^{-2}]t^2$ = integral. For thin crystals the value of t can be determined from measurements of s at subsidiary fringes either in the image or in the diffraction patterns,[7] provided that the values of ξ_g are known.

From the foregoing, contrast effects in an otherwise perfect crystal are expected to be due to the following:

1. Changes in t—wedge fringes, fringes at inclined defects.
2. Changes in s—Bragg contour fringes.
3. Changes in orientation—changes in s and g.

Thus in polycrystals the intensity varies from grain to grain because of differences in diffracting conditions due to the different orientation of each grain. In general, therefore, contrast from crystals is not limited by resolution except in special cases. However, since the contrast is very orientation sensitive, it is essential to use a goniometric specimen stage, preferably with as large a tilting range as possible so that the diffracting conditions can be varied in a systematic matter. Without such a stage, quantitative characterization of microstructure is almost impossible.

3 Contrast in Imperfect Crystals

3.1 General Comments

In practical materials there are complicated microstructures; and the aim of electron microscopy is to recognize and characterize them. Important microstructure features include:

1. Changes in orientation with or without change in structure or composition, such as grains, twins, precipitates, structure of boundaries.
2. Lattice defects: point defects, line defects, planar defects, volume defects (effects due to elastic displacements).
3. Multiple-phase systems (*a*) changes in composition but not structure (e.g., spinodals), (*b*) changes in composition and structure (general precipitation), (*c*) changes in structure but not composition (e.g., martensites), (*d*) interphase interfaces (coherent, partially coherent, incoherent).

Contrast from these features will arise from such effects as changes in the local diffracting conditions {changing s and g (d-spacings)}, phase changes on crossing interfaces, structure factor changes, and changes in effective thickness (changing ξ_g). The situation can become quite complex especially when the defect density is high and strain fields overlap as in heavily deformed crystals, or crystals containing large volume fractions of particles.

The combination of bright and dark field imaging techniques and diffraction pattern analysis is essential in the characterization procedure. Analysis should always start from the diffraction pattern and most of the interpretation will be carried out *at the microscope.* For contrast work it is recommended that two-beam orientations be used (as in Fig. 2.9*a*). For this reason it becomes essential to recognize orientations by inspection so that particular reflections of interest can be brought into operation. It is convenient to start by tilting the foil into a recognizable symmetry orientation and then proceed from there with the tilting. The use of Kikuchi maps greatly facilitates this process, as will be shown later. It cannot be emphasized too strongly how important it is to have a sound working knowledge of diffraction patterns and three-dimensional crystallography. Thus all the information developed in Chapters 1 and 2 is needed for analysis.

3.2 Information Requirements for Analysis (Two-Beam Conditions)

Generally it is necessary to know the foil orientation, direction of the diffraction vector, sign of s, and foil thickness.

3.2.1 *Precise Orientation.* Because of the 180° ambiguity in spot patterns, the spot pattern by itself does not give the unique foil orientation, and hence the geometry of defects in the foil is not known since the image is a two-dimensional projection of the object. Thus the top of the foil is not distinguishable from the bottom, nor up from down. However, a Kikuchi pattern can be indexed uniquely (provided that at least two poles are present), and this is facilitated by comparison with the appropriate Kikuchi map. If Kikuchi patterns are not obtainable, several other methods can be utilized. One example is the use of special absorption contrast effects, such as the asymmetry in the dark field image when s is not quite zero, namely, for $s > 0$ the bottom of the foil is in stronger contrast than the

top; the reverse is true for $s < 0$ (see, e.g., Fig. 3.15 and ref. 8). Alternatively, large angle tilting experiments, in which the change in projected size of an object in the foil is observed, can be performed.

Once the orientation is known, the geometry of the foil is known, and hence so also is the sense of slope of planes and directions. This information is needed for quantitative analyses involving determination of the sense of strain fields (e.g., vacancy or interstitial loops or faults).

Stereomicroscopy is useful also, because by this technique one can obtain information on the depth distribution of defects. Stereo pairs can be obtained by tilting 8 to 10° along a Kikuchi band so that the diffracting conditions are not altered (see Chapter 1, Section 4.10).

In orienting the pattern with the image, because of the inversion between the image and the diffraction pattern, the diffraction pattern should be rotated 180° (plus magnification rotation) with respect to the image, with both negatives emulsion side up, as described in Chapter 1, Section 4.2.* Crystallographic data can then be transferred directly from the pattern to the image. In this way the geometry is preserved with minimum confusion, and the correct sense of the diffraction vector **g** in the image is retained. Of course **g** is identified from the diffraction pattern, and the region in the specimen corresponding to this **g** will reverse contrast in the dark field image of this reflection.

3.2.2 *Sign of* s. The sign of the deviation parameter is important in several instances. In a two-beam absorbing case the intensity in bright field is maximum at $s > 0$ and in dark field at $s = 0$. These conditions are thus readily seen directly in the image. In the diffraction pattern for $s > 0$ the Kikuchi line will lie outside the corresponding spot since the reciprocal lattice point will lie inside the reflecting sphere. For $s < 0$ the Kikuchi lines lie inside the spot, for example, in symmetrical orientations (Fig. 2.9*b*).

3.2.3 *Foil Thickness.* This can be found in several ways, such as (*a*) from trace analysis of projected defects that go completely through the foil (faults, twins, precipitates); or (*b*) from measurements of subsidiary fringes either in the convergent beam pattern (Fig. 2.33) or from Bragg contour fringes.[7] In a general method developed by von Heimendahl[9] one or two latex balls of known diameter are applied to both foil surfaces in the area viewed. Changes in dimensions are observed after a known tilt, and from the geometry, the thickness can be calculated to within about 4% accuracy.

*The sense of this rotation (which can be zero also) depends on the type of instrument, the number of the lenses, and the way in which the lenses are assembled. Each instrument therefore requires its own rotation calibrations.

4 Visibility of Lattice Defects: General Criteria

Defects can be described in terms of translational vectors which represent displacements of atoms from their regular positions in the lattice. If the general displacement vector is **R**, the kinematical amplitude scattered from the crystal (see eq. 2.10) as a whole becomes

$$\psi \sim \int_{\text{crystal}} \{\exp [2\pi i(\mathbf{g} + \mathbf{s}) \cdot (\mathbf{r}_i + \mathbf{R})]\} \, dt \tag{3.3}$$

or

$$\psi \sim \int_{\text{crystal}} [\exp (2\pi i \mathbf{s} \cdot \mathbf{r}_i) \exp (2\pi i \mathbf{g} \cdot \mathbf{R})] \, dt \tag{3.4}$$

since $\mathbf{g} \cdot \mathbf{r}_i$ = integer, and $\mathbf{s} \cdot \mathbf{R}_n$ may be neglected.

Thus the amplitude scattered by the perfect crystal is modified by the phase factor $2\pi \mathbf{g} \cdot \mathbf{R} = n2\pi$, and n can be integral, zero, or fractional. The case $\mathbf{g} \cdot \mathbf{R} = 0$ is particularly important in contrast work. It has a simple physical meaning, as can be seen from Fig. 2.4. If **R** lies in the reflecting plane, d (and thus $|\mathbf{g}|$) is unaltered, so that the path difference between transmitted and diffracted waves is unaffected by **R**. Since **g** is normal to (hkl), $\mathbf{g} \cdot \mathbf{R} = 0$ is the condition for no contrast due to a displacement **R** (Figs. 3.1 and 3.2). It should be pointed out that for a general defect **R** varies with position. However, for a stacking fault **R** is a constant equal to the displacement vector for the fault (Section 8.1). In this case no contrast arises when $\alpha = 2\pi \mathbf{g} \cdot \mathbf{R} = n \cdot 2\pi$, n an integer.

The magnitude of $\mathbf{g} \cdot \mathbf{R}$ must be sufficient to change the intensity from background so that contrast is detectable (about 10% is enough). For example, for dislocation line defects in crystals $\mathbf{g} \cdot \mathbf{b}$ should be greater than $\frac{1}{3}$ if the lines are to be detectable. Nonintegral values of $\mathbf{g} \cdot \mathbf{b}$ for dislocations mean that **b** is not a lattice translational vector, and such dislocations must therefore always be associated with faults, **b** is the Burgers vector.

From a visibility viewpoint a general displacement **R** can be resolved along the principal axes:

$$\mathbf{R} = \mathbf{R}_x + \mathbf{R}_y + \mathbf{R}_z. \tag{3.5}$$

Examples of defects that are of general interest in the study of crystals and can be described in terms of such displacements are dislocations, dislocation loops, coherent volume defects (e.g., point defect clusters, voids, coherent particles),

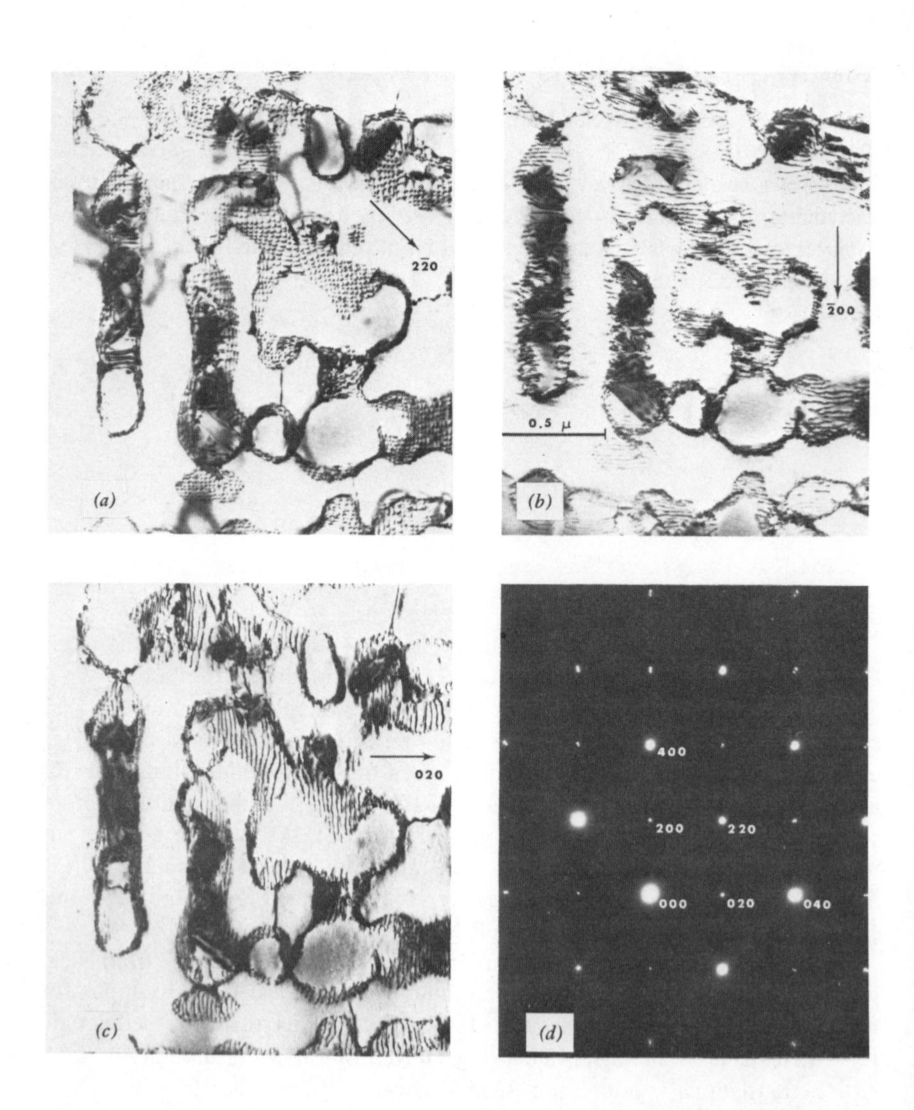

Fig. 3.2 Similar to Fig. 3.1 (different foil) "conventional" bright field images with **g** ($s > 0$) excited. Notice that the image width of these dislocations is much greater than in Fig. 3.1. Since those dislocations are visible when the Burgers vectos are parallel to **g**. They are all pure edge character in Figs. 3.1 and 3.2, with **b** in ⟨100⟩ directions. The diffraction pattern (*d*) shows the alloy to be ordered (*L*21) structure. Courtesy North-Holland.[26]

planar defects such as stacking faults, domain boundaries (chemical, magnetic, electric, order), twin and grain boundaries, and interphase interfaces.

The *visibility* of these defects can all be understood in terms of the simple $\mathbf{g} \cdot \mathbf{R}$ criterion. However, the detailed interpretation of contrast behavior, such as intensities, variation in contrast with depth in the foil, behavior in the exact Bragg case, and influence of other reflections, as well as the analysis of the sense of the displacements associated with the defects, requires the application of the dynamical theory, as described in Chapters 4 and 5.

A truly kinematical situation arises when large s values are used, for example, when imaging in $\mathbf{g}$ with $ngs > 0$ ($n = 2$, 3, or larger) excited. Such images are called weak beam images (because the intensity in $\mathbf{g}$ is low) and have high resolution.[1] An example for pure edge dislocations is shown in Fig. 3.1, which can be compared to Fig. 3.2. The method of obtaining weak beam images is given in Chapter 1, Section 4.7.

Since, from Eq. 3.5, $\mathbf{R}_z \cdot \mathbf{g}$ is always zero, only displacements lying in the plane of the foil are of importance in producing contrast. For screw dislocations the displacements $\mathbf{R}$ are always parallel to the Burgers vector $\mathbf{b}$; hence when $\mathbf{g} \cdot \mathbf{b} = 0$ screws are invisible. In the case of an edge dislocation, the principal components of $\mathbf{R}$ are $\mathbf{R}_b$ and $\mathbf{R}_n$ (displacements parallel to and normal to $\mathbf{b}$ respectively). For an edge dislocation with its half plane parallel to the beam, $\mathbf{g} \cdot \mathbf{b} = n$ (including zero), whereas $\mathbf{g} \cdot \mathbf{R}_n = 0$. On the other hand, if the dislocation is oriented with its half plane normal to the beam, $\mathbf{g} \cdot \mathbf{b} = 0$, but $\mathbf{g} \cdot \mathbf{R}_n = m$ (including zero). Thus because of the displacements $\mathbf{R}_n$ edge dislocations do not necessarily go out of contrast completely when $\mathbf{g} \cdot \mathbf{b} = 0$, except under conditions when $\mathbf{g} \cdot \mathbf{R}_n$ also goes to zero. For this reason it is possible to see edge dislocations when their Burgers vectors are parallel to the incident beam. For example, pure edge prismatic dislocation loops are visible (by so-called residual contrast) when they are parallel to the foil plane, as shown in Fig. 3.3*a*. Notice in this case that loop segments are invisible where $\mathbf{g} \cdot \mathbf{R}_n = 0$, so the loops have a line of no contrast for the parts that are normal to $\mathbf{g}$. The geometry of the situation may be seen even more clearly for the sides of the interstitial Frank (faulted) loop of hexagonal shape shown in Fig. 3.3*b*. The sides are only truly invisible when $\mathbf{g}$ is parallel to the dislocation line direction, even though $\mathbf{g} \cdot \mathbf{b} = 0$ for the whole loop. A similar contrast occurs for spherical strain fields, as in the case of coherent particles, because the plane normal to $\mathbf{g}$ is unaffected by the strain; hence $\mathbf{g} \cdot \mathbf{R}_n = 0$–a well-known example is Cu–Co (Fig. 3.4). In both cases therefore, as the direction of $\mathbf{g}$ is changed, the line of no contrast shifts so as to always lie normal to $\mathbf{g}$. This behavior allows such defects to be distinguished from perfect loops, which also exhibit arc contrast (independently of $\mathbf{g}$, $\mathbf{g} \cdot \mathbf{b} \neq 0$) when inclined to the foil plane (Fig. 3.9).

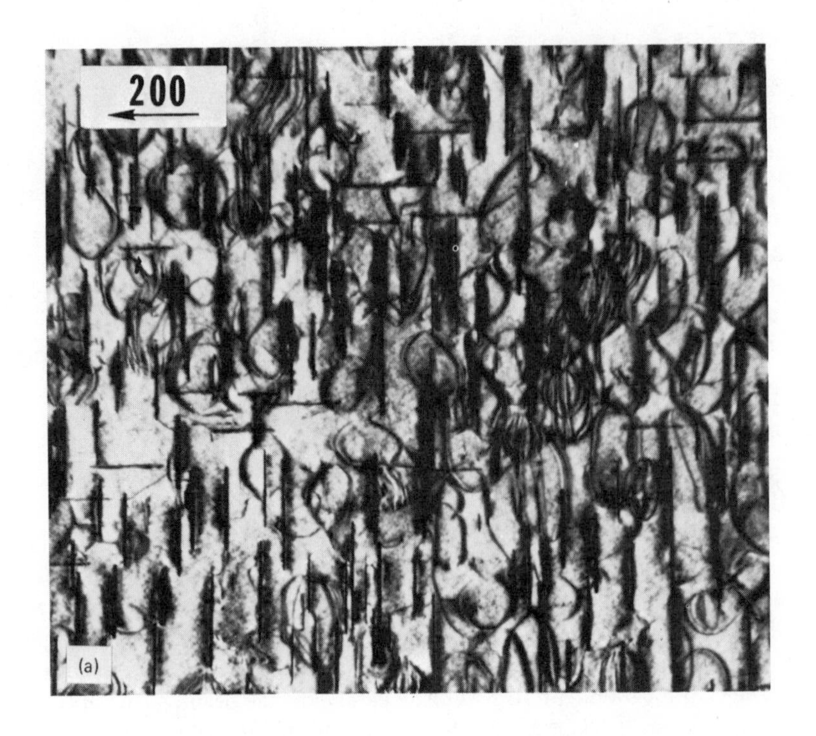

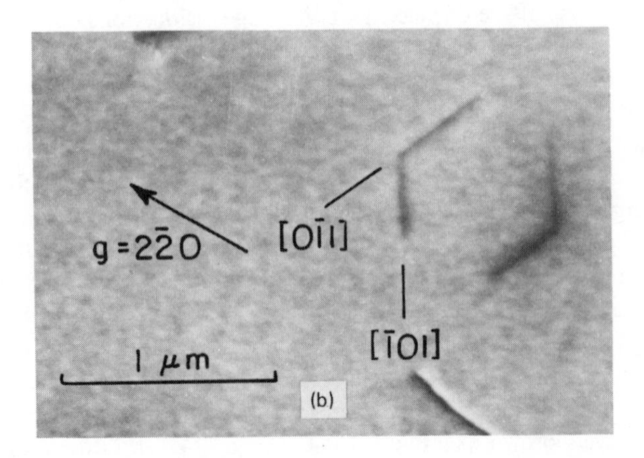

Fig. 3.3 (*a*) Contrast from dislocation loops around θ' plates in aged Al-4% Cu alloy. The loops lie in $\{100\}$ and are pure edge with **b** in $\langle 100 \rangle$. For plates normal to the beam $\mathbf{g} \cdot \mathbf{b} = 0$ and $\mathbf{g} \cdot \mathbf{R}_n$ is radial and zero along directions normal to **g**. This gives rise to "residual" arced contrast. The dislocations around (010) plates are invisible since $\mathbf{g} \cdot \mathbf{b}_{(010)} = 0$. Courtesy North-Holland.[29] (*b*) Hexagonal Frank loops in copper for which $\mathbf{g} \cdot \mathbf{b} = 0$. Note residual contrast for the directions for which $\mathbf{g} \cdot \mathbf{b} \wedge \mathbf{u}$ is nonzero (**u** is the dislocation line direction).

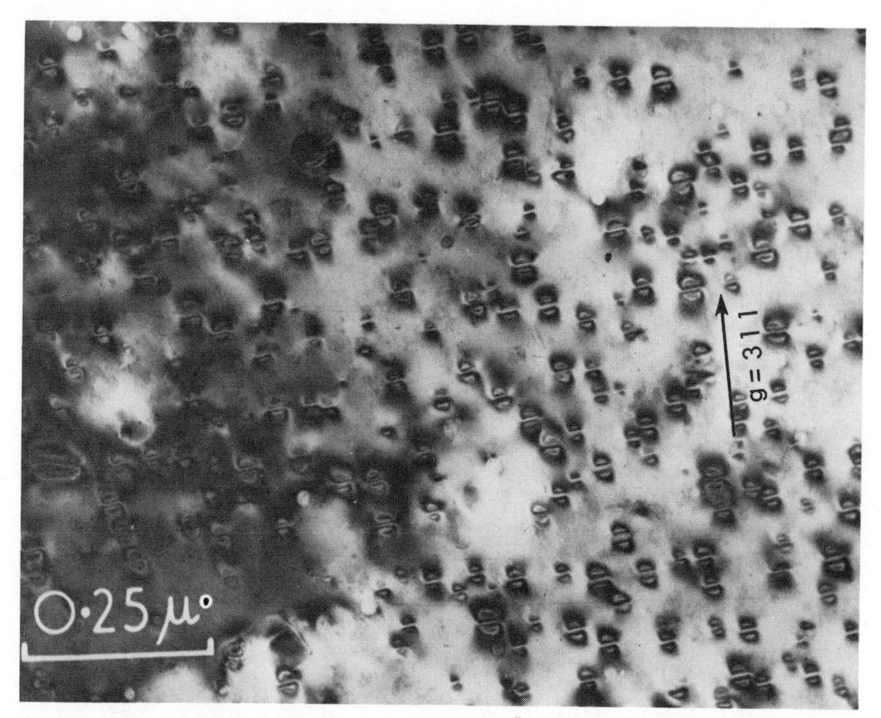

Fig. 3.4 Contrast from spherically symmetrical strain fields due to precipitation in Cu-Co alloy. Notice similarity to arc contrast in Fig. 3.3*a*. Courtesy M. F. Ashby and L. M. Brown; see *Philosophical Magazine*, **8**, 1083, 1649 (1963).

5 Burgers Vector Analysis

Table 3.1 gives examples of various **g** · **b** values for perfect dislocations in fcc, bcc, and hcp structures, and Tables 3.2 and 3.3 list **g** · **b** values for imperfect dislocations in fcc and hcp lattices. By considering the various **g** · **b** possibilities, it is possible to determine the most useful orientations needed for analysis (orientations [*uvw*] such that [*uvw*] · **g** = 0, Fig. 2.1) and the changes in orientation needed to arrive at the range of reflections required. These must lie within the capability of the specimen-tilting device. For this purpose the use of Kikuchi maps (see Chapter 2, Section 5.4, and refs. 10, 11, 13) greatly facilitates the required procedures. Several different reflections will normally be needed to obtain unique identifications [e.g., disappearances (**g** · **b** = 0) for two nonlinear values of **g** are sufficient to define the direction of **b**]. For partial dislocations **g** · **b** = $\frac{1}{3}$ does not produce enough contrast to be detected, so that this condition is one for invisibility.[12]

Table 3.1 Values of g · b for Perfect Dislocations

a. Fcc

Plane of Dislocation	b	g · b, g = $1\bar{1}1$	g · b, g = $\bar{1}11$	g · b, g = $11\bar{1}$
$(1\bar{1}1)$ or $(1\bar{1}\bar{1})$	$\frac{1}{2}[110]$	0	0	1
$(1\bar{1}\bar{1})$ or $(11\bar{1})$	$\frac{1}{2}[101]$	1	0	0
$(1\bar{1}1)$ or $(11\bar{1})$	$\frac{1}{2}[011]$	0	1	0
(111) or $(11\bar{1})$	$\frac{1}{2}[1\bar{1}0]$	1	$\bar{1}$	0
(111) or $(1\bar{1}1)$	$\frac{1}{2}[10\bar{1}]$	0	$\bar{1}$	1
(111) or $(\bar{1}11)$	$\frac{1}{2}[0\bar{1}1]$	1	0	$\bar{1}$

b. Bcc

Plane of Dislocation	b	g · b, g = $0\bar{1}1$	g · b, g = $1\bar{1}0$	g · b, g = 110
$(0\bar{1}1)$, $(1\bar{1}0)$, $(\bar{1}01)$	$\frac{1}{2}[111]$	0	0	1
$(0\bar{1}1)$, (110), (101)	$\frac{1}{2}[\bar{1}11]$	0	$\bar{1}$	0
$(\bar{1}01)$, (110), (011)	$\frac{1}{2}[1\bar{1}1]$	1	$\bar{1}$	0
(101), $(\bar{1}10)$, (011)	$\frac{1}{2}[\bar{1}\bar{1}1]$	1	0	$\bar{1}$

c. Hcp

Perfect dislocations in the hcp lattice are $\frac{1}{3}\langle 2\bar{1}\bar{1}0\rangle$ (three in number), [0001] (one in number), and $\frac{1}{3}\langle 11\bar{2}3\rangle$ (six in number).

The table below illustrates for orientation near $[1\bar{1}05]$ the **g · b** conditions necessary to distinguish dislocations invisible in **g** = $1\bar{1}00$.

b	g · b, g = $2\bar{2}00$	g · b, g = $\bar{2}3\bar{1}1$	g · b, g = $\bar{3}211$
$\frac{1}{3}[\bar{1}\bar{1}20]$	0	-1	1
$\frac{1}{3}[\bar{1}\bar{1}23]$	0	0	2
$\frac{1}{3}[11\bar{2}3]$	0	2	0

Table 3.2 Values of g · b for Imperfect Dislocations in the Fcc Lattice[a]

Fault Plane	b	g · b							
		g =							
		200	$0\bar{2}0$	$2\bar{2}0$	220	111	$1\bar{1}\bar{1}$	$4\bar{2}\bar{2}$	311
(111)	$\frac{1}{6}[\bar{1}\bar{1}2]$	$-\frac{1}{3}$	$\frac{1}{3}$	0	$-\frac{2}{3}$	0	$-\frac{1}{3}$	-1	$-\frac{1}{3}$
	$\frac{1}{6}[2\bar{1}\bar{1}]$	$\frac{2}{3}$	$\frac{1}{3}$	1	$\frac{1}{3}$	0	$\frac{2}{3}$	2	$\frac{2}{3}$
	$\frac{1}{6}[\bar{1}2\bar{1}]$	$-\frac{1}{3}$	$-\frac{2}{3}$	-1	$\frac{1}{3}$	0	$-\frac{1}{3}$	-1	$-\frac{1}{3}$
$(11\bar{1})$	$\frac{1}{6}[2\bar{1}1]$	$\frac{2}{3}$	$\frac{1}{3}$	1	$\frac{1}{3}$	$\frac{1}{3}$	$\frac{1}{3}$	$\frac{4}{3}$	1
	$\frac{1}{6}[\bar{1}\bar{1}\bar{2}]$	$-\frac{1}{3}$	$\frac{1}{3}$	0	$-\frac{2}{3}$	$-\frac{2}{3}$	$\frac{1}{3}$	$\frac{1}{3}$	-1
	$\frac{1}{6}[\bar{1}21]$	$-\frac{1}{3}$	$-\frac{2}{3}$	$-\frac{1}{3}$	$\frac{1}{3}$	$\frac{1}{3}$	$-\frac{2}{3}$	$-\frac{5}{3}$	0
$(1\bar{1}1)$	$\frac{1}{6}[\bar{1}\bar{2}\bar{1}]$	$-\frac{1}{3}$	$\frac{2}{3}$	$\frac{1}{3}$	-1	$-\frac{2}{3}$	$\frac{1}{3}$	$\frac{1}{3}$	-1
	$\frac{1}{6}[\bar{1}12]$	$-\frac{1}{3}$	$-\frac{1}{3}$	$-\frac{2}{3}$	0	$\frac{1}{3}$	$-\frac{2}{3}$	$-\frac{5}{3}$	0
	$\frac{1}{6}[21\bar{1}]$	$\frac{2}{3}$	$-\frac{1}{3}$	$\frac{1}{3}$	1	$\frac{1}{3}$	0	$\frac{4}{3}$	1
$(\bar{1}11)$	$\frac{1}{6}[\bar{2}\bar{1}\bar{1}]$	$-\frac{2}{3}$	$\frac{1}{3}$	$-\frac{1}{3}$	-1	$-\frac{2}{3}$	0	$-\frac{2}{3}$	$-\frac{4}{3}$
	$\frac{1}{6}[1\bar{1}2]$	$\frac{1}{3}$	$\frac{1}{3}$	$\frac{2}{3}$	0	$\frac{1}{3}$	0	$\frac{1}{3}$	$\frac{2}{3}$
	$\frac{1}{6}[12\bar{1}]$	$\frac{1}{3}$	$-\frac{2}{3}$	$-\frac{1}{3}$	1	$\frac{1}{3}$	0	$\frac{1}{3}$	$\frac{2}{3}$
(111)	$\frac{1}{3}[111]$	$\frac{2}{3}$	$-\frac{2}{3}$	0	$\frac{4}{3}$	1	$-\frac{1}{3}$	0	$\frac{5}{3}$
$(11\bar{1})$	$\frac{1}{3}[11\bar{1}]$	$\frac{2}{3}$	$-\frac{2}{3}$	0	$\frac{4}{3}$	$\frac{1}{3}$	$-\frac{1}{3}$	$\frac{4}{3}$	1
$(1\bar{1}1)$	$\frac{1}{3}[1\bar{1}1]$	$\frac{2}{3}$	$\frac{2}{3}$	$\frac{4}{3}$	0	$\frac{1}{3}$	$\frac{1}{3}$	$\frac{4}{3}$	1
$(\bar{1}11)$	$\frac{1}{3}[\bar{1}11]$	$-\frac{2}{3}$	$-\frac{2}{3}$	$-\frac{4}{3}$	0	$\frac{1}{3}$	-1	$-\frac{8}{3}$	$-\frac{1}{3}$
(111)	$\frac{1}{6}[1\bar{1}0]$	$\frac{1}{3}$	$\frac{1}{3}$	$\frac{2}{3}$	0	0	$\frac{1}{3}$	1	$\frac{1}{3}$
	$\frac{1}{6}[01\bar{1}]$	0	$-\frac{1}{3}$	$-\frac{1}{3}$	$\frac{1}{3}$	0	0	0	0
	$\frac{1}{6}[10\bar{1}]$	$\frac{1}{3}$	0	$\frac{1}{3}$	$\frac{1}{3}$	0	$\frac{1}{3}$	1	$\frac{1}{3}$
$(1\bar{1}1)$	$\frac{1}{6}[\bar{1}01]$	$-\frac{1}{3}$	0	$-\frac{1}{3}$	$-\frac{1}{3}$	0	$\frac{1}{3}$	-1	$\frac{1}{3}$
	$\frac{1}{6}[110]$	$\frac{1}{3}$	$-\frac{1}{3}$	0	$\frac{2}{3}$	$\frac{1}{3}$	0	$\frac{1}{3}$	$\frac{2}{3}$
	$\frac{1}{6}[011]$	0	$-\frac{1}{3}$	$-\frac{1}{3}$	$\frac{1}{3}$	$\frac{1}{3}$	$-\frac{1}{3}$	$-\frac{2}{3}$	$\frac{1}{3}$
$(11\bar{1})$	$\frac{1}{6}[101]$	$\frac{1}{3}$	0	$\frac{1}{3}$	$\frac{1}{3}$	$\frac{1}{3}$	0	$\frac{1}{3}$	$\frac{2}{3}$
	$\frac{1}{6}[1\bar{1}0]$	$\frac{1}{3}$	$\frac{1}{3}$	$\frac{2}{3}$	0	0	$\frac{1}{3}$	1	$\frac{1}{3}$
	$\frac{1}{6}[011]$	0	$-\frac{1}{3}$	$-\frac{1}{3}$	$\frac{1}{3}$	$\frac{1}{3}$	$-\frac{1}{3}$	$-\frac{2}{3}$	$\frac{1}{3}$
$(\bar{1}11)$	$\frac{1}{6}[110]$	$\frac{1}{3}$	$-\frac{1}{3}$	0	$\frac{2}{3}$	$\frac{1}{3}$	0	$\frac{1}{3}$	$\frac{2}{3}$
	$\frac{1}{6}[0\bar{1}1]$	0	$\frac{1}{3}$	$\frac{1}{3}$	$-\frac{1}{3}$	0	0	0	0
	$\frac{1}{6}[101]$	$\frac{1}{3}$	0	$\frac{1}{3}$	$\frac{1}{3}$	$\frac{1}{3}$	0	$\frac{1}{3}$	$\frac{2}{3}$

[a] $\frac{1}{6}\langle 112\rangle$ are Shockley partials; $\frac{1}{3}\langle 111\rangle$ are Frank partials; and $\frac{1}{6}\langle 110\rangle$ are stair rod dislocations.

Table 3.3a Values of g · b for Imperfect Dislocations, Hcp Crystals, Using [0001] Orientation

	g · b		
	b =		
g	$\frac{1}{3}[0\bar{1}10]$	$\frac{1}{6}[0\bar{2}23]$	$\frac{1}{6}[02\bar{2}3]$
$20\bar{2}0$	$\frac{2}{3}$	$-\frac{2}{3}$	$\frac{2}{3}$
$2\bar{2}00$	$\frac{2}{3}$	$\frac{2}{3}$	$-\frac{2}{3}$
$02\bar{2}0$	$-\frac{4}{3}$	$-\frac{4}{3}$	$\frac{4}{3}$
$2\bar{1}\bar{1}0$	0	0	0
$\bar{1}2\bar{1}0$	-1	-1	1
$\bar{1}\bar{1}20$	1	1	-1

Table 3.3b Values of g · b for Partial Dislocations, Hcp Crystals, Using [0001]-[$4\bar{2}\bar{2}9$] Orientations

	g · b		
	b =		
g	$\frac{1}{3}[0\bar{1}10]$	$\frac{1}{6}[0\bar{2}23]$	$\frac{1}{6}[02\bar{2}3]$
$2\bar{1}\bar{1}0$	0	0	0
$\bar{3}032$	1	2	0
$\bar{3}302$	-1	0	2

Table 3.3c Values of g · b for Partial Dislocations, Hcp Crystals, Using [0001]-[$1\bar{1}03$] Orientations

	g · b		
	b =		
g	$\frac{1}{3}[0\bar{1}10]$	$\frac{1}{6}[0\bar{2}23]$	$\frac{1}{6}[02\bar{2}3]$
$2\bar{1}\bar{1}0$	0	0	0
$\bar{2}111$	0	$\frac{1}{2}$	$-\frac{1}{2}$
$\bar{3}302$	-1	0	2

As an illustration consider the case of hcp crystals, for which Kikuchi maps are very helpful for several reasons,[10,11] especially since the spot diffraction patterns from such crystals are in general much more difficult to analyze than those from cubic crystals. In addition to the complexities due to double diffraction (see Chapter 2, Section 4.1) and the fact that the c/a ratios differ from material to material, the d-spacings of certain planes in hcp crystals are so close together that obtaining two-beam orientations (e.g., in $\langle 11\bar{2}0 \rangle$ foils) and unambiguously identifying the foil orientation may be impossible in certain cases. The above problems can be circumvented, however, if foils thick enough to produce Kikuchi reflections are used. In general, Kikuchi patterns become increasingly useful as the symmetry of the crystal system decreases, as well as for recognizing orientations under two-beam conditions. Furthermore, for hcp structures the basal-plane orientation and the orientations within about a 20° tilt from [0001] are all that are required for solving problems such as Burger's vector determinations. This tilting range is within the capability of most commercial tilting stages.

5.1 Perfect Dislocations

Perfect dislocations must vanish for one of the $20\bar{2}0$ Kikuchi bands (see Fig. 2.31). These, rather than $10\bar{1}0$, are used because of their higher intensities. If a particular reflection is designated as $2\bar{2}00$, then **b** must be one of the following: $\pm\frac{1}{3}$ $[\overline{11}20]$, $\pm\frac{1}{3}$ $[\overline{11}23]$, or $\pm\frac{1}{3}$ $[11\bar{2}3]$. Since a dislocation image vanishes for any Kikuchi band that converges to the pole of the Burgers vector ($\mathbf{g} \cdot \mathbf{b} = 0$), any two Kikuchi bands converging to the $[\overline{11}23]$ and $[11\bar{2}3]$ poles can be used to distinguish between the three possibilities. Reference to the Kikuchi map (Fig. 2.31) shows that the nearest of such bands are the $\bar{2}3\bar{1}1$ and $\bar{3}211$, which intersect at the $[1\bar{1}05]$ pole. This pole can be reached from [0001] by a direct tilt along the $11\bar{2}0$ Kikuchi band (e.g., for Titanium this involves about a 14° tilt). Table 3.1c lists the $\mathbf{g} \cdot \mathbf{b}$ values for the three reflections used in the procedure and clearly shows that they are sufficient to determine the Burgers vector.

5.2 Imperfect Dislocations

Imperfect dislocations must vanish for one of the $11\bar{2}0$ Kikuchi bands. If this reflection is designated as $2\overline{11}0$, then **b** must be one of the following: $\pm\frac{1}{3}$ $[0\bar{1}10]$, $\pm\frac{1}{6}$ $[0\bar{2}23]$, or $\pm\frac{1}{6}$ $[0\bar{2}23]$. Table 3.3a shows that other $11\bar{2}0$ and $20\bar{2}0$ reflections cannot be used to distinguish between the three possibilities. In this case the poles of interest are the $[0\bar{2}23]$ and $[02\bar{2}3]$, and the corresponding Kikuchi bands are the $\bar{3}302$ and $\bar{3}032$, which intersect at the $[4\overline{22}9]$ pole. This pole can be reached from [0001] by tilting (~24° for titanium) along the $01\bar{1}0$ Kikuchi band. Table 3.3b shows the $\mathbf{g} \cdot \mathbf{b}$ values for the three reflections used in the procedure.

Since the tilt (~24°) needed to reach the $[4\bar{2}\bar{2}9]$ pole from the [0001] orientation is very near the limit of many specimen-tilting devices, an alternative procedure involving somewhat smaller tilts uses the $\bar{2}111$ forbidden reflections resulting from double diffraction. The procedure is the same as the one described above except that the foil is tilted into the $[1\bar{1}03]$ rather than the $[4\bar{2}\bar{2}9]$ orientation. (For titanium this involves about a 20° tilt.) Since, it will be recalled, extinction of imperfect dislocations occurs when $|\mathbf{g} \cdot \mathbf{b}| \leqslant \frac{1}{3}$, Table 3.3*c* shows that the $\bar{2}111$ and $\bar{3}302$ reflections accessible in the $[1\bar{1}03]$ orientation provide sufficient information for a unique determination of the Burgers vectors.

In many problems the object is to distinguish between dislocations having Burgers vectors of the same type. For example, one may wish to distinguish between dislocations of the primary and secondary slip systems having Burgers vectors of the form $\frac{1}{3}\langle 2\bar{1}\bar{1}0\rangle$. The lowest order reflections needed for $\mathbf{g} \cdot \mathbf{b} = 0$ conditions are of the form $10\bar{1}0$ and $\bar{2}201$; $10\bar{1}0$ reflections cannot distinguish between $\frac{1}{3}\langle 2\bar{1}\bar{1}0\rangle$ and $\frac{1}{3}\langle 2\bar{1}\bar{1}3\rangle$. Furthermore, the tilting device often tilts further in one direction than in another. If the [0001] orientation does not lie within the range of the tilt, all three $10\bar{1}0$ Kikuchi bands may not be accessible. On the other hand, the $\bar{2}201$ Kikuchi bands intersect in such a way as to form six equivalent triangles situated symmetrically about the [0001] zone axis. The six triangles are centered on the $\langle 1\bar{1}03\rangle$ poles and have vertices composed of poles of the forms $\langle 1\bar{2}16\rangle$ and $\langle 0\bar{1}12\rangle$. Any one of the six triangles can be used to distinguish between the three $\frac{1}{3}\langle 2\bar{1}\bar{1}0\rangle$ Burgers vectors. For example, if the foil is initially in some high index orientation such as the $[0\bar{3}34]$, the [0001] zone axis may lie outside the range of the specimen-tilting device. In this case the triangle centered on the $[0\bar{1}13]$ pole can be used where $\bar{2}201$, $02\bar{2}1$, and $20\bar{2}1$ Kikuchi bands are readily accessible. An example of this application is shown in Fig. 3.5 for dislocations in Ag_2Al.[14] The dislocations occur in the form of perfect loops clustered in slip bands lying parallel to the trace of the primary slip plane.

Figures 3.5*a–c* show a series of micrographs of a typical slip band. The corresponding reflections are indicated on the Kikuchi map, Fig. 3.5*d*. The micrographs have been oriented with respect to the Kikuchi map so that the slip band lies parallel to the trace of the $(01\bar{1}0)$ plane, that is, parallel to the $01\bar{1}0$ Kikuchi band. As shown in Section 7.2, in fcc metals Burgers vectors of double-arc loops can be uniquely determined from the direction of the line of no contrast;[15] however, in hcp metals the line of no contrast can be used only to eliminate all but three of the nine possible Burgers vectors of the forms $\frac{1}{3}\langle 2\bar{1}\bar{1}0\rangle$ and $\frac{1}{3}\langle 2113\rangle$. As indicated in Fig. 3.5*a*, the line of no contrast is parallel to the $[1\bar{1}00]$ direction; hence the double-arc loops can have any of the following Burgers vectors: $\pm\frac{1}{3}[\bar{1}\bar{1}20]$, $\pm\frac{1}{3}[\bar{1}\bar{1}23]$, or $\pm\frac{1}{3}[1123]$. To distinguish between these, the foil was tilted to the nearest $\bar{2}201$ triangle centered on the $[0\bar{1}13]$ pole. Figure 3.5*a*

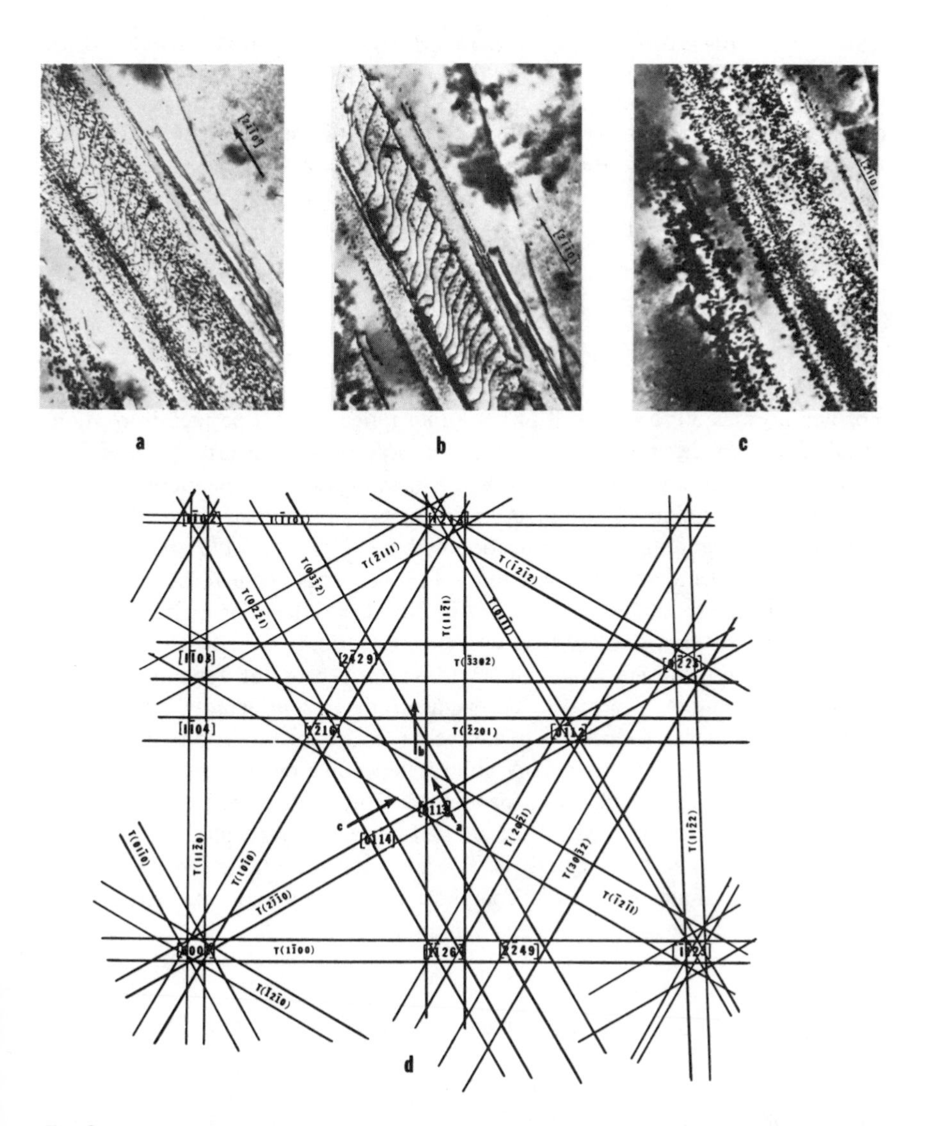

Fig. 3.5 Illustrating the use of the Kikuchi map for analyzing the slip band structure in deformed Ag_2Al. Courtesy *Journal of Applied Physics.*[10]

was taken with $2\bar{1}\bar{1}0$ so that all three possible types of dislocations could be in contrast. Figure 3.5*b* shows that all loops are out of contrast with $\bar{2}201$; therefore they have the same Burgers vector, $\frac{1}{3}$ $[\bar{1}\bar{1}20]$. By tilting to $\mathbf{g} = 02\bar{2}1$, the pile-up of dislocations within the band and the helical dislocations near the edges vanish (Fig. 3.5*c*). Therefore these dislocations have the Burgers vector $\frac{1}{3}$ $[2\bar{1}\bar{1}0]$.

6 Magnitude of g · b

For hcp crystals, $\mathbf{g} \cdot \mathbf{b}$ can be 2 in low order reflections for perfect dislocations (Table 3.1*c*). In these conditions the image is doubled. This effect is useful for distinguishing different Burgers vectors. Unless high order reflections are used, $\mathbf{g} \cdot \mathbf{b} > 1$ is not commonly observed in fcc or bcc crystals. For example, in the $[\bar{1}10]$ fcc orientation a screw dislocation with $\mathbf{b} = \frac{1}{2}$ [110] would show a double image in $\mathbf{g} = 220$ and a single image in $\mathbf{g} = 111$, and would be invisible in $\mathbf{g} = 002$. Thus, in principle, studies of the image in different reflections make it possible to determine the *magnitude* of **b**, as well as its direction.

For partial dislocations some complexities arise because of dynamical effects. In the fcc structure, especially when anisotropy is considered,[15] it has been shown that, whereas $\mathbf{g} \cdot \mathbf{b} = \pm\frac{1}{3}$ is always an invisibility criterion, the case of $\mathbf{g} \cdot \mathbf{b} = \mp\frac{2}{3}$ leads to visibility or invisibility, depending on the sign of s and the position in the foil (thickness dependence). These difficulties have been discussed by Clarebrough,[17] who suggests that computations will be necessary to determine Burgers vectors of partials. This approach is facilitated by using image simulation techniques for direct comparisons of computed data with experimental results as discussed in Chapter 5 and ref. 16.

7 Image Position

The position of the dislocation image with respect to the actual line depends on the sign of s. This can be seen geometrically by considering Fig. 3.6. In Fig. 3.6*a*, s is set positive and an edge dislocation is oriented as shown. If **g** points to the right, the sense of tilt of the planes on the left-hand side causes greater deviation, $s \gg 0$, from diffraction, whereas on the right-hand side the reverse is true. In this case the image will thus appear to the right-hand side. Similarly, in Fig. 3.6*b*, if the foil is tilted slightly to make s negative, the image flips to the left-hand side. Thus on crossing an extinction contour (s changing sign) the image changes position. The same effect is true when the sign of **g** is changed (for s invariant).

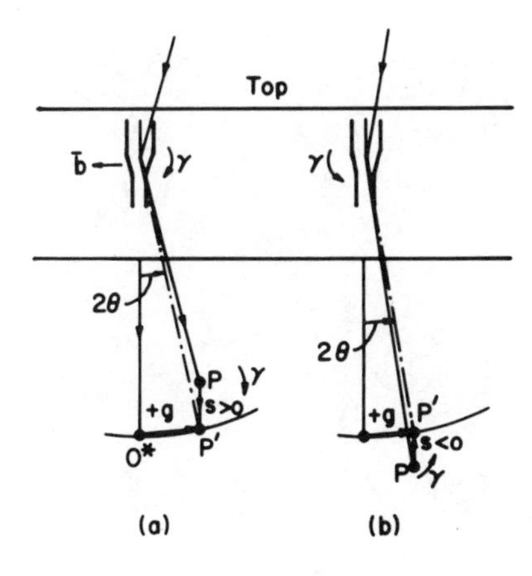

Fig. 3.6 Scheme showing that a dislocation locally tilts reflecting planes closer to or away from the Bragg condition on opposite sides of the extra half plane. Notice that (*a*) and (*b*) would be reversed if the dislocation was inverted (extra half plane down). The images are therefore to one side of the true position of the dislocation.

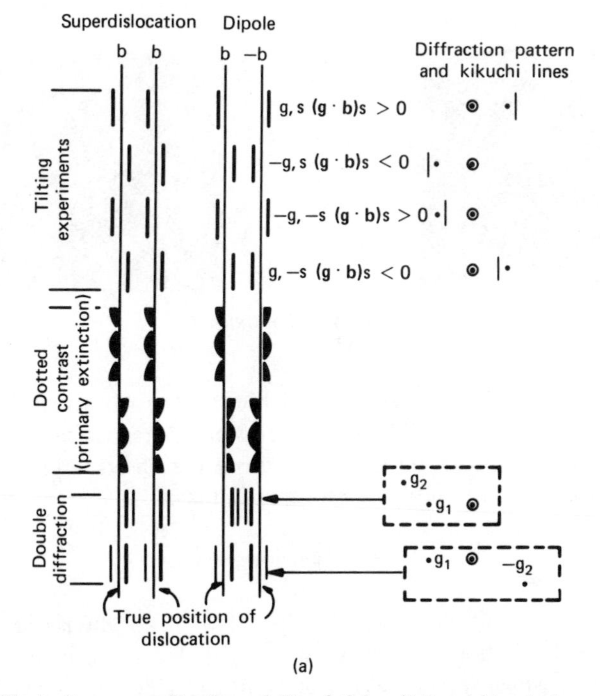

Fig. 3.7 (*a*) Illustrating an application of Fig. 3.6 for distinguishing between dipoles and superdislocations; three different methods, all involving tilting, are shown. Fig. 3.7(*b*, *c*) shows examples for stainless steel. The pairs at *A* do not change in width upon changing the sign of *g*, hence these must be super-dislocations. Courtesy Pergamon Press.[18]

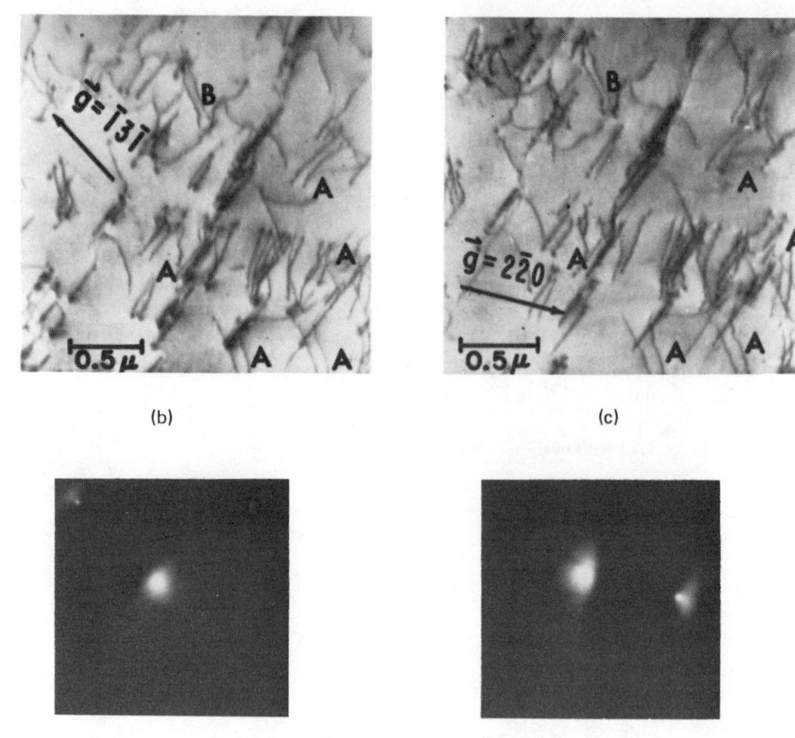

Fig. 3.7 (*Continued*)

Thus, without predefining the sense of Burgers vectors, as long as the sign of s and the direction of g are known it is possible to reconstruct in the foil the orientation of the dislocation that gives rise to the observed image shift when g or s changes sign. This result has useful applications.

7.1 Dislocation Pairs

Dislocation pairs are of two types: the dipole (two dislocations of opposite sign), and the superdislocation (two of the same sign). These can be distinguished by the manner in which the images change with changing **g** or s, as is seen in Fig. 3.7*a*. The example of Figs. 3.7*b* and *c* for austenitic stainless steel, obtained by changing the sign of **g**, shows that the pairs at A are superdislocations, indicating the existence of order. Dipoles are seen at B. These experiments are particularly valuable for studying order in alloys of similar elements for which possible superlattice reflections cannot be detected in the electron diffraction patterns, for example, brass (Cu-Zn) and austenitic stainless steel (Fe-Cr-Ni).[18] Pairs can be

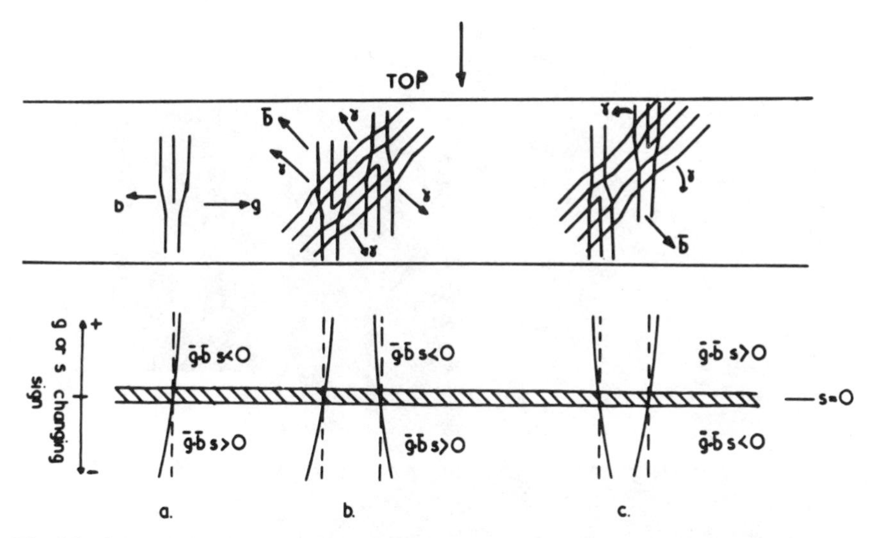

Fig. 3.8 Scheme showing image behavior of large, perfect dislocation loops in crystals when g or s changes sing. From G. Thomas, *Thin Films*, copyright 1964, American Society of Metals.

distinguished from double images since the latter occur only when more than one strong reflection operates on opposite sides of a single dislocation, or when $\mathbf{g} \cdot \mathbf{b} = 2$.

7.2 Loops

The fact that the dislocation image changes position with a change in sign of **g** or s can be used to determine whether dipoles or loops are vacancy or interstitial in character, again without prior definition of the sign of **b**. Figure 3.8 illustrates how the images vary with a change in sign of **g** or s. If the defect lies along the plane inclined from bottom left to top right, the vacancy loop will always be in outside contrast for **g** to the right-hand side and $s > 0$.

It is important to notice that, if the defects in Fig. 3.8 were inclined in the opposite sense (top left to bottom right), the vacancy loop would be in outside contrast at $s > 0$ with **g** to the left-hand side. Figure 3.9 shows an actual example of this case, for loops in quenched aluminum. Thus it is essential to know the sense of inclination of the defect and to correctly orient the direction of **g** on the micrograph, as discussed earlier in this chapter. The example of Fig. 3.9 also shows that for perfect loops the Burgers vector is perpendicular to the line of no contrast dividing the loop image into arcs.[15] For fcc crystals this line is a $\langle 110 \rangle$, and so **b** can be uniquely found once the foil is correctly oriented. This double-

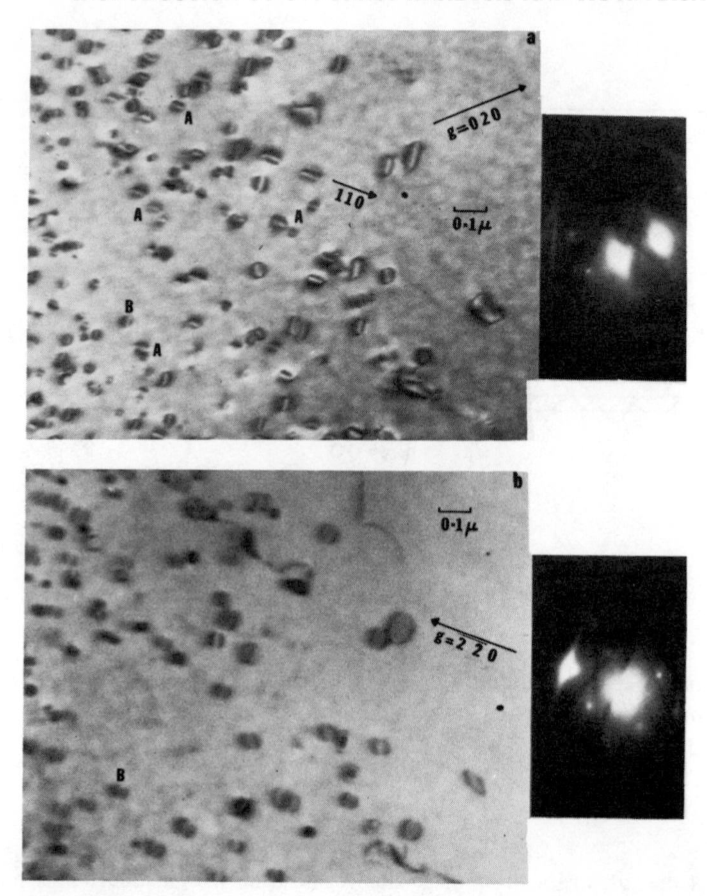

Fig. 3.9 Quenched aluminum containing perfect loops on $\{111\}$: loops B increase in size as the sign of $\mathbf{g}$ is changed from (a) to (b) ($s > 0$, as shown by the diffraction patterns); loops A go out of contrast in $\mathbf{g} = \overline{2}20$, so $\mathbf{b} = \frac{1}{2}\,[1\overline{1}0]$. Notice that this $\mathbf{b}$ is normal to the [110] no-contrast lines of the double-arc images of loops A in (a). Courtesy Taylor and Francis.[15]

arc contrast arises because the strain fields tend to cancel at the strong edge components of the loops. For two-beam imaging conditions at $s \sim 0$ this rule appears to hold generally for large perfect loops in fcc, bcc, and hcp crystals, and is thus quite useful when large numbers of defects are to be analyzed.[19]

8 Planar Defects

8.1 Stacking Faults

For purposes of illustration consider stacking faults in the fcc structure. Two possibilities exist, namely, formation of a stacking fault by the splitting of a whole dislocation under a shear stress:

$$\tfrac{1}{2}[\bar{1}01] = \tfrac{1}{6}[\bar{1}\bar{1}2] + \tfrac{1}{6}[\bar{2}11] \quad \text{on } (111),$$

or creation of a fault by growth or point defect aggregation, forming imperfect edge prismatic loop dislocations of the $\frac{1}{3}\langle 111\rangle$ type.

The stacking fault is the least complicated of planar defects since only a displacement of the crystal across the fault plane is involved (i.e., no change in s or **g**). The contrast was first considered by Whelan and Hirsch.[20] A wave crossing a faulted region of a crystal suffers a phase change $\alpha = 2\pi\mathbf{g}\cdot\mathbf{R}$, and, similarly to the wedge case, fringes occur when the fault is inclined in the foil. Since α is a phase factor, everything is invariant under changes of 2π (see, eq. 3.4); conventionally α is quoted in its principal range, $-\pi < \alpha < \pi$. If **R** is a lattice translation vector, the phase contrast is zero since $\mathbf{g}\cdot\mathbf{R}$ is integral. For stacking faults, however, **R** cannot be a lattice translation vector, although particular values can occur for which $\mathbf{g}\cdot\mathbf{R}$ is integral. This result affords a means for studying the faults in a manner similar to that adopted for dislocations, putting **R** equal to the fault vector.

In fcc materials $\mathbf{R} = \frac{1}{6}\langle 112\rangle$ or $\frac{1}{3}\langle 111\rangle$. Therefore α can take the values

$$\alpha = \frac{2\pi(h+k+2l)}{6} \quad \text{or} \quad \frac{2\pi}{3}(h+k+l).$$

In either case $\alpha = \pm 2\pi/3$ (fault visible), or $n2\pi$ (fault invisible); hence the shear $\frac{1}{6}\langle 112\rangle$ fault cannot be distinguished from the $\frac{1}{3}\langle 111\rangle$ fault purely from fringe contrast, as expected, since physically they are identical.

The possible $\mathbf{g}\cdot\mathbf{R}$ values for faults in fcc or dc crystals can be found from Table 3.2. The values of $\alpha = 2\pi\mathbf{g}\cdot\mathbf{R}$ change along each parallel $\langle 111\rangle$ row of the reciprocal lattice (see Fig. 2.13*a*). Thus faults are visible in reflections along the 2nd, 3rd, 5th, 6th, etc., $\langle 111\rangle$ rows, but are invisible in the 1st, 4th, etc.

It is instructive to consider various examples of contrast to be expected from faults bounded by partials. Consider the (111) fault plane with the operating partials $\frac{1}{6}[1\bar{2}1]$ and $\frac{1}{6}[2\bar{1}\bar{1}]$. It is assumed also that $\mathbf{g}\cdot\mathbf{b} = \pm\frac{2}{3}$ will always be visible (remembering that this is not necessarily true in all situations).

1. $\mathbf{g} = 200$

$$\tfrac{1}{2}[1\bar{1}0] = \tfrac{1}{6}[1\bar{2}1] + \tfrac{1}{6}[2\bar{1}\bar{1}] \quad \text{on } (111)$$

$$\mathbf{g}\cdot\mathbf{b} = \quad 1 \qquad +\tfrac{1}{3} \qquad +\tfrac{2}{3}$$

α for fault $= 2\pi/3$; fault visible, one partial visible.

2. $\mathbf{g} = 200$

$$\tfrac{1}{2}[01\bar{1}] = \tfrac{1}{6}[\bar{1}2\bar{1}] + \tfrac{1}{6}[11\bar{2}] \quad \text{on } (111)$$

$$\mathbf{g}\cdot\mathbf{b} = \quad 0 \qquad -\tfrac{1}{3} \qquad +\tfrac{1}{3}$$

α for fault $= -2\pi/3$; fault visible, no partials visible.

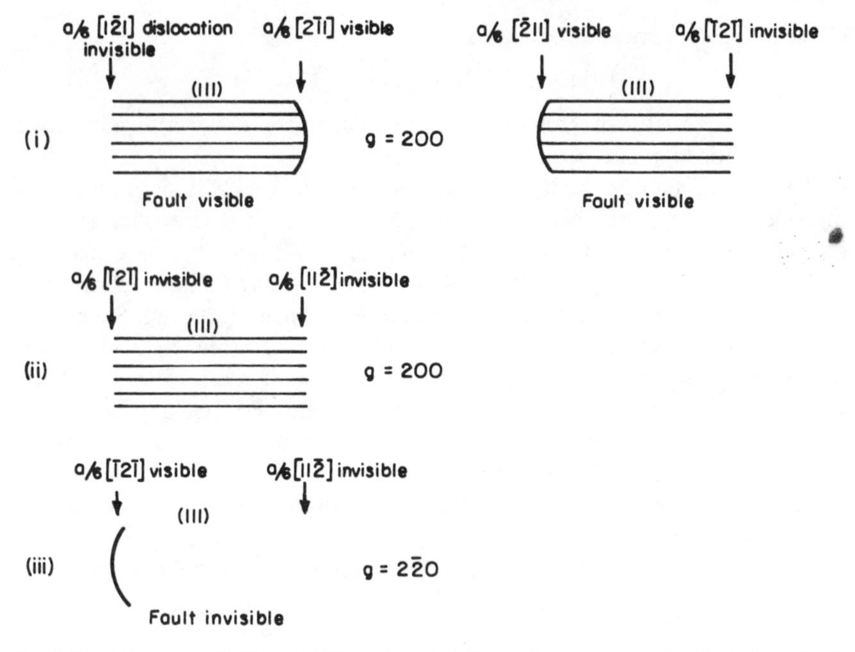

Fig. 3.10 Scheme predicting visibility of partial dislocations and associated stacking faults in fcc crystals.

3. $\mathbf{g} = 2\bar{2}0$

$$\tfrac{1}{2}\,[01\bar{1}] = \tfrac{1}{6}\,[\bar{1}2\bar{1}] + \tfrac{1}{6}\,[11\bar{2}] \qquad \text{on } (111)$$

$$\mathbf{g}\cdot\mathbf{b} = \quad -1 \qquad\quad -1 \qquad\quad 0$$

α for fault = 0; fault invisible, only one partial visible. This case could be mistaken for a perfect dislocation.

These cases are illustrated by the sketches in Fig. 3.10. Figure 3.11 shows an example of case 1 in which faults of opposite sense are present at A and B.

If one wishes to distinguish between a $\frac{1}{3}\langle 111\rangle$ and a $\frac{1}{6}\langle 112\rangle$ fault, it is necessary to determine the Burgers vector for the bounding dislocations. Reference to Table 3.2 for fcc crystals shows the reflections needed to make this distinction. For a fault on a $\{111\}$, use of $\mathbf{g} = 2\bar{2}0$ and $\mathbf{g} = 200$ is sufficient and a foil in $\langle 001\rangle$ will be required (e.g., Fig. 3.12; the partial at A on $(1\overline{11})$ must be the Shockley, with $\mathbf{b} = \frac{1}{6}\,[12\bar{1}]$).

8.1.1 *Determination of Type of Stacking Fault.* The dynamical theory (Chapters 4 and 5) predicts the dependence of the intensity of the fringes on the sign of α. In bright field, for α positive the first fringe is light, whereas for α negative

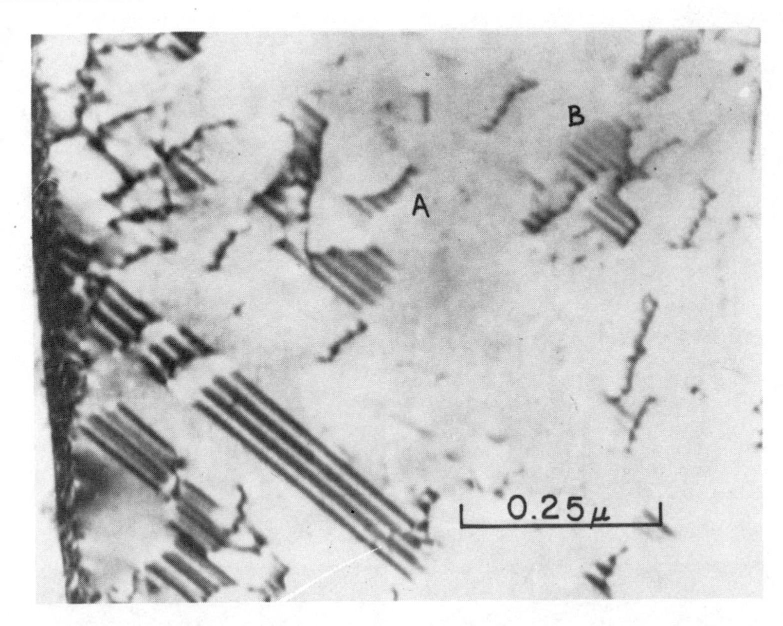

Fig. 3.11 Partial dislocations and stacking faults in austenitic stainless steel (cf. Fig. 3.10).

the first fringe is dark. The reverse is true for dark field images, although the effect of absorption modifies the symmetry so that the fringes are complementary only at the lower surface of the foil (Fig. 3.13). This dependence of the color of the first fringe on α can also be obtained intuitively by considering phase advance or retardation on an amplitude phase diagram.

The above rules can be used to determine whether a fault or thin slab of precipitate is intrinsic or extrinsic,[21] but it is essential to know the sense of slope of the fault plane with respect to **g**. Since an extrinsic fault may be thought of as one due to the insertion of an extra plane of atoms, and an intrinsic fault as one formed by removal of a plane, the sign of **R** will be opposite for extrinsic and intrinsic faults, and hence the sign of $\alpha = 2\pi \mathbf{g} \cdot \mathbf{R}$ will be reversed for these cases when the same **g** operates. The two types of faults may be formally defined as follows. Consider the top half of the crystal to be fixed, and the bottom half displaced by the fault displacement vector, as shown in Fig. 3.14. Now, knowing the direction of **g**, after allowing for optical rotations, and observing the color of the first fringe, one may discover whether β is acute or obtuse and hence determine the nature of the fault. For example, in the fcc case the value of $\alpha = 2\pi|\mathbf{g}|\,|\mathbf{R}| \cos \beta = \pm 2\pi/3$, where $|\mathbf{R}| = \sqrt{3}/3$; then suppose that $\mathbf{g} = 200$ and is to the right-hand side of the fault. Also, $\cos \beta = \pm 1/\sqrt{3}$, since $\mathbf{R} = \pm\frac{1}{3}$ [111]. If the first fringe is white, α is positive, that is, $\alpha = 4\pi/3(\sqrt{3}) \cos \beta = 2\pi/3$ if $\cos \beta =$

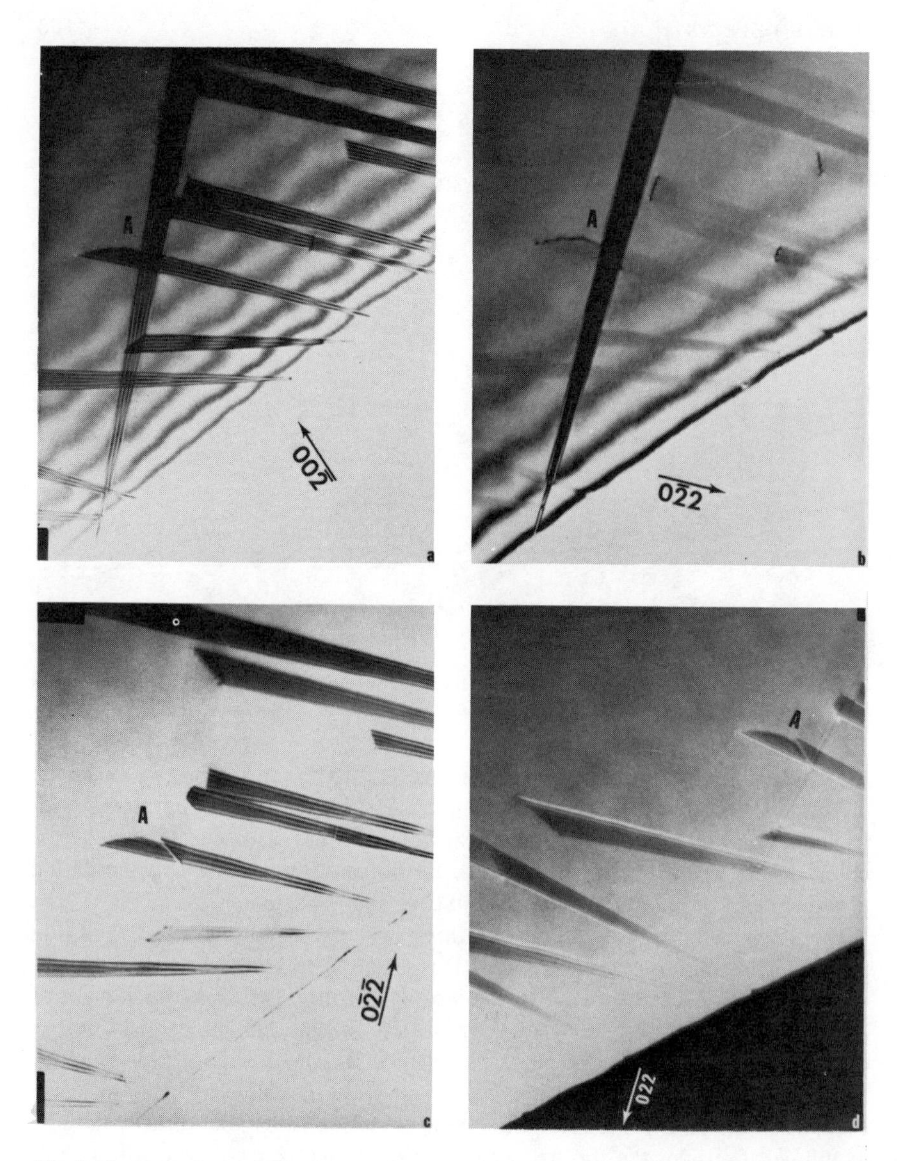

Fig. 3.12 Contrast experiments for faults in $TaC_{0.8}$. (*a–c*) Bright field images showing the shear nature of the faults $\{\mathbf{b} = \langle 112 \rangle\}$. The faults do not completely vanish in (*b–d*), indicating local changes in composition (structure factor contrast). The dark field image in (*d*) shows that (111) and $(1\bar{1}\bar{1})$ are intrinsic and extrinsic, respectively.

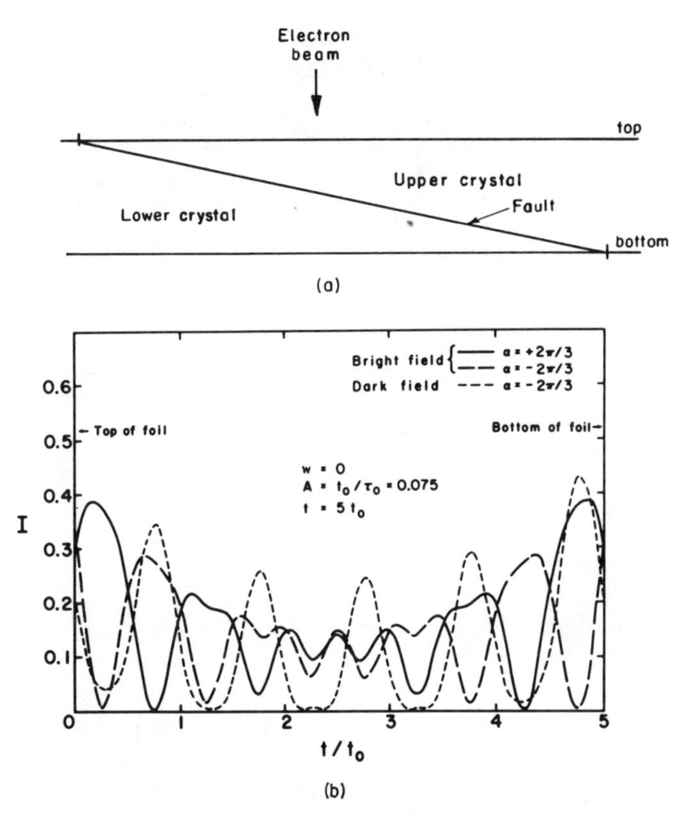

Fig. 3.13 (*a*) Geometry used to describe an inclined fault. (*b*) Intensity distribution for $\alpha = \pm 2\pi/3$ contrast as predicted by two-beam dynamical theory with absorption for the geometry shown in (*a*).

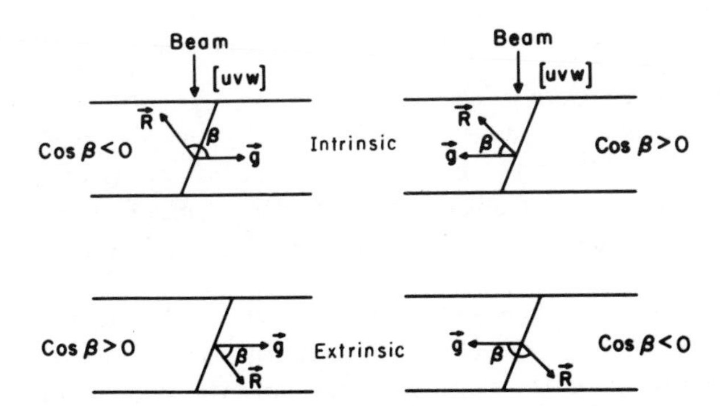

Fig. 3.14 Scheme showing orientation of intrinsic (top row) and extrinsic (lower row) stacking faults with respect to **g** for determining fault character (see Table 3.4). After R. Gevers, A. Art, and S. Amelinckx, *Physica Status Solidi*, **3,** 1563 (1963).

Table 3.4 Determination of Nature of Stacking Faults in Fcc Crystals[a]

g	Orientation of g (Fig. 3.14)	Color of First Fringe[b]	
		Bright	Dark
200	L	E	I
	R	I	E
400	L	I	E
	R	E	I
220	L	I	E
	R	E	I
111	L	I	E
	R	E	I

[a] Reference 21.
[b] Bright field.
In this table, L = left; R = right; E = extrinsic; I = intrinsic.

$-1/\sqrt{3}$. In other words, β must be obtuse and the fault must be intrinsic. For the same situation an extrinsic fault would show a black first fringe. These rules are summarized in Table 3.4 for faults oriented as shown in Fig. 3.14. In practice the micrograph can always be oriented so that the fault plane slopes as shown in these sketches.

Figure 3.15 shows an example of the use of these rules, and illustrates how rapidly one can analyze the fault inclination by dark field imaging at $s \neq 0$. The first fringe is the most intense one when $s < 0$ in dark field.

8.1.2 *Overlapping Faults.* Consider two overlapping faults of the same kind (Fig. 3.16*a*). If these are close together, the phase factors add, giving a net phase shift of $2\pi/3 + 2\pi/3 \equiv -2\pi/3$, that is, the color of the first fringe changes at the point of overlap. If three faults overlap, the phase change is $2\pi \equiv 0$ and no contrast occurs. If the faults are far apart, the outer fringes will be of the same color (Fig. 3.16*b*).

Similarly, if two overlapping intrinsic-extrinsic faults exist, if they are close together $\alpha = +2\pi/3 + (-2\pi/3) = 0$ and no contrast occurs. However, if they are far enough apart, the outer fringes will be of opposite color and the center part of the overlap will have weak or zero contrast where the two phase shifts cancel (Fig. 3.16*c*).

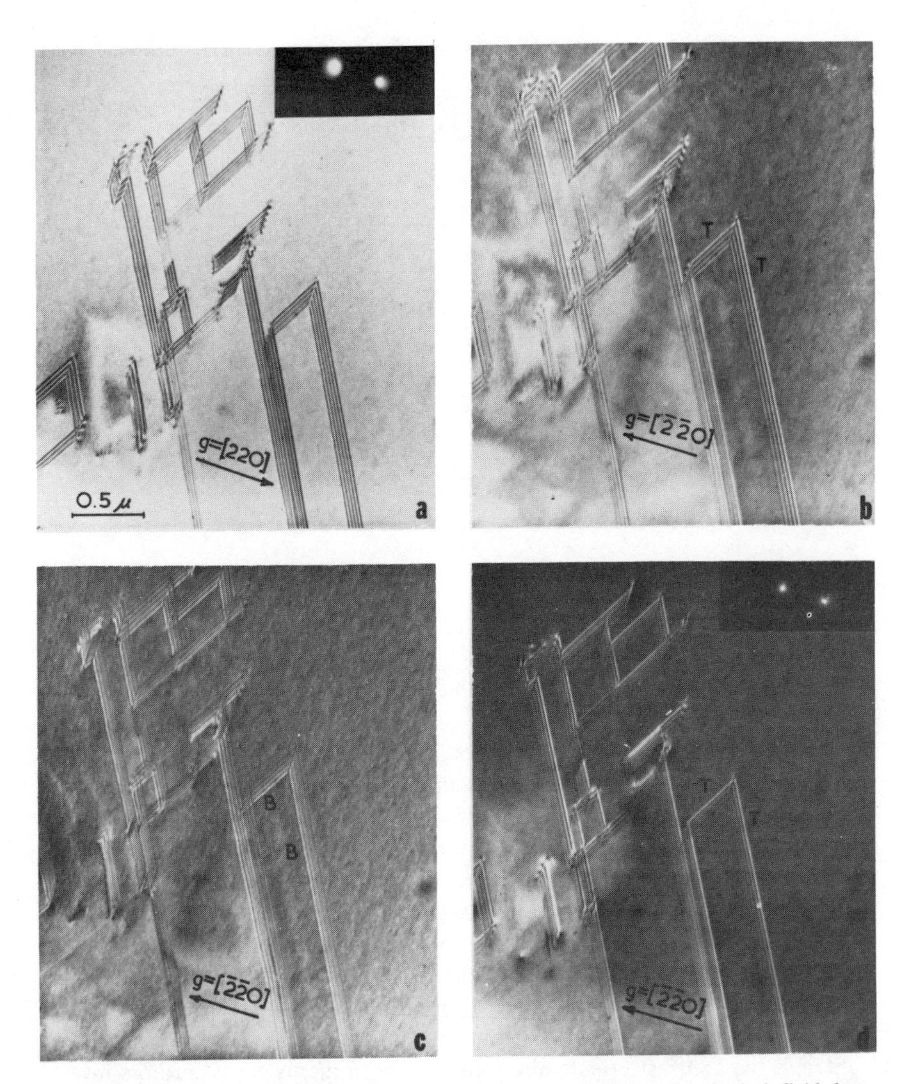

Fig. 3.15 Contrast from intrinsic faults in silicon: (*a*) bright field and (*b*) dark field showing the predicted behavior of Fig. 3.13; (*c*, *d*) show the asymmetry properties of dark field for $s \neq 0$, enabling the first, or last, fringe to be immediately identified. Courtesy *Physica Status Solidi.*[8]

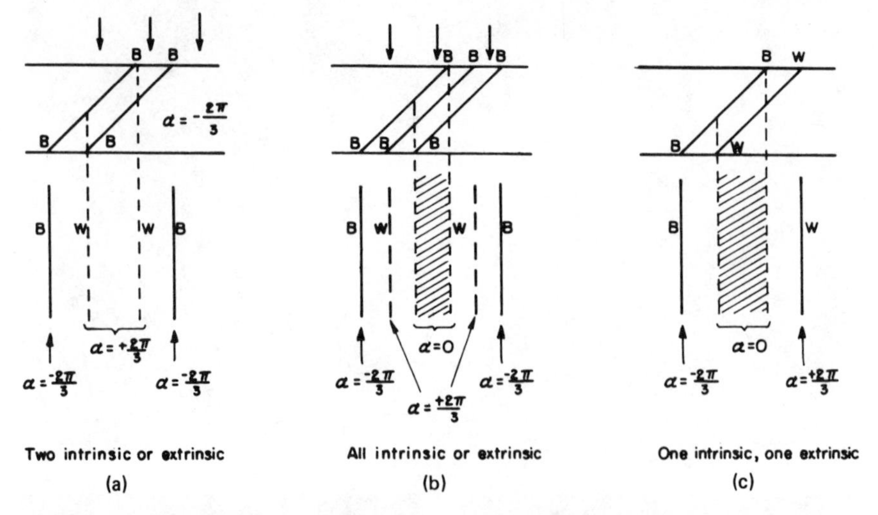

Fig. 3.16 Scheme showing contrast expected from overlapping faults in fcc crystals with $\alpha = \pm 2\pi/3$. B and W are dark and light fringes, respectively. Courtesy Springer-Verlag.

8.2 Orientation Faults

So far, the contrast considered at faults has been produced by displacement of the lower crystal (see Fig. 3.13*a*) without change of orientation. This is exactly true for stacking faults but is not generally true for planar faults or for crystals of lower symmetry. The present section concentrates on the next simplest special case, change of orientation only; subsequent sections will deal with lower symmetry. The physical situation considered is again illustrated by Fig. 3.13*a*, but now the upper and lower crystals have slightly different orientations, although basically the same structure, that is, s is slightly different in the two crystals. Such a situation may arise at coherent twin boundaries in tetragonal crystals and in materials such as barium titanate and the superconducting alloy V_3Si, and the contrast effects observed have been discussed by authors investigating these materials.[22] The principal effects are as follows:

1. Fringes much like stacking fault fringes are produced across the fault; such fringes are referred to as δ-fringes.[22]
2. The perfect crystal on either side of the region of overlap in general transmits different intensities, in both bright and dark field (in contrast to the stacking fault) because of the slight differences in orientation.
3. In sufficiently thick crystals the bright field image is definitely asymmetrical, while the dark field image is more nearly symmetrical (again in contrast to the stacking fault).
4. The nature of the tilt determines the color of the outermost fringes.

5. A pair of overlapping boundaries of opposite sign may, if moderately closely spaced, reverse the symmetry noted in item 3.
6. The fringe pattern, like the stacking fault pattern, is dependent on the specimen thickness.

Most of these effects may be seen in Fig. 3.31, where the orientation differences are produced by the magnetostrictive effects arising from the differing magnetization directions in different parts of the specimen.

8.2 Antiphase Boundaries in Ordered Alloys

A fault in the periodic array of ordered planes of atoms gives rise to an antiphase boundary. If the displacement of the crystal associated with this type of fault is $\mathbf{R} = [uvw]$, the phase shift $\alpha = 2\pi\mathbf{g} \cdot \mathbf{R} = 2\pi(hu + kv + lw)$. Ordered alloys are often characterized by having primitive symmetry in that all values of hkl are allowed. The superlattice reflections are those that would not exist in the disordered alloy, for example, in the $B2$ superlattice, based on a body-centered structure (CsCl):

$$F = f_A + f_B \qquad h + k + l \text{ even} \quad \text{(fundamental)}$$

$$= f_A - f_B \qquad h + k + l \text{ odd} \quad \text{(superlattice)}.$$

In this superlattice $\mathbf{R} = \frac{1}{2}\langle 111 \rangle$; hence

$$\alpha = \pi(h + k + l) = 0 \qquad \text{for } h + k + l \text{ even}$$

$$= \pm\pi \qquad \text{for } h + k + l \text{ odd}.$$

in other words, only superlattice reflections (e.g., 100, 111) can produce phase contrast in this case. This follows because $\frac{1}{2}\langle 111 \rangle$ is a **T** vector for the disordered alloy. In the case of π boundaries the first fringe is always dark in bright field for $s > 0$.

Another type of superlattice is the $L1_2$ superlattice based on a face-centered structure. In this case

$$F = 3f_A + f_B \qquad \text{for } hkl \text{ unmixed}$$

$$= f_A - f_B \qquad \text{for } hkl \text{ mixed}.$$

One type of antiphase boundary vector $\mathbf{R}$ is $\frac{1}{2}\langle 110 \rangle$; hence $\alpha = \pi(h + k)$. Thus for fundamental reflections α is always zero ($\frac{1}{2}\langle 110 \rangle$ is a **T** vector for the disordered alloy) and is zero or $\pm\pi$ for superlattice reflections.* Again, in bright

*In Figs. 2.18*a–c* faint antiphase domain boundary contrast is visible in the fundamental reflection–this is so because there is a faint contribution from the superlattice reflection (e.g., if the fundamental is 2**g**, **g** is also excited to some extent).

field the first fringe is always black. Another possible antiphase vector is $\frac{1}{6}\langle 112 \rangle$, a partial vector in the disordered alloy. Hence for fundamental reflections $\alpha = 0$ or $\pm 2\pi/3$, as for a stacking fault. In the case of superlattice reflections α can be $\pm\pi/3, \pm 2\pi/3, \pm\pi$, or zero.

The periodicity of fringes for superlattices is quite different from that of stacking faults in nonordered materials because of the fact that the extinction distances for superlattice reflections are much greater than those for fundamental reflections. Therefore few fringes are visible in the case of domain boundaries. Antiphase domain boundaries (APBs) can be formed either thermally by growth

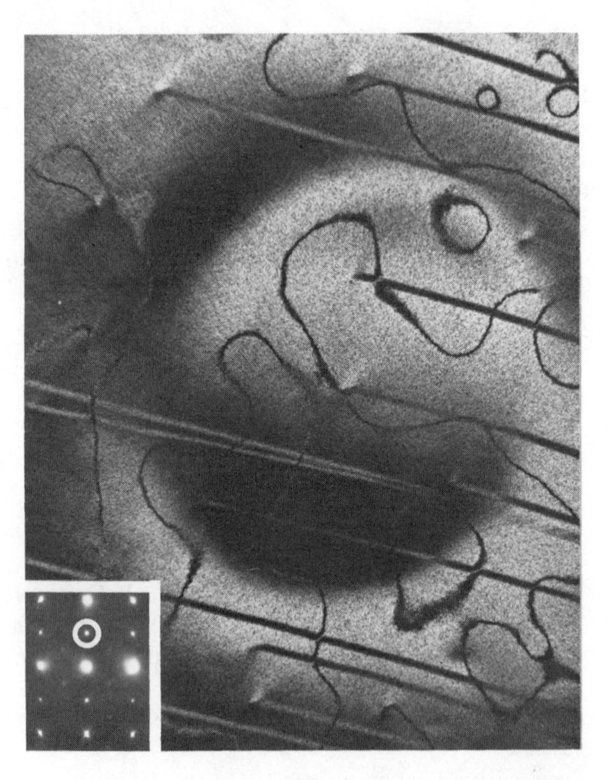

Fig. 3.17 Showing curved thermal and straight slip-produced antiphase boundaries in Fe-22.5 at % Al cooled from 660°C. The micrograph was taken with a 111 superlattice reflection (encircled on SAD in $[1\bar{1}0]$ orientation). The superstructure is of the $B2$ (CsCl) type, and both kinds of antiphase boundaries have displacement vectors equal to body-centering translations ($a/2\ \langle 111 \rangle$). Note the diffuse spots at $\frac{1}{2}, \frac{1}{2}, \frac{1}{2}$ positions in the SAD and the fine mottled background structure in the micrograph, due to the beginning of the $B2 \rightarrow DO_3$ order-order transformation. The residual contrast of the dislocations terminating antiphase boundaries is a dynamical effect. At points of intersection thermal and slip boundaries can be seen to react and move apart. The two large dark semicircles are extinction contours arising from the elastic buckling of the foil as a result of surface contamination. Courtesy U. Dahmen.

and impingement of superlattice domains that are out of phase, or by slip of a dislocation, which restores the periodicity of the fundamental lattice but creates a discontinuity in the superlattice.

Figure 3.17 shows a set of parallel APBs that have been produced by slip of $\frac{1}{2}\langle 111 \rangle$ dislocations in Fe-Al (CsCl type). They intersect a set of randomly distributed thermal APBs. Note the reversal of contrast at the points of intersection. Only the first fringe is visible because of the large extinction distance (2710 Å) of the 111 superlattice reflection.

8.4 Inversion Boundaries in Enantiomorphic Systems

When the symmetry operations, which constitute the space group of a structure, do not include an inversion or a reflection operation, the structure can exist in two forms, a right-handed and a left-handed one, called enantiomorphs. The presence of the two enantiomorphs coexisting within a sample can be verified in the electron microscope by imaging in dark field in a multibeam orientation, with the electron beam parallel to a zone axis (Fig. 3.18), along which the crystal does not show a center of symmetry in projection.[23] One takes advantage here

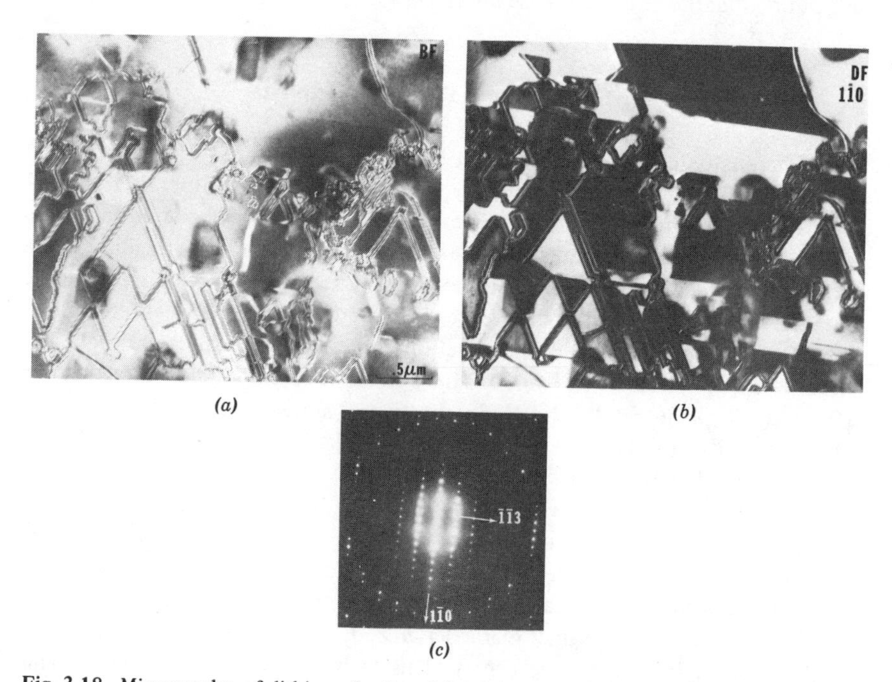

(a) (b) (c)

Fig. 3.18 Micrographs of lithium ferrite: (*a*) bright field, (*b*) dark field, (*c*) diffraction pattern taken under many-beam nonsystematic conditions. Note that the bright field shows no contrast changes from adjacent enantiomorphs, while in the dark field case pronounced differences are observed in agreement with the calculations of Fig. 4.14. Courtesy *Acta Crystallographica*.[86]

of violations in Friedel's law,[24] which may cause a difference in background intensity between the two structures. The calculated image intensities for these situations and another example will be discussed in Chapter 4.

8.5 Additional Effects

Other factors also can give rise to contrast at planar defects. For example, the values of both s and **g** can change across a fault, giving rise to moiré (Δg) and other fringes, sometimes called δ-fringes (see Fig. 3.31), discussed in detail by Amelinckx.[25] An example of moiré fringes across coherent interfaces in Cu-Mn-Al was given in Fig. 2.12. Measurements of fringe spacings enable one to calculate the mismatch across the interface, especially very small mismatches [Δg small, D (fringe spacing) large]. Imaging at conditions where $s \gtrsim\!> 0$ (e.g., dark field weak beam) is also useful for distinguishing moiré contrast from interface dislocations, as has been shown for coarsened alloys by Bouchard et al.[26]

Analysis of contrast changes across interfaces is necessary if the nature of the interface is to be determined. Recently, considerable interest has been generated in grain boundaries and interphase interfaces (see, e.g., refs. 26–28).

9 Small Volume Defects—Precipitates

For coherent defects the contrast can be considered by use of the appropriate displacements (eq. 3.5) in the intensity calculations. The visibility criterion $\mathbf{g} \cdot \mathbf{R}$ makes it possible to establish the direction of **R**, as already shown in Figs. 3.3*a* and 3.4, so that plates, spheres, and rods of precipitates can be distinguished (see ref. 29 for review). Contributing to contrast in the case of precipitates will be intensity changes due to the differences in structure factor between precipitate and matrix. Figure 3.19*a* shows diffraction contrast from small $\{100\}$ coherent plates in a $[011]$ foil of Al-4% Cu. These plates have strain fields normal to the $\{001\}$ habit plane, so $\mathbf{R} = \langle 001 \rangle$. In the figure, since $\mathbf{g} = 020$, only plates on (020) are visible ($\mathbf{g} \cdot [100] = \mathbf{g} \cdot [001] = 0$); in Fig. 3.19*b*, which is a dark field image of the $[010]$ streak from θ'' occurring in the diffraction pattern (Fig. 3.19*c*), "structure factor" contrast is obtained for the (010) plates. In Fig. 3.19*d*, which is an example of the powerful technique of lattice imaging, small GP zones of thickness ~8 Å and diameter 110 Å are actually resolved. Here the strains from each GP zone can be measured directly from the displacements of the (002) planes along the lines marked *a*, *b*, *c*, *d*. The strains fall off to near zero at about 30 Å from the zone center. This technique is very valuable for analyzing the complex situation existing when there is a very large density of defects (e.g., GP zones in Al-4% Cu can exist in densities $\geqslant 10^{17}\ \mathrm{m}^{-3}$) so that strain fields overlap. The resultant strain patterns in amplitude contrast images are very complex and in many cases completely mask the individual particles. Such contrast, referred to as tweed or basket weave, often occurs in ⟨110⟩ in many alloys undergoing different types of phase transformations.[29] Progress is being made in understanding these images by computational techniques[30] and lattice imaging, see Section 12 and refs. 31 and 32.

Another example of the use of lattice imaging to make possible the interpretation of the complicated strain field contrast present in steels[33] is shown in Fig. 3.20. Here the conventional bright field (Fig. 3.20*a*) shows the presence of γ'' precipitates, while in the high magnification strong beam image (Fig. 3.20*b*) the precipitates are masked by strain contrast. They are better defined in the dark field "structure factor" image (Fig. 3.20*c*), taken using one of the weak super-

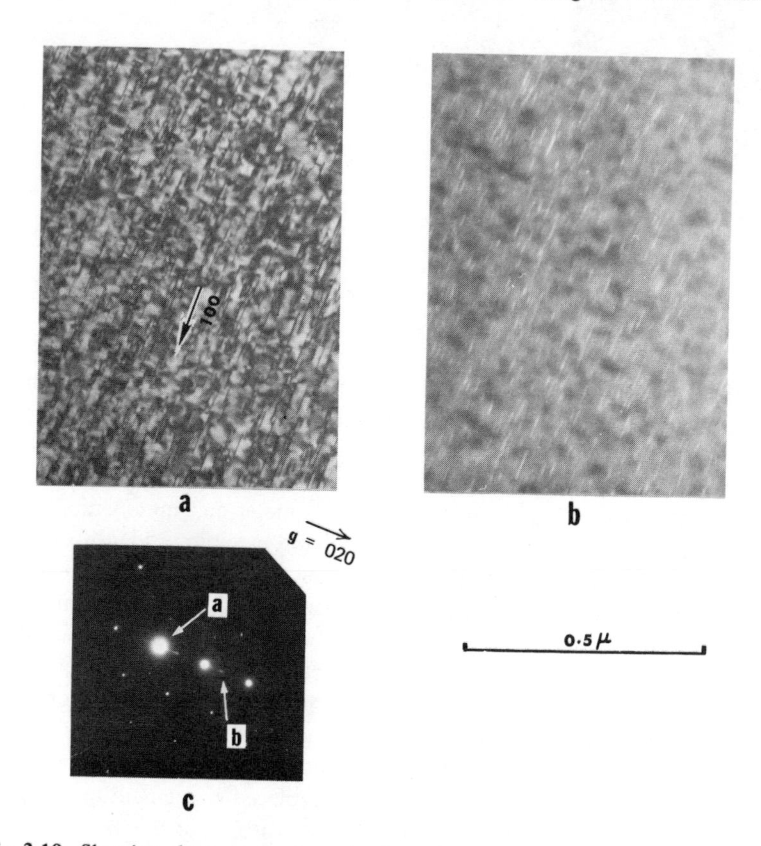

Fig. 3.19 Showing three contrast mechanisms for imaging small coherent plates of GP zones or θ'' in Al-Cu alloys: (*a*) bright field diffraction contrast, (*b*) structure factor contrast dark field image of the streak in (*c*) (Al-4% Cu-0.5% Sn), (*d*) two-beam tilted (002) lattice images of GP zones (Al-3% Cu). Courtesy North-Holland[29] and Pergamon Press.[32]

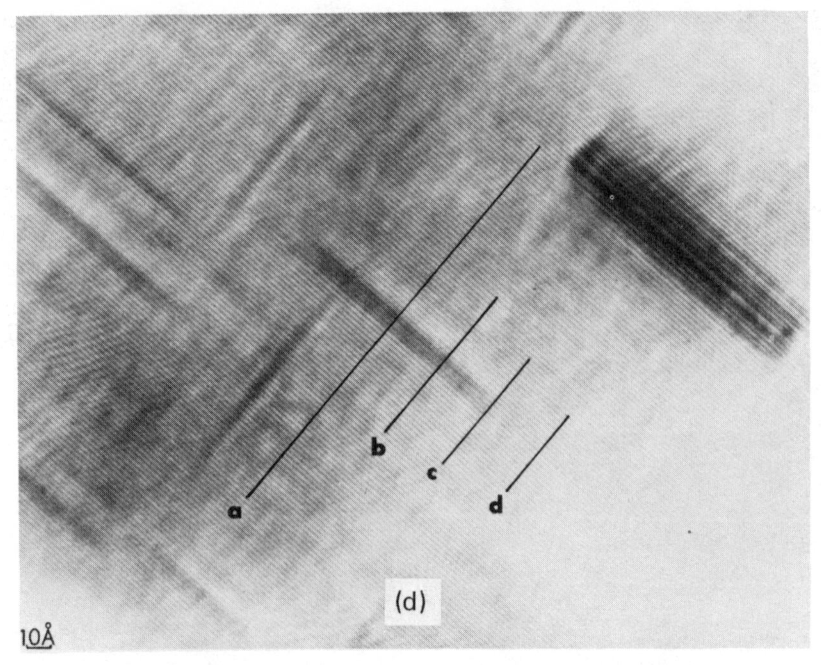

Fig. 3.19 (*Continued*)

lattice diffraction spots, but best defined in the lattice image (Fig. 3.20*d*), where the superlattice and matrix periodicities are clearly seen in the precipitate and its surroundings, respectively.

Martensitic steels are characterized by even more complex microstructures containing $\sim 10^{16}$ to 10^{18} dislocations per square meter and innumerable interfaces and boundaries. For this reason small volume precipitate identification is extremely difficult. First, the strain contrast often interferes with the identification of any small precipitates that may be present; and if it is not minimized, it masks all the other structural details. Second, there are two carbide types of interest in martensitic plain carbon steels: ϵ-carbide (hexagonal) and cementite (orthorhombic), which may be precipitated either during quenching from high temperature (autotempering) or during subsequent tempering. Normally ϵ-carbide exists (in the absence of silicon, aluminum, and nickel) in steels subjected to low tempering temperatures ($\leqslant 200°C$) and is replaced by cementite at higher tempering temperatures.[34] The precipitation of ϵ-carbide in the early stages is characterized by a high degree of coherence, and this precipitate retains some coherency even during precipitate growth. To recognize its presence, it is necessary to tilt the foil so as to minimize the strain contrast both in the matrix (due

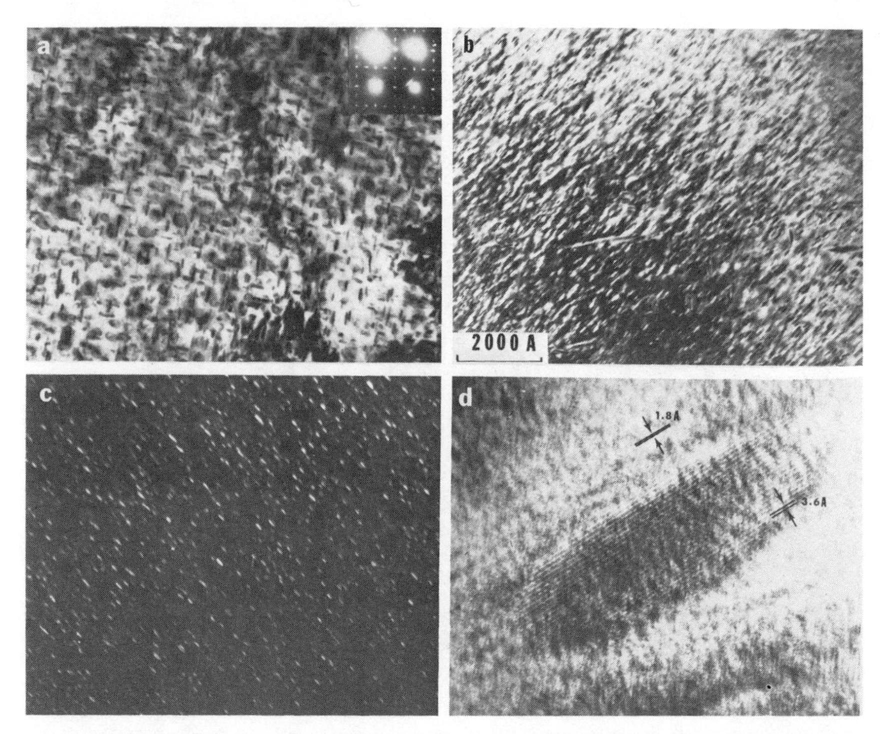

Fig. 3.20 Micrographs of a specimen of an iron-nickel steel (Fe-36Ni-3Ti-3Ta-2.5Nb-1Mo-0.3V-0.01B) aged at 750°C for 4 hours and water quenched, showing γ'' precipitates: (*a*) bright field (*b*) high magnification strong beam bright field, (*c*) superlattice dark field, (*d*) lattice image. Sections (*a–c*) are all at the same magnification. Courtesy J.-Y. Koo.

to dislocations) and at the carbide-matrix interface (coherency strains). By doing so, one may enhance the structure factor contrast by bringing the carbides into a better diffracting condition. Even so, it is rarely possible to recognize the weak carbide spots on the screen and thus be able to take dark field micrographs. Similarly, it is very rare that one can obtain single crystal patterns from these fine carbides. Therefore the distinction between ϵ-carbide and cementite becomes a difficult task, particularly in their early stages of precipitation (Fig. 3.21).

It is important to emphasize here that one cannot depend on the *d*-spacing measurements to distinguish the two carbides in their early stages for the following reasons: (*a*) their equilibrium *d*-spacings obtained from X-ray analysis are too close to permit distinction, and (*b*) the use of equilibrium *d*-spacings is questionable in view of the fact that in the initial stages the carbides may be characterized by nonequilibrium compositions. Fortunately, there is an alternative way of distinguishing them—by the use of habit plane analysis. It is well

Fig. 3.21 Structure of martensite in Fe-5Ni-0.26C steel double tempered 2 + 2 hours at 400°F, showing fine precipitates of ϵ-carbides; arrow marks indicate cementite $\{110\}$ habit, which has also begun to form. Courtesy American Institute of Mining, Metallurgical and Petroleum Engineers.[36]

established[34–36] that the ϵ-carbide has a $\{100\}_\alpha$ habit and grows in $\langle 100 \rangle_\alpha$ directions, whereas cementite forms with a $\{110\}$ habit. Also, the ϵ-carbide interfaces are somewhat wavy (Fig. 3.21), as opposed to the more or less straight interfaces of cementite,[36,37] and this information, together with habit planes, is used to identify the presence of these carbides. In Fig. 3.21 this information serves to identify the presence of both ϵ-carbide and cementite. Figure 3.22 shows the carbide precipitation at a later stage[34] where the carbides have grown sufficiently large to yield single-crystal patterns. They may therefore be identified as cementite by bright and dark field imaging and the corresponding selected area diffraction analysis.

Analysis is also difficult when large densities of small defect clusters are produced by irradiation. The contrast is very depth dependent, making identification in terms of vacancy or interstitial character rather tedious. These difficulties have been discussed in detail elsewhere,[38] and it is clear that computer simulation and image processing techniques are essential also for these problems; see, for example, Fig. 5.18.

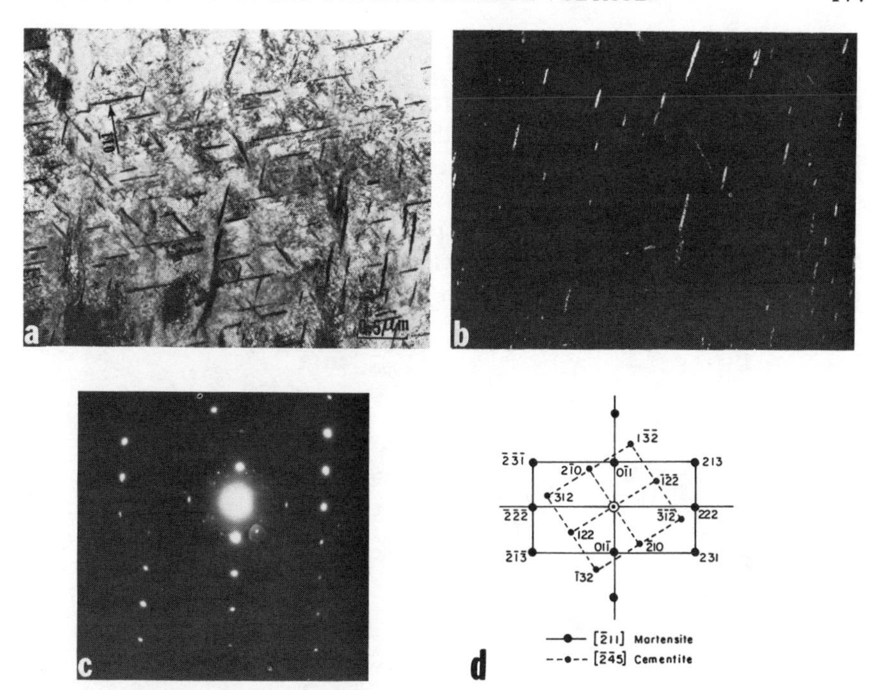

Fig. 3.22 (*a*) Bright field, (*b*) dark field, (*c*) selected area diffraction pattern, and (*d*) corresponding indexed pattern, revealing Widmanstatten cementite in 200°C tempered specimens of Fe-4Cr-2Mn-0.3C steel. Courtesy B. V. N. Rao.

10 Many-Beam Effects and Contrast at High Voltages

At high voltages the sphere of reflection becomes much flatter (Table 1.1), so that many beams are excited. This means that two-beam theory no longer applies, and it is more difficult, or even impossible, to predict contrast from simple geometrical arguments. However, if many-beam theory (see Chapters 4 and 5) is used, it appears that useful practical applications become possible. A particularly interesting effect is that of critical voltage, where intensities of second-order reflections go through a minimum at certain voltages (e.g., 430 kV for 222 in aluminum). Since this particular phenomenon is discussed in Chapter 5, two other examples of many-beam effects are described here.

10.1 Dislocation Contrast

For a defect with a continuously varying strain field of long range, for example, a dislocation, it is found that the width of the image in both bright field and dark field is approximately $\xi_g^{\text{eff}}/3$, where ξ_g^{eff} is the effective extinction distance for the important interactions occurring. Discussion of the theoretical basis of this statement is deferred until Chapter 5. For the present is is sufficient to note that many-beam effects make the effective extinction distance ξ_g^{eff} different from the calculated two-beam value ξ_g. Thus, although extinction distances increase with increasing voltage (and hence two-beam image widths increase), the overall effect is often a decrease in the effective extinction distance, and this effect is particularly marked for higher order reflections of a strong systematic row. Systematic orientation means that a reciprocal lattice row along $n\mathbf{g}$ is excited; in simultaneous orientation several *different* rows are excited.

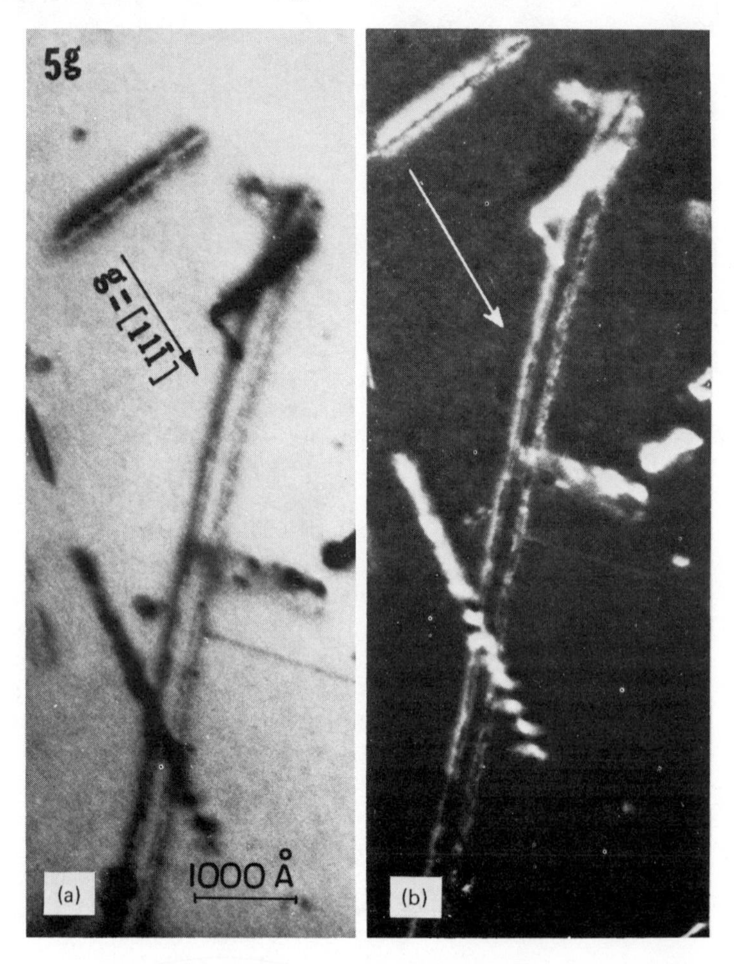

Fig. 3.23 (*a, b*) Dissociated dislocation loops in boron-implanted silicon imaged in high order bright field (BF) and weak beam dark field (WB). (*a*) 5g BF for $\mathbf{b} = \frac{1}{2}\,[10\bar{1}]$. (*b*) 5g WB. (*c, d*) Computed dissociated dislocation image profiles corresponding approximately to (*a*) and (*b*). (*c*) WB. (*d*) BF. Courtesy *Physica Status Solidi*.[40]

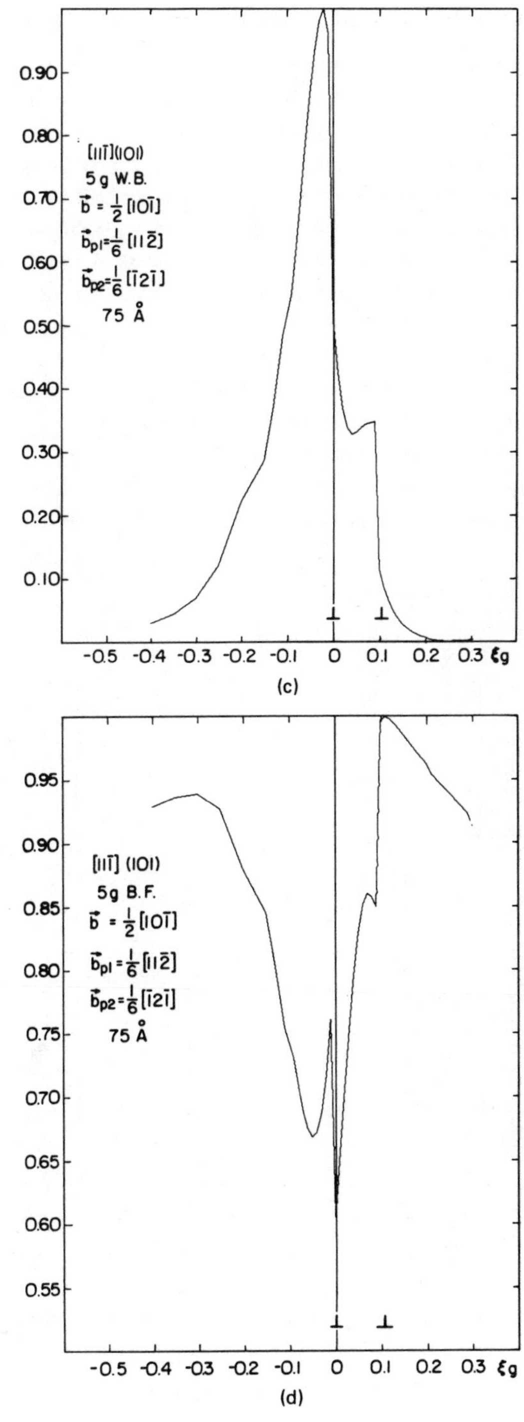

Fig. 3.23 (*Continued*)

The technique is to orient the crystal to excite the reflection $n\mathbf{g}$, usually with $s_{ng} > 0$, where $n\mathbf{g}$ is the nth order of the first reflection of the systematic row. The method of utilizing this effect is described in Chapter 1, Section 4.7.2. A typical example is shown in Fig. 3.23*a*, together with a weak beam dark field image (Fig. 3.23*b*) and the corresponding 12-beam calculated image profiles (Figs. 3.23*c* and *d*). Under suitable conditions[3] such high order bright field images compare favorably with weak beam dark field images (see, e.g., Figs. 3.23*b* and *c*, and also Fig. 3.1) and offer the advantage of high intensity for focusing and exposing the picture. Comparison of the theoretical bases of the two methods of reducing the image width is deferred until Chapter 5. Both have been successfully used to study narrow dislocation separations in dissociation, dipole formation, superdislocation structure, and so on.

Dislocation image characteristics under systematic diffracting conditions have also been found useful for determining the magnitude of the product $\mathbf{g}_1 \cdot \mathbf{b}$ where $\mathbf{g}_1$ is the first-order reflection of the systematic set and thus can be of

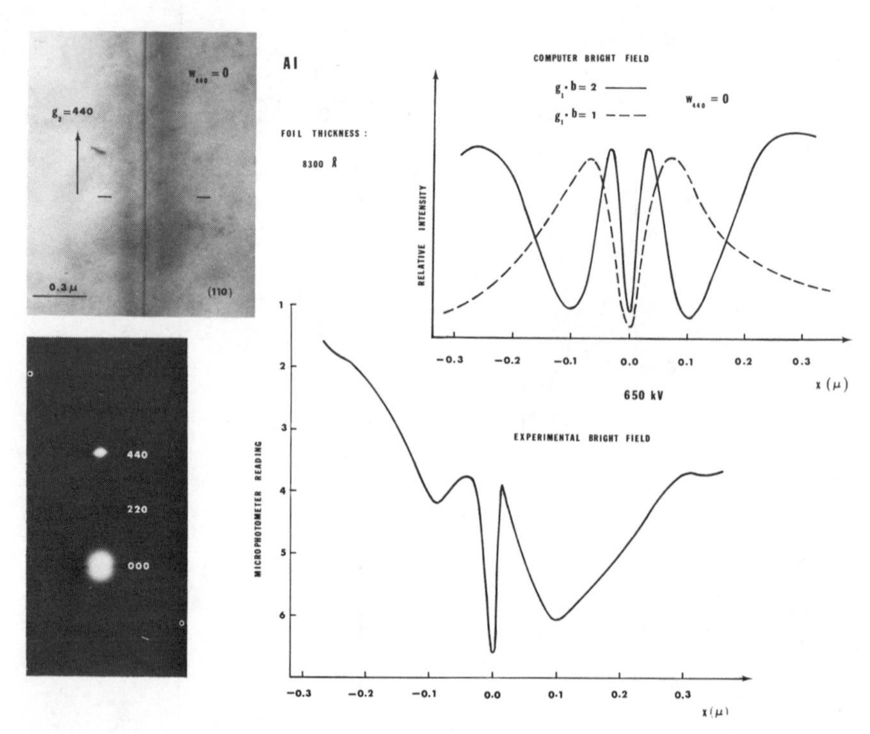

Fig. 3.24 Comparison between observed and calculated image contrast for $\frac{1}{2}\langle 110\rangle$ screw dislocation in aluminum. The magnitude of the Burgers vector is found from the fringe spacings, not from their intensities. Calculations for 650 kV, systematic 12-beams. Courtesy L. C. deJonghe.

tremendous help when it is necessary to choose between a number of different possibilities, as well as in examining more complex crystals. The example cited here is a simpler case of comparison between a known result and theory. In the fcc system, for an undissociated dislocation with Burgers vector of the type $\frac{1}{2}\langle 110\rangle$, when a 220 reflection is used the only three possibilities are $\mathbf{g}_1 \cdot \mathbf{b} = 0$, 1, and 2. Figure 3.24 shows the image of a screw dislocation in aluminum compared to the image profile and multiple-beam bright field profiles obtained by computer methods for the possibilities $\mathbf{g}_1 \cdot \mathbf{b} = 1$ and 2. Comparing the trace of the image with the theoretical profiles, it is clear that there is very good correspondence for the width of the central minimum and the positions of all extrema for the $\mathbf{g}_1 \cdot \mathbf{b} = 2$ case. Variations in relative intensities in the experimental image are probably due to thickness variations and discrepancies in dislocation depth.

10.2 Stacking Fault Contrast

In Section 8.1 rules were discussed for determining the nature of stacking faults in fcc and dc crystals; these rules rely on observation of the outermost fringes when the stacking fault intersects the foil surface. Subsequent calculations using the systematic many-beam dynamical equations have shown that the same information can be obtained without as many restrictions. Bell[39] suggested a many-beam method suitable for high voltage microscopy that would determine the nature of faulted loops without requiring faults to intersect the foil surface. The method depends upon the fact that different contrast is observed in bright field images if the phase angle change $\alpha = 2\pi\mathbf{g} \cdot \mathbf{R}$ at the fault is positive or negative. Further work[40] has shown that the relative contrast of bright field fringes is dependent on the thickness of the foil, but that if an image is formed in 1**g** dark field, the contrast is uniquely determined by $\alpha = 2\pi\mathbf{g} \cdot \mathbf{R}$ when the 2**g** Bragg condition is exactly satisfied. The dark field stacking fault image exhibits higher contrast when the phase change $\alpha = 2\pi\mathbf{g} \cdot \mathbf{R}$ at the fault is $-2\pi/3$ and an observably lower contrast when $\alpha = +2\pi/3$, independently of foil thickness. Similar contrast effects can be found when the symmetry diffraction condition is satisfied, but its application is less practical.

By use of the systematic eight-beam dynamical approximation (see Chapters 4 and 5) the bright and dark field image profiles of stacking faults in silicon have been calculated for different diffraction conditions, accelerating voltages, and thicknesses of the foil. The results are shown in Fig. 3.25, where the maximum intensity has been normalized to unity to show the details of fringe contrast. Figure 3.25*a* shows the dark field image profiles of two stacking faults. The foil thickness varies from four to five extinction distances ξ_g with increments of $\frac{1}{4}\xi_g$. It can be seen clearly that the visibility of the fringes for $\alpha = -2\pi/3$ is

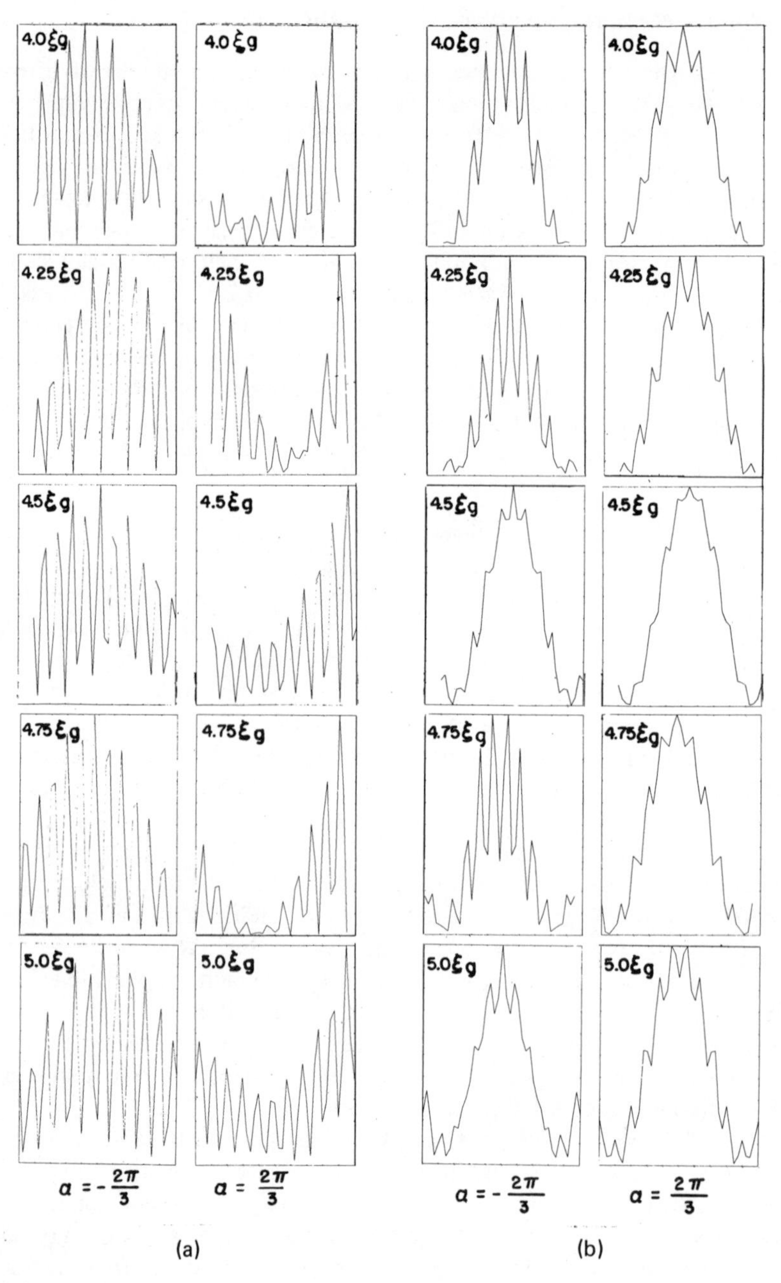

Fig. 3.25 (*a*) Computed profiles of dark field images for $\alpha = \pm 2\pi/3$ when second-order reflection conditions are satisfied. The thickness of the foil varies from $4\xi_g$ to $5\xi_g$ in $\frac{1}{4}\xi_g$ increments. (*b*) Bright field image profiles corresponding to (*a*). Courtesy *Physica Status Solidi.*[40]

generally much higher than that for $\alpha = +2\pi/3$ at all thicknesses. For $\alpha = +2\pi/3$ it is only at the thickness of minimum background intensity that the visibility of even a few fringes has about the same value as that for $\alpha = -2\pi/3$, but this is unlikely to be observed in practice because of the low intensity. Figure 3.25*b* shows the bright field image profiles corresponding to those in Fig. 3.25*a*. The contrast difference of fringes for $\alpha = \pm 2\pi/3$ are large at foil thicknesses t of $4.25\xi_g$ and $4.75\xi_g$. At all other thicknesses the contrast in both cases is much lower. Calculations for situations where 2g is not quite exactly satisfied, that is, for $s > 0$ or <0, show that the effect is thickness dependent for *both* bright and dark field images. From a practical viewpoint, therefore, the foil must be oriented at exactly $s = 0$ for 2g if the value of α is to be found; this can be readily obtained if sharp Kikuchi lines are present.

The features of the bright and dark field images predicted by theory and shown in Fig. 3.25 are confirmed experimentally, as illustrated in Figs. 3.26 and 3.27. Figure 3.26*a* shows a normal dark field image of a stacking fault for $\mathbf{g} = [\bar{2}\bar{2}0]$. The fault is readily identified to be extrinsic since, if **g** is placed at the center of the fault, it points toward the light fringe and the method of Amelinckx et al. applies.[12,22] The top edge of the fault is determined to be at the light fringe edge in Fig. 3.26*a* (see Section 8.1). The fault plane is therefore (111), and the geometrical configuration is shown in Figs. 3.26*b* and *c*. Since the fault is extrinsic, $\mathbf{R} = \frac{1}{3}[111]$, $\alpha = 2\pi\mathbf{g}\cdot\mathbf{R} = 2\pi \times \frac{4}{3} = +2\pi/3$ (modulo 2π). From

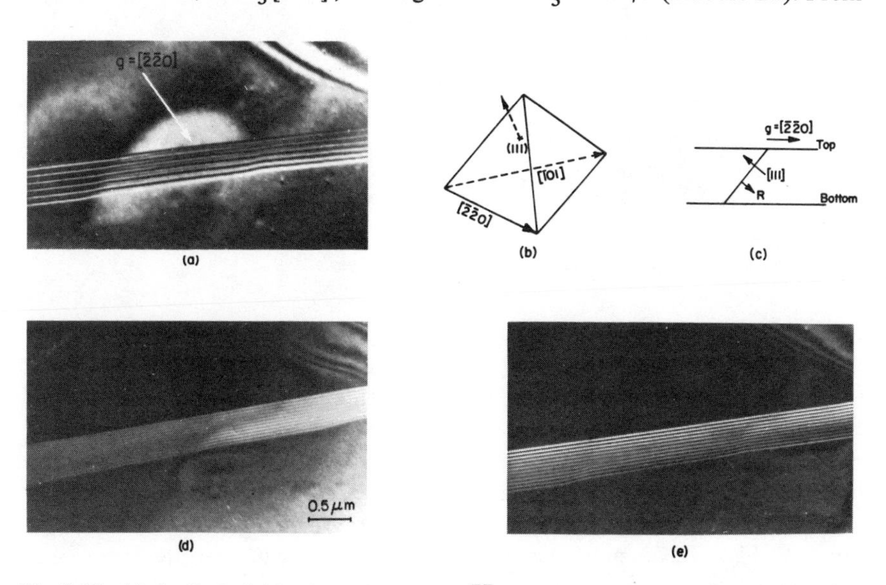

Fig. 3.26 (*a*) 1g Dark field micrograph for $\mathbf{g} = \bar{2}\bar{2}0$. (*b*, *c*) Geometrical configuration. (*d*) 2g Dark field micrograph, $\alpha_\mathbf{g} = +2\pi/3$. (*e*) $-2\mathbf{g}$ Dark field micrograph, $\alpha_\mathbf{g} = -2\pi/3$. Courtesy *Physica Status Solidi.*[40]

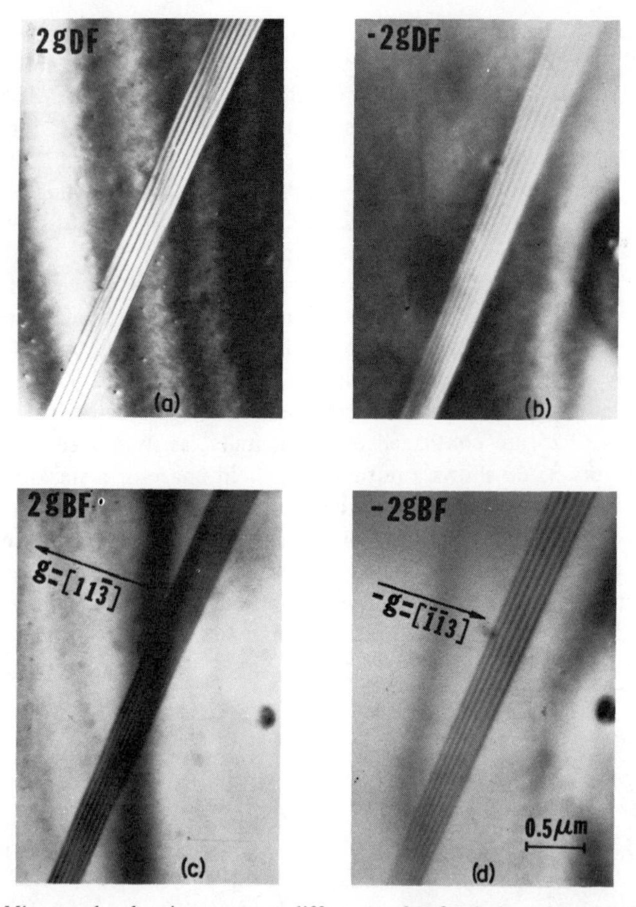

Fig. 3.27 Micrographs showing contrast differences for 2g dark and bright fields for $\alpha = \pm 2\pi/3$ at various thicknesses. Courtesy *Physica Status Solidi.*[40]

Fig. 3.25 the 1g dark field stacking fault images with 2g satisfied should show observably lower contrast for $\alpha = +2\pi/3$ than for $\alpha = -2\pi/3$. Figures 3.26*d* ($\alpha_g = +2\pi/3$) and 3.26*e* ($\alpha_g = -2\pi/3$) clearly demonstrate this is indeed the case. Figures 3.27*a* and *b* show 1g(2g, $s = 0$) dark field images of a stacking fault for $\alpha = -2\pi/3$ and $+2\pi/3$, respectively, when 2g is satisfied for a foil of varying thickness. The fringe contrast for the $\alpha = -2\pi/3$ case is much stronger than that for $\alpha = +2\pi/3$ at all thicknesses. However, the corresponding bright field images shown in Figs. 3.27*c* and *d* clearly show the oscillations in intensity with changing foil thickness.

Since the high or low contrast nature of the dark field fault fringes is determined by the sign of α only and is independent of the thickness, the method is particularly useful for analyzing the nature of faulted loops that *do not intersect* the foil surfaces. Thus it can be useful for analyzing small loops inside the foil, provided that the correct dark field conditions are satisfied.

The experimental procedure can be summarized as follows:

1. Obtain one systematic set of reflection conditions, for example, $n\mathbf{g} = n\langle 111 \rangle$.
2. Take $1\mathbf{g}$ dark field micrographs when the $\pm 2\mathbf{g}$ condition is satisfied.
3. Determine the slope of the fault plane by the usual method (the actual fault plane indices are not required).

For analysis, from step 2 the sign of the phase angle change α at a fault can be determined by applying the rule that the fringe contrast is much stronger for $\alpha = -2\pi/3$ than for $\alpha = +2\pi/3$. When this is combined with knowledge of the slope of the fault plane, the nature of the fault can be readily determined. Since this rule of high and low contrast is strictly thickness independent only when the exact $2\mathbf{g}$ conditions are satisfied, the presence of Kikuchi lines is most helpful to ensure this is achieved.

11 Microscopy of Magnetic Materials

11.1 Lorentz Microscopy and Its Applications

The trajectories of the electrons through the specimen may also be affected by defect features that are *not* structural in the sense of being "distorted or displaced lattice planes," which have been considered until now. An example of such a feature is a magnetic domain in a ferro- or ferrimagnetic material. On passing through such a magnetized specimen, the electron beam is deflected by the field in the specimen, through an angle ϕ as shown in Fig. 3.28. This deflection is due to the Lorentz force, which inevitably acts on any moving charged

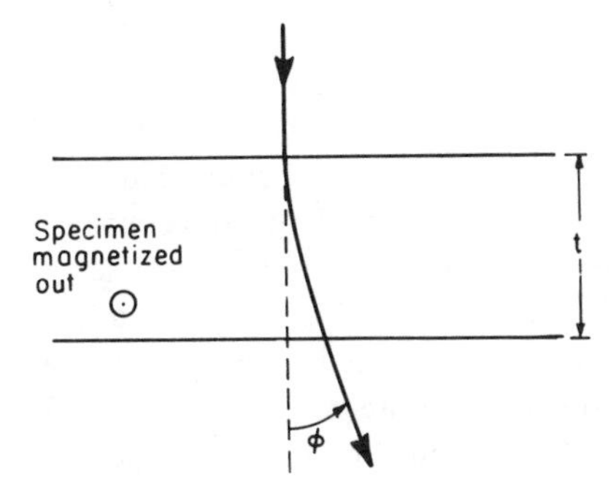

Fig. 3.28 Schematic diagram showing how the electron beam is deflected by the Lorentz force on passing through a magnetized specimen.

Table 3.5 Magnetic Data for Some Materials

Material	Magnetization M_s(A m^{-1})	Saturation Induction B_s(Wb m^{-2})	Deflection[a] ϕ (rad)
Co	1.43×10^6	1.79	3.2×10^{-4}
Ni	4.80×10^5	0.60	1.1×10^{-4}
Fe	1.71×10^6	2.15	3.9×10^{-4}
$Gd_{20}Co_{80}$	$\sim 2 \times 10^5$	~0.2	$\sim 4 \times 10^{-5}$
Cd-Co-Au	$\sim 1 \times 10^5$	~0.1	$\sim 2 \times 10^{-5}$

[a] Deflection for a foil of thickness 200 nm and electrons of energy 100 keV.

particle. The force **F** is given by

$$\mathbf{F} = q\mathbf{V} \wedge \mathbf{B},$$

where **B** is the magnetic induction, q the charge on the particle, and **V** its velocity. For the geometry of Fig. 3.28 the small deflection ϕ after the electron beam has passed through a specimen of thickness t may be shown to be

$$\phi = \frac{\dot{q}Bt}{mV}, \tag{3.6}$$

where m is the relativistically corrected mass. Inserting typical figures for the saturation induction of, say, cobalt (B_s = 1.79 Wb m^{-2}, thickness t = 200 nm, and electron energy = 100 keV) yields a typical deflection of $\phi \sim 3 \times 10^{-4}$ rad. Data for cobalt and other materials are listed in Table 3.5, including the materials of smallest induction yet studied.[42]

Such a small deflection (less than one tenth of a typical Bragg angle) will only have a slight effect on the standard in-focus image (see below), being effectively equivalent to a small change in the incident beam direction or to a very small tilt of the specimen. Nevertheless it is sufficiently large for significant contrast effects to be detected, as shown in Fig. 3.29, which illustrates the imaging for 180° domains separated by either Bloch or Neel walls.

A magnetic specimen (e.g., the top of Fig. 3.29), typically consists of adjacent oppositely magnetized domains. The diffraction pattern from such a specimen consists of spots split by ϕ either side of the "nonmagnetic" position (all spots being similarly split). The in-focus image (Fig. 3.29*a*) (which for convenience is a bright field image here) shows little magnetic contrast; standard bend contours, slip traces, and so on are easily visible as usual. However, if the specimen is viewed with the objective lens either slightly overfocused (Fig. 3.29*b*) or underfocused (Fig. 3.29*c*), electrons from oppositely magnetized domains produce regions of excess and deficient brightness in the image at the positions of the walls that separate the domains. The "excess" and "deficient" walls interchange con-

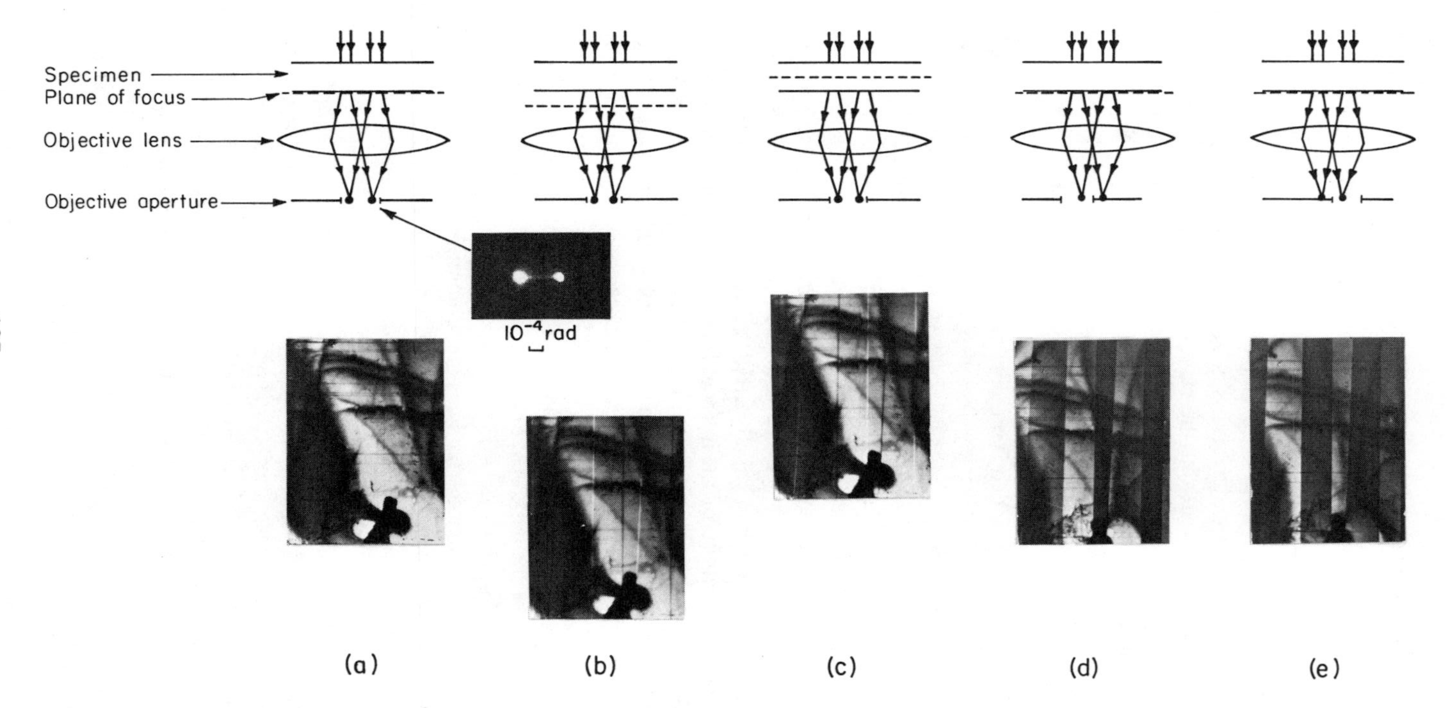

Fig. 3.29 Schematic illustration of the ways in which the presence of a magnetic domain structure can be made visible in the electron microscope, together with the diffraction pattern obtained and the micrographs, which are all of the same area of specimen. Micrographs courtesy J. P. Jakubovics.

trast on going through focus (compare Figs. 3.29*b* and *c*), and for many materials (see Table 3.5) the degree of defocus required is sufficiently small that the contrast produced by other features is not too blurred. Hence interactions between magnetic domain walls and structural defects may be studied. If higher resolution of structural defects is simultaneously required, the objective aperture may be displaced to cut out the electrons deflected by one set of domains only (see Figs. 3.29*c* and *d*) and the specimen viewed in focus. Since the angles involved are very small, this requires accurate positioning of the objective aperture, which must be exactly in the focal plane and must also be very clean to avoid aberration of the beam passing through close to its edge.

An example of the application of two of these observational techniques is shown in Fig. 3.30, which represents an Fe-Cr-Co-V ductile permanent magnet alloy, consisting of a weakly ferromagnetic chromium-rich matrix containing strongly ferromagnetic iron-cobalt-rich particles. The domain walls in the weakly ferromagnetic phase are shown in Fig. 3.30*a* by the out-of-focus method and in Fig. 3.30*b* by the displaced aperture technique, and may be seen to bend around the strongly ferromagnetic particles in some instances, for example, at *A*. Figures such as these provide evidence to suggest that the magnetization reversal process of this alloy is controlled by domain wall pinning (as opposed to single-domain hardening) with the iron-cobalt particles acting as pinning sites.

More careful study of the in-focus image shows that even the small deflections involved produce detectable contrast effects in both bright and dark fields and that the changes produced are not identical in the two cases. The effects of the oppositely magnetized domains on certain contours are (*a*) to displace the bend contours in both bright and dark field (by $\pm\phi/2$), as if the beam were *incident* at the angle of trajectory obtaining at the middle of the foil; and (*b*) to change the intensity of the dark field only (see ref. 43 for a discussion of the orientation relationships for which the effect is most visible).

The imaging techniques and magnetic contrast effects noted above may be given the generic name "Lorentz microscopy," since all depend on the Lorentz deflection. For successful Lorentz microscopy the specimen magnetization must have a component in the orientation shown in Fig. 3.28, that is, perpendicular to the electron beam. Specimens must be thin and parallel to the beam for reasons of penetration, and this also favors the magnetization arrangement that is suitable for reasons of magnetostatic energy (free pole suppression). Thus interesting specimens are readily obtained which, however, may not be representative of the bulk material in either scale of domain structure or even direction of magnetization. However, provided that this reservation, which is common to a greater or lesser extent to all electron microscopy, is accepted, the technique is of considerable value. In the case of Lorentz microscopy, though, one further limitation must be overcome. The magnetic field of the objective lens in the standard specimen position is usually sufficiently large to drastically modify the domain arrangement. Also, any thicker pieces of the specimen (e.g., the rim of a

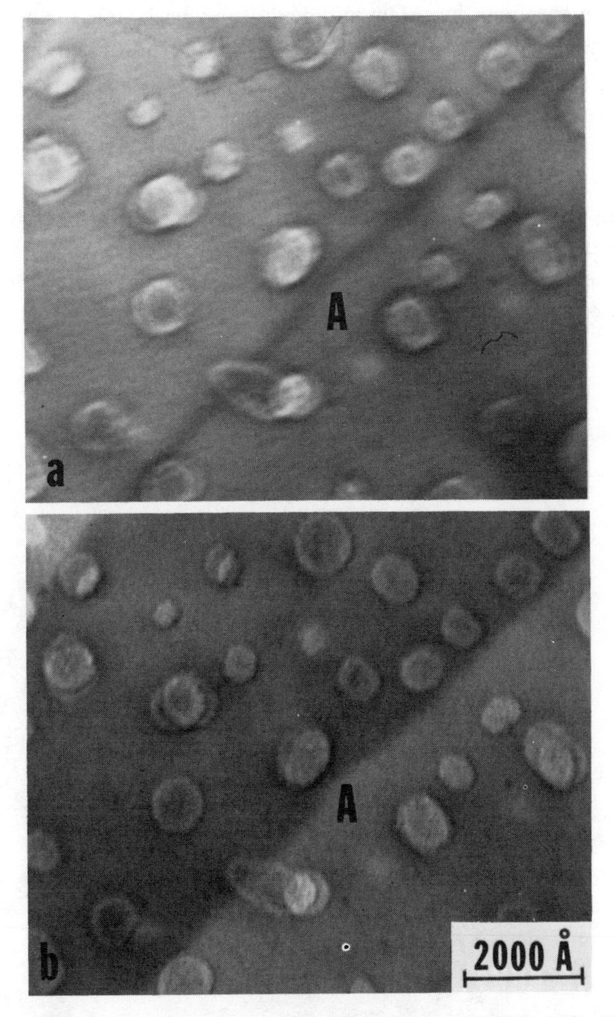

Fig. 3.30 Micrograph of a ductile permanent magnet alloy [Fe-23Cr-15Co-5V (wt %) aged for 50 hours at 650°C] with the magnetic domains imaged by (*a*) out-of-focus method, (*b*) displaced aperture technique. Courtesy M. Okada.

disk specimen) tend to distort the objective lens field and hence the whole image. Both effects may be minimized by placing the specimen in a position of small magnetic field, and a number of methods of achieving this have been successfully devised:

1. Switching the objective lens off and operating at low resolution and/or low magnification, using the other lenses.

2. Raising the specimen by a few millimeters and weakening the objective for a compromise of reasonable resolution (~5 nm) and magnification.
3. Designing special objective lens pole pieces with small fields at the specimen position (usually with the additional feature of placing the objective aperture exactly in the back focal plane to achieve good discrimination in the displaced aperture mode).
4. Operating at high voltages where, although the deflections are smaller, so are the effects of distortion produced by the "bulk" parts of the specimen; thicker specimens are penetrated, and the lower level of chromatic aberration produces an improvement in the contrast.

11.2 Structural Effects

When a material is magnetized, certain dimensional changes take place, the phenomenon usually being termed magnetostriction. Although for metals the effects

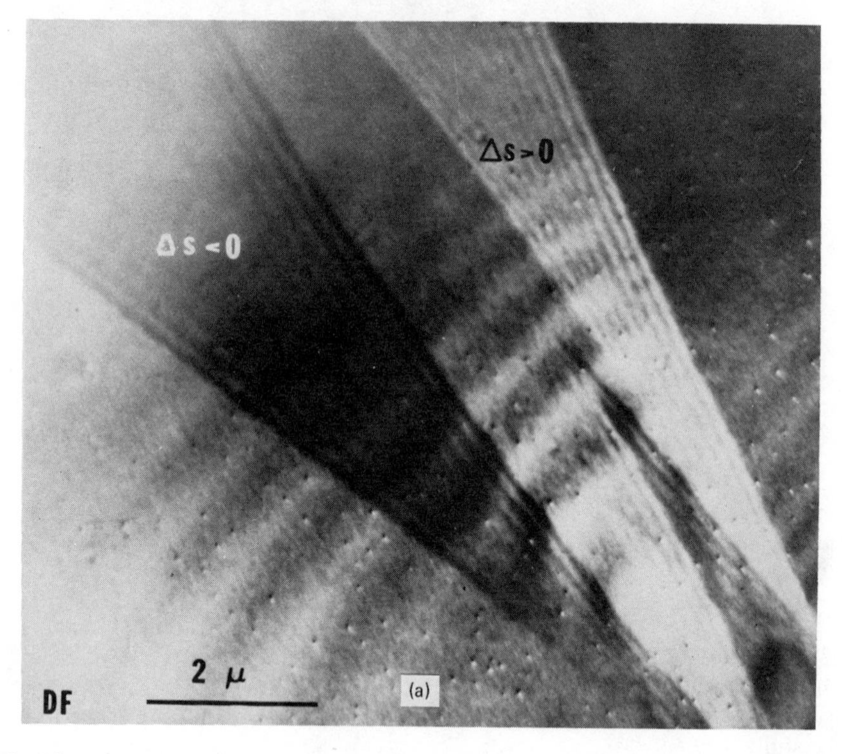

Fig. 3.31 Magnetic domains in a cobalt ferrite producing "standard" diffraction contrast through the small changes of orientation between the differently magnetized regions: (*a*) dark field, (*b*) bright field. Courtesy L. C. de Jonghe.[44]

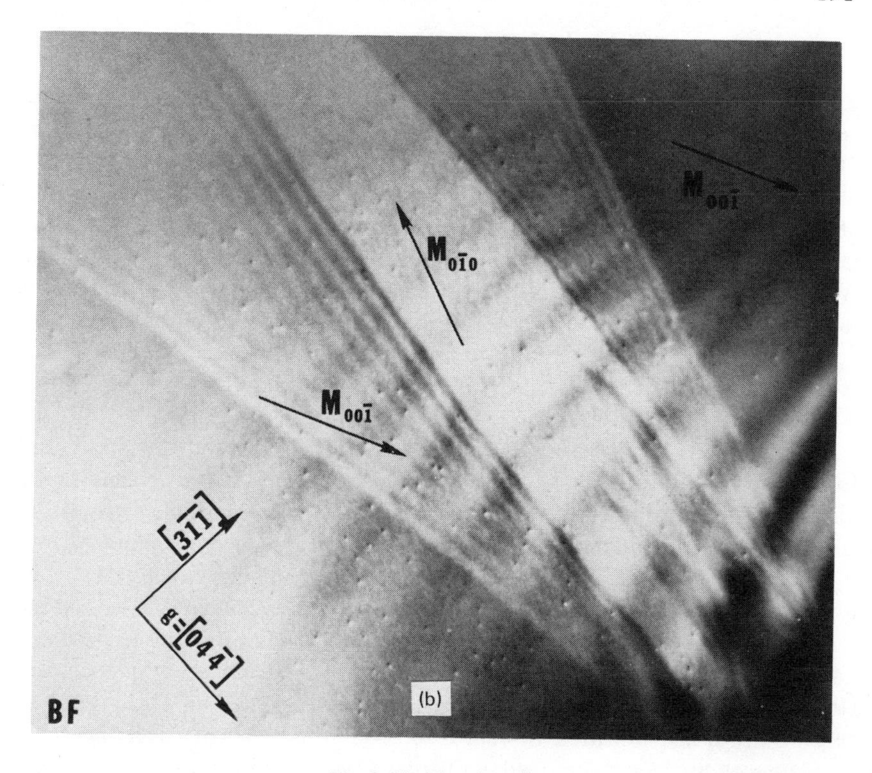

Fig. 3.31 (*Continued*)

are usually too small to be detected in the electron microscope, this restriction often does not apply in the case of ferrimagnetic oxides such as cobalt ferrites. An example showing magnetic domains and domain boundaries imaged by conventional diffraction contrast is presented in Fig. 3.31. The magnetization directions indicated in Fig. 3.31*b* produce contrast through the small differences in orientation Δs indicated in Fig. 3.31*a*; see Section 8.2.

PHASE CONTRAST IMAGING

The images discussed in the preceding sections of this chapter have all, with minor exceptions, been amplitude contrast images, that is, the intensities in the images have been the square of the amplitude of the single beam allowed through the aperture. In the following sections the results of allowing a number of beams to form the image by interfering with each other will be considered. The contrast is therefore a composite of interference effects and amplitude changes produced by defects.

12 Lattice and Structure Imaging

As discussed in Chapter 1, modern 100 kV transmission electron microscopes possess the capability of 1.5 Å line resolution and 3 Å point-to-point resolution. Thus by suitable choice of imaging conditions it is possible to image directly the crystalline lattice itself. This provides an approach to studying details that offers an alternative to the amplitude phase contrast so far discussed and was in fact one of the earliest ways in which the electron microscope was used to study crystalline materials,[45] namely, lattice imaging in phthalocyanine crystals at resolutions of approximately 12 Å. An increasing amount of research has been devoted over the past few years to establishing the conditions under which lattice images can be satisfactorily and readily interpreted in terms of the real lattice of the specimen.

The conditions for direct resolution of crystal lattices are very stringent, and for close-packed structures, as those in most metals, alloys, and ceramics, only one-dimensional or lattice imaging can be done at the present time with reasonable facility. In this method (Chap. 1, Sec. 4.8) thin specimens are required, and the microscope is operated marginally out of focus to minimize spherical aberration (e.g., for 5 Å resolution the defocus may be approximately 500 Å, and for 3 Å resolution about 1500 Å—the Scherzer[46] focus condition). Thus to optimize image contrast it is essential to choose the optimum conditions for specimen (thickness, orientation) and instrument (defocus, illumination, objective aperture with high magnification viewing $\geqslant$400,000×). Methods of calculating these optimum conditions are described in Chapter 5. In the following discussion some examples of recent applications of lattice and structure imaging are given.

12.1 Examples of Lattice Imaging

At 100 kV it is rarely possible to form a lattice image of metals and ceramics using any higher than first-order reflections, and usually no more than two beams at a time. The resulting images consist of a set(s) of fringes normal to the operating diffraction vector(s), which represent a particular set of atomic planes.

12.1.1 *Early Developments.* Many of the early studies utilizing lattice imaging for quantitative analysis of metallurgical problems involved crystal lattice dislocations.[47–49] In these investigations, measurements were made of spacing perturbations surrounding terminating fringes and were related through isotropic elasticity theory to the core width, core strain, and long-range matrix strains associated with dislocations. Such studies were also applied to crossed lattice images[50] to examine the effects of lattice strain on atomic positions. However, it was later demonstrated,[51] both by experiment and through contrast calculations, that an inclined lattice dislocation may produce either one or three terminating fringes in a two-beam image, depending upon the direction (+**g** or −**g**) of the operating reflection. Only when the dislocation line was oriented end-on with respect to the electron beam was a single-fringe termination produced. These results demonstrated that extreme care must be taken to assure a one-to-one correspondence between terminating fringe profiles and matrix dislocations.

Subsequent research on dislocation strain fields was carried out using specimens prepared under rigid geometrical constraints[52] in order to assure the required end-on dislocation configuration. In other defect studies[53] Phillips successfully used lattice imaging to reveal microtwins of three-atom plane thicknesses in silicon, as well as the occurrence of highly regular periodic structures in incoherent twin boundaries. It was also observed[54] that in ion-bombarded germanium damaged regions exhibited no fringe contrast, suggesting a lack of crystallinity, while similar defects in copper were found to have a complex strain field detected only in the lattice image mode.[55]

By providing detail that exceeds the resolution capabilities of other experimental techniques, lattice imaging has considerably enhanced the characterization of phase transformation behavior. Following the direct observation of GP zones in Al-Cu by this method (see e.g., Fig. 3.19*d*), a quantitative assessment of zone size, density, and displacement field was subsequently performed at the atomic level and more recently extended through the precipitation sequence.[56] The latter study revealed a continuous structure progression from GP zones through θ' precipitation, allowing identification of each phase by its characteristic fringe spacing even when such evidence in the electron diffraction patterns was questionable. When utilized in an examination of the omega transformation in Zr-Nb, lattice imaging furthermore confirmed the existence of three subvariants within a previously identified single omega variant[57] and led to the formulation of a new model for a linear, omega-like defect in these materials.[58]

As shown by these studies, the most obvious advantage of high resolution electron microscopy is the enhanced structural detail which it can provide, making it possible to perform real space crystallography on even the most complex materials. Furthermore, because the microscope is operated in a transmission mode, this type of information can be extracted from the vicinity of internal interfaces without the need for fracturing the specimen to expose the appropriate boundaries. Hence it is possible to study directly the atomic configurations at intact interfaces which are responsible for both the physical and mechanical behavior of materials, as is shown in Section 12.2.2 for germanium. Such information is vitally important in the analysis of materials behavior, for example, phase transitions, the deformation behavior of polycrystals, the structure at and near grain boundaries, etc. Examples of recent developments are given in the following subsections.

12.1.2 *Ordering Reactions.* A fruitful area of application for high resolution electron imaging has been the study of short-range order[59] and ordering[60] in a number of alloy systems. For these investigations, experiments are designed to permit the imaging of the ordered superlattice planes as well as the fundamental lattice planes, when they coexist within the specimen. The most successful imaging technique has been to combine the transmitted, first-order superlattice and the first-order fundamental reflections along a systematic row within the objective aperture.[61] The illumination is also tilted to align the superlattice reflection on the optic axis. An example is shown in Fig. 3.32. Because this situation results in a complicated dependence of the image on focus, calculations must also be performed to ensure accurate image interpretation.[61-63]

A comparison of the lattice and conventional dark field image is shown in Fig. 3.32 for partially ordered Mg_3Cd. The bright areas in the superlattice dark field image represent ordered regions, and the dark regions are disordered, as

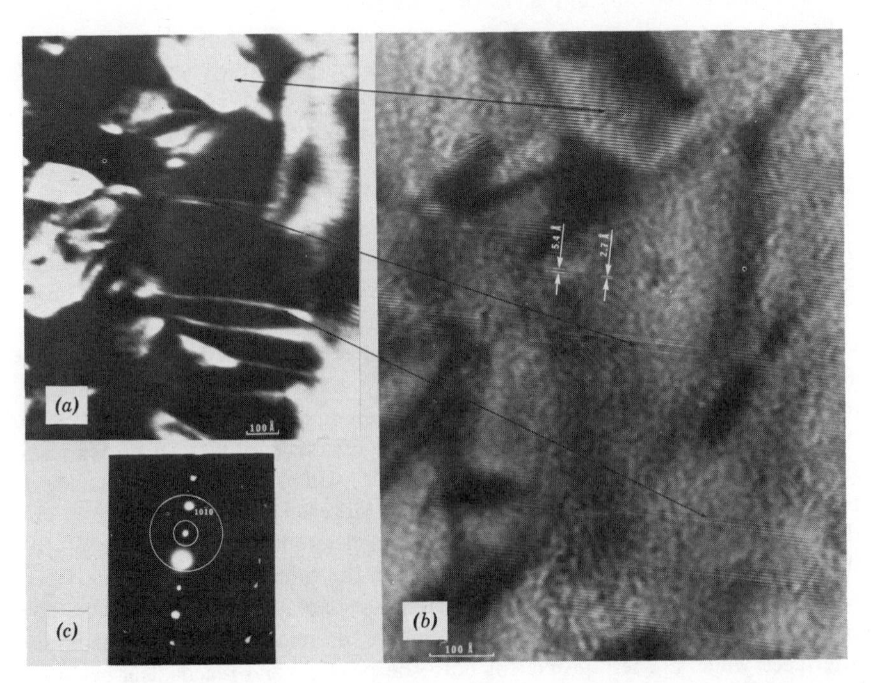

Fig. 3.32 Comparison of the lattice image (*b*) and superlattice dark field image (*a*) of partially ordered Mg_3Cd. Ordered regions possess twice the finge periodicity of disordered regions in the lattice image. An exact correspondence between the two is clear, with more detailed information present in the former (see Fig. 3.33). The corresponding diffraction pattern is shown in (*c*), with the smaller circle showing the objective aperture used for the dark field image (*a*), and the larger circle that for the lattice image (*a*). Courtesy J. Dutkiewicz and R. Sinclair.

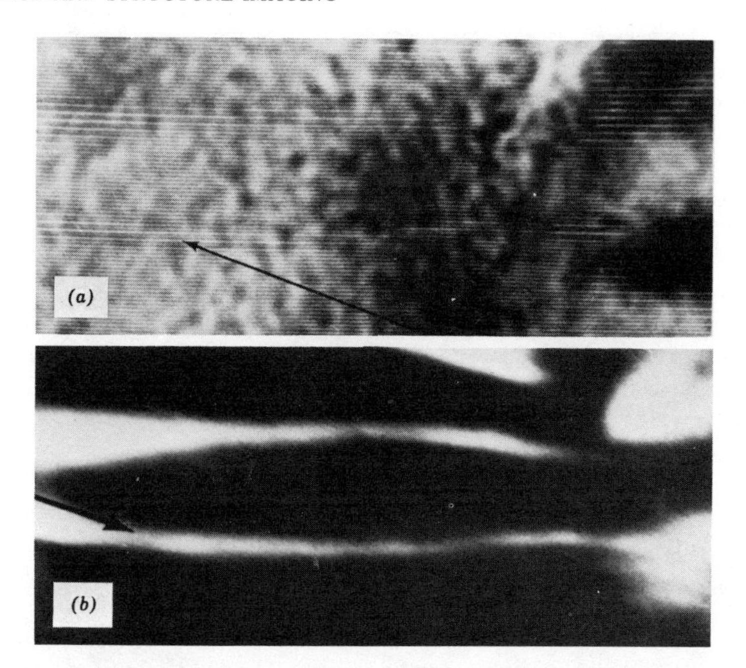

Fig. 3.33 A detail of the lattice (*a*) and dark field (*b*) images shown in Fig. 3.32. The dark field image reveals thin ribbons of ordered material that show some curvature. The lattice image indicates that this is due to unit cell high steps in the interface between ordered and disordered material. In addition to the higher resolution, information about local degree of order can also be obtained from the lattice image, but not from the dark field image. Courtesy J. Dutkiewicz and R. Sinclair.

revealed by differences in fringe spacings, with double periodicity in the ordered material.[60, 62] This gives confidence in the interpretation of dark field images of diffuse reflections. The additional information in the lattice image explains such features as bending of the interface, seen in dark field (see Fig. 3.33). There are discrete unit cell high steps at the interface which are not revealed in the dark field image, suggesting that the mechanism of ordering in this alloy is one of movement of these steps or ledges across the interface; this would account for the flat domain boundaries found in the fully ordered alloy (see Fig. 1.2*c*).

The superior information available from lattice imaging of alloys, over that gained in conventional imaging modes, is illustrated by Fig. 3.34, which compares the lattice image and superlattice dark field image of an antiphase boundary (APB) in ordered Ni_4Mo. In this material the boundary is revealed as a black line on a white background, in some regions approximately 20 Å thick. The translation at the APB must be determined by a series of various dark field micrographs. However, the lattice image can be taken as representing the posi-

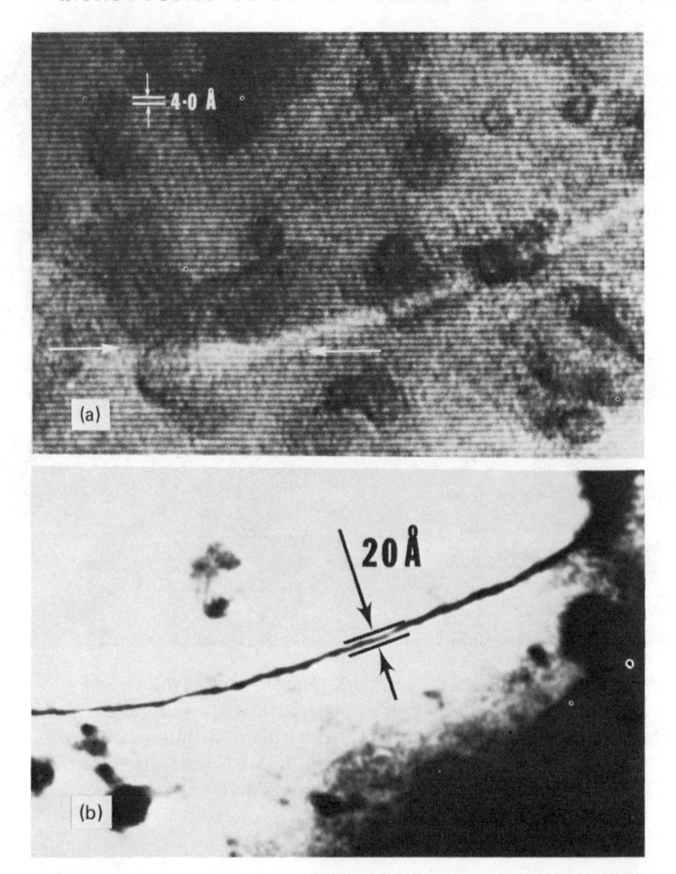

Fig. 3.34 Comparison of the lattice image (*a*) and conventional superlattice dark field image (*b*) of an antiphase boundary in ordered Ni_4Mo. In the lattice image information is given concerning the positions of atomic planes right up to the boundary, the degree of order in the vicinity of the boundary, the nature of the fault, and any local compositional changes (as in the position indicated). Such detailed information is not available in the dark field image. Courtesy MTM Association for Standards and Research; G. Thomas, *Journal of Metals*, **29**, 31 (1977).

tions of the molybdenum plane of atoms, which occur every fifth plane in the superlattice, the other planes being all nickel. The continuity of the fringes (atomic planes) is shown to within an atomic diameter of the fault. From calculations of the fringe visibility with degree of long-range order it is established that little change in degree of order occurs up to the APB. It can be seen directly that the left-hand side is shifted upward by two fifths of the superlattice spacing (i.e., two fundamental lattice planes) with respect to the right and that there is a local change of spacing (and hence composition) where the boundary becomes parallel to the imaging planes (arrowed). This is expected in this region, where the boundary is nonconservative.[65] Thus considerably more information is present in the lattice image than is available from the conventional micrograph.

12.1.3 *Interfaces and Boundaries.* The conditions necessary for high resolution imaging of interfaces and boundaries are also very stringent. The interface must be parallel to the incident beam, and the crystals on either side of the boundary should be simultaneously in a strong diffracting condition. In this way lattice images can be obtained simultaneously, and therefore details of boundary structure and possible intergranular phases can be revealed to approximately 2 Å resolution. Figure 3.35 shows a schematic of these conditions.

An example of a situation where resolution and characterization of intergranular phases is essential occurs in ceramic materials, where processing of polycrystals involves additives that can form glassy films at the boundaries. One such example, typical of many, is the nitrogen ceramics, which have very attractive

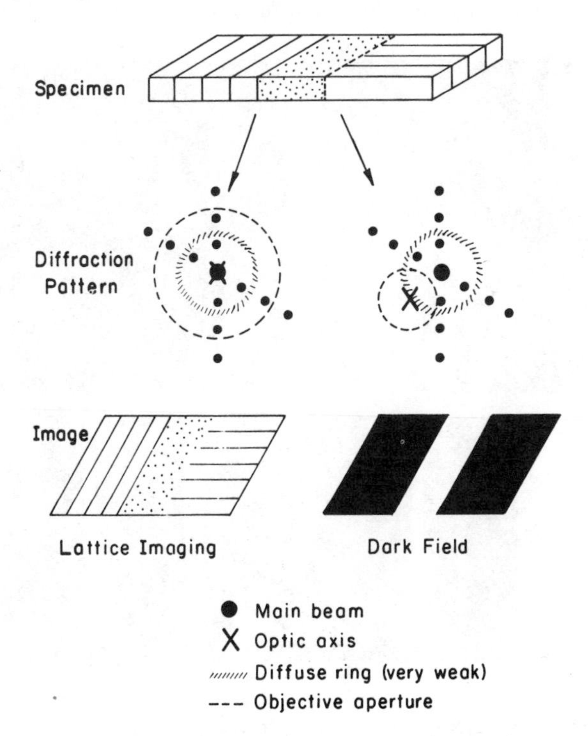

Fig. 3.35 Necessary contrast conditions for analyzing grain boundary interfaces and for detecting intergranular phases. The boundary must be parallel to the incident beam for simultaneous lattice imaging in both grains 1 and 2. Courtesy O. L. Krivanek.

properties (high modulus-to-density ratios, high melting points, oxidation resistance, etc.). However, because of fabrication difficulties a hot-pressing additive such as MgO or Y_2O_3 is needed, and the properties at high temperatures are impaired. It has been proposed that the impairment is due to the formation of an intergranular phase, probably glassy, as a result of the formation of silicates or crystalline oxynitrides. Attempts to prove this, using high resolution TEM, have been successful.[66,67] The problem of resolving intergranular phases and determining whether or not they are amorphous is not trivial. From an electron microscopy viewpoint, therefore, characterization at grain boundaries requires attention to the following features: (*a*) detecting the intergranular phases and their distribution, (*b*) determining whether these phases are crystalline, and (*c*) determining their chemical compositions.

From a morphological viewpoint it is essential to choose the proper orientation conditions as sketched in Fig. 3.35. The contrast from intergranular phases, if present and resolved, depends on whether they are amorphous or crystalline. Amorphous phases (most probably silicates) generally appear in light contrast (low atomic number-mass thickness contrast), as shown in Figs. 3.36 and 3.37*a*.

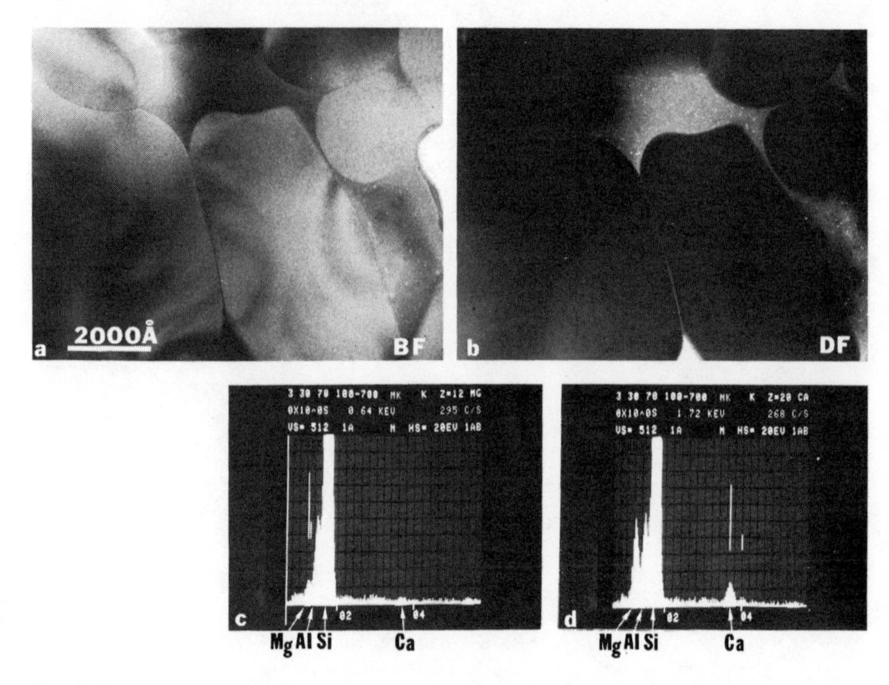

Fig. 3.36 MgO-Fluxed Si_3N specimen. (*a*) Bright field showing ionization damage in glassy phase. (*b*) Dark field image using diffuse scattering from glassy phase. (*c*) STEM X-ray analysis of Si_3N_4. (*d*) STEM X-ray analysis of glassy phase; notice impurities, especially calcium. Courtesy T. M. Shaw and O. L. Krivanek.

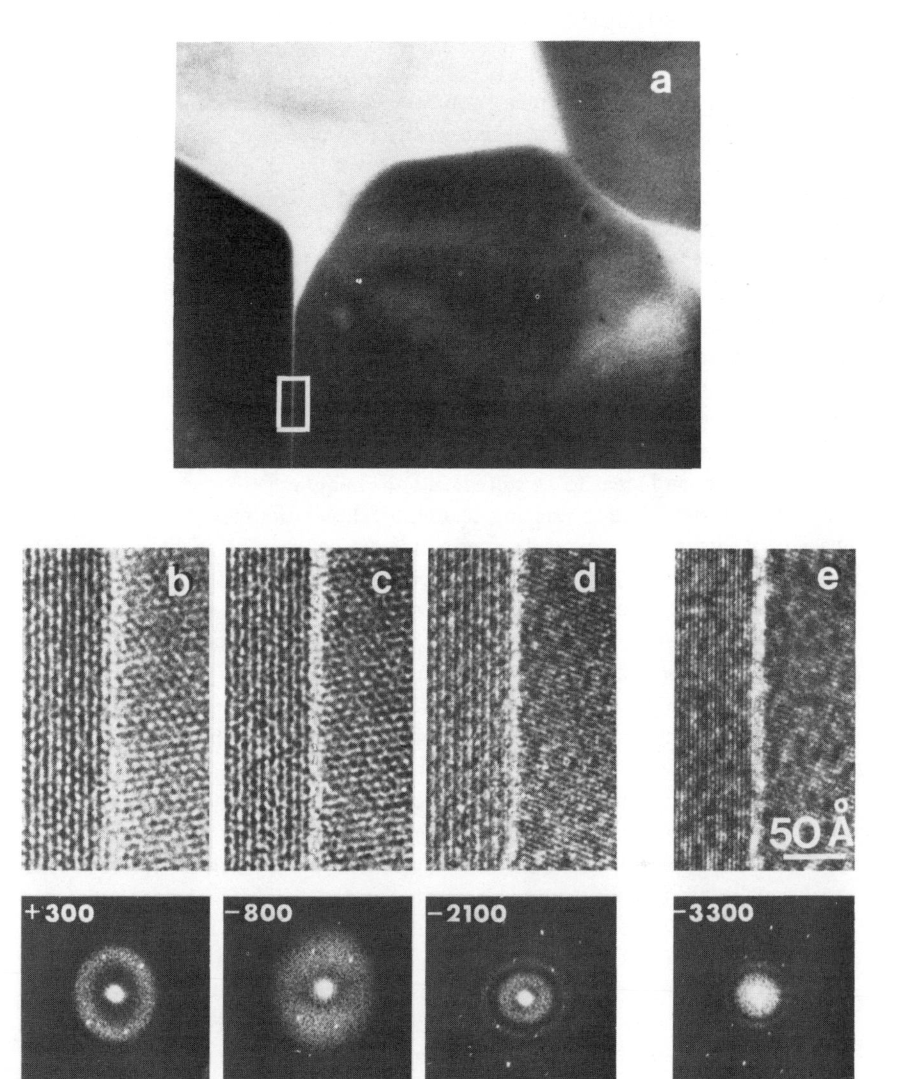

Fig. 3.37 (*a*) Dark field image of the same material as Fig. 3.36. (*b-e*) Through-focal lattice image series of the boxed area in (*a*), together with optical diffractograms. (*c*) is at the Scherzer defocus. Courtesy O. L. Krivanek.

Dark field imaging of the grains (Fig. 3.35), although difficult, is preferable to enhance grain boundary interphase contrast. One can take advantage of the well-known ionization radiation damage sensitivity of glasses in order to detect this phase. At 100 kV, exposure to the electron beam results in such damage, and so it very quickly becomes apparent which parts of the microstructure are glassy. This may be seen in Fig. 3.36. The incoherent scattering from these regions can then be quickly identified by placing the SAD aperture over them and locating the area in the diffraction pattern. Dark field analysis of this area reveals the glassy phases. However, it is still not certain that all the dark field contrast, especially in very narrow intergranular regions, can be claimed to be due exclusively to glassy phases. Also, these materials are very complex; both the amount and the composition of glassy phase vary (e.g., Figs. 3.36*c* and *d*). Thus the task of characterization is not trivial and must be done with meticulous care and with enough samples and areas to be statistically meaningful.[66,68]

The bright field lattice imaging technique shows the glassy grain boundary phase very directly, but careful experiments are essential. A slight mistilt of the grain boundary away from the edge-on configuration can obscure a 10 Å wide glassy phase, and not even a perfect edge-on orientation can guarantee the detection of the phase. This point is illustrated in Figs. 3.37*b–e*, where a through-focal series of lattice images is shown, together with optical diffractograms of the Si_3N_4 image and the amorphous carbon film present on the foil surface. Only the Scherzer defocus image shows the amorphous phase with any clarity; in the first contrast transfer *overfocus* image the two silicon nitride grains appear to join up. The dark field image of this region, on the other hand, shows the glassy phase quite clearly. The dark field approach is thus seen to offer two advantages: (*a*) defocus and specimen tilt are not very critical, and (*b*) large areas of the material can be examined at once. However, the dark field image does not reveal the details of the interface (presence of ledges, etc.).

Recent work on Y_2O_3 hot-pressed Si_3N_4[66] has shown that with proper care lattice imaging can reveal amorphous phases approximately 10 Å wide between crystalline Si_3N_4 and the crystalline yttrium oxynitride (see Fig. 3.38). Here again STEM X-ray analysis reveals considerable impurity segregation (Fig. 1.37*b*). Thus the composition of the glassy phase, which controls its melting point, is of critical significance in determining its mechanical properties.

One of the problems with chemical analysis (spectroscopy) of nitrogen ceramics (e.g., sialons) is that they contain light elements. Since for most practical purposes X-ray analysis is limited to elements for which $Z \geqslant 11$, electron energy loss spectroscopy becomes particularly important. Qualitatively this point is well illustrated by Fig. 1.38.

12.1.4 *Amorphous Materials.* Interpretation of the structural nature of amorphous materials has been ambiguous since it was found that the diffraction

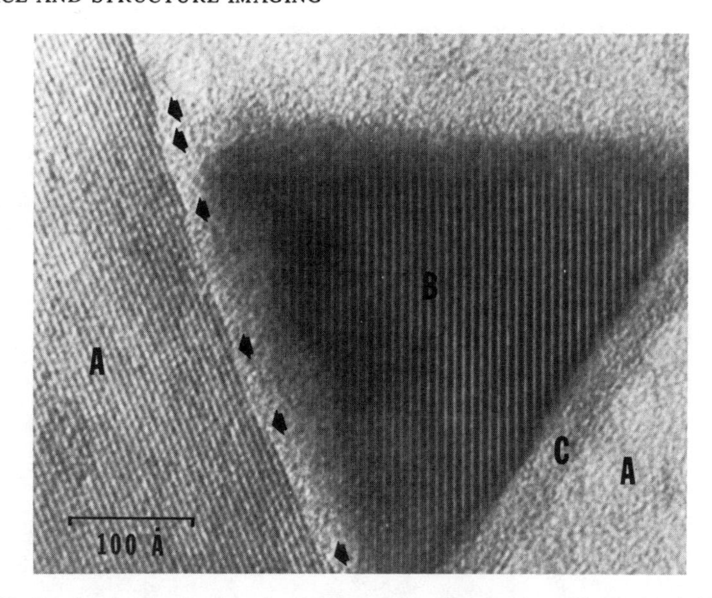

Fig. 3.38 Lattice image of a junction between crystalline yttrium-silicon oxynitride grain (*B*) and two Si_3N_4 grains (*A*). There is a narrow amorphous region at *C*, and microledges are resolved at the arrows; these may be sites for crystallization initiation (see Fig. 1.37). Courtesy D. R. Clarke.

pattern showed diffuse peaks at well-defined angular positions. Renewed interest in this problem has come from lattice imaging of such materials. Rudee and Howie[73] formed images using the transmitted beam plus part of the first diffuse ring and found regions approximately 14 Å in diameter in amorphous silicon and germanium films possessing regular fringes. These they interpreted as small crystallites existing in the predominantly irregular structure. However, more recent experiments[74] on a range of amorphous films has indicated that their result was an imaging artifact arising from lens astigmatism on tilting the electron illumination, and that the extent of coherently scattering regions is ⩽6 Å. Overlap of the image due to small crystallite size and the images of atoms from the remaining thickness of the specimen directly determines the smallest size of any ordered (crystalline) region detectable by electron microscopy. Recent experiments indicate that optical diffraction may be utilized to sort out random noise from meaningful structure.[75] It is obvious that great care must be taken in the interpretation of fringe images, and for detecting crystallites approximately 14 Å in diameter foil thicknesses must be less than 50 Å in amorphous specimens.[75]

Figure 3.39 is an example of a lattice image of "glassy" carbon obtained using the imaging conditions shown (recombining all the diffuse reflections in the

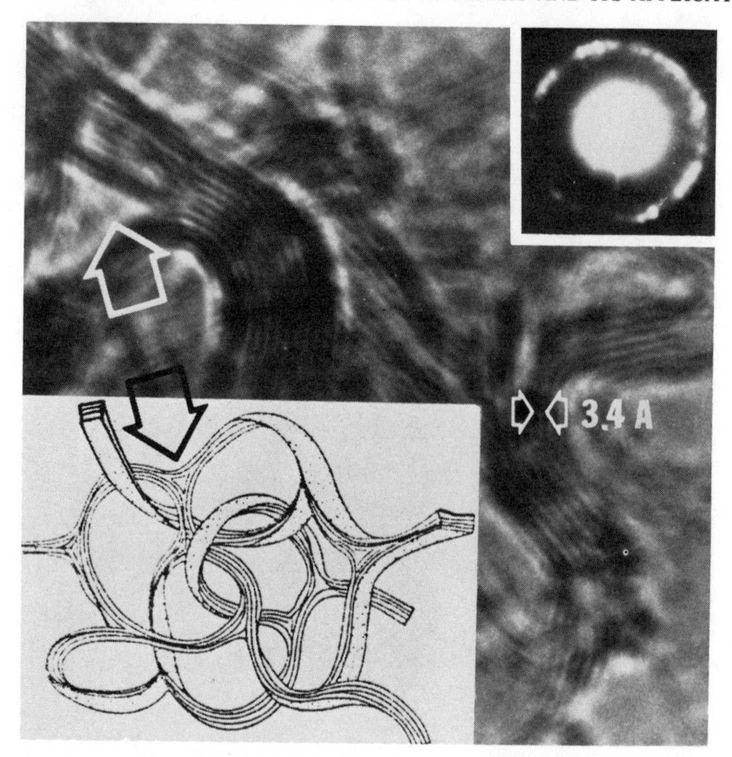

Fig. 3.39 Lattice image of "LMSC" glassy carbon with inset diffraction pattern and structure model; the first spotty ring ($d \sim 3.4$ Å) falls inside the objective lens aperture, and hence all orientations are imaged. Courtesy *J. Am. Cer. Soc.*, **61**, 174, (1978).

first diffuse ring with the transmitted beam). In the phase contrast mode an aperture was used to allow imaging in the two-beam "0.00–00.2" situation to occur, and lattice fringes corresponding to the (00.2) planes of graphite were obtained over large, thin areas. The lattice images show fringes but with no preferred orientation, indicating the isotropic nature of glassy carbon. The fringe spacing is 3.4 Å, and the fringes are continuous, usually over 50 Å. The fringe pattern resembles the "Jenkins nightmare" model,[72] and there are no definite crystallite boundaries. The layers show extensive bending; stacking disorders are encountered in some areas. The stacked layers bifurcate occasionally, indicating that pores are enclosed among interweaving layers.

High resolution electron microscopy has clearly shown in this case that "glassy" carbon is not, in fact, amorphous but is composed of a complex mixture of small crystallites of large aspect ratio having preferred orientations of (00.1) planes parallel to the long axis of the crystallite.

12.2 Examples of Structure Imaging

12.2.1 *Complex Oxides.* In spite of the stringent resolution limitations, when viewing crystals with large unit cells it is still possible to retain on the order of 100 beams within the objective aperture and produce correspondingly detailed "structural images."[73, 74] The extent to which these images actually represent specimen structure is determined on the basis of image computations, using the dynamical theory of electron scattering.[75] Agreement between experiment and theory in these studies has been excellent and has fostered confidence in the interpretation and extended application of structural imaging.[76]

At the appropriate underfocus the structure image can be interpreted in terms of a projection of the charge density of the material,[73] that is, the contrast is roughly proportional to the atomic number (see Chapter 5, Section 9). Thus in metallic oxides the projection of metal atom rows appears dark in the image, while rows of oxygen atoms appear light. Metal atoms too close to one another for resolution appear as darker regions than those that can be resolved by the microscope. An example of the correlation between the electron structure image of $Ti_2Nb_{10}O_{29}$ and the projection of the structure as established by X-ray diffraction is shown in Fig. 3.40. Other materials so imaged include various titanium, niobium, and tungsten oxides, silicates, wüstite perovskite polytypes, and tourmaline. The correlation between the structure and the projected images indicates that structure determination is possible if a reasonable basis of the structure is known. Examples whereby ambiguities of X-ray structures were resolved in various tetragonal tungsten bronze-type materials in the Nb_2O_5-WO_3 system have been described by Iijima and Allpress.[77]

The advantage of using a direct technique such as microscopy to view defects is that their nature and distribution can be observed directly on a picture. The further advantage of structure and fringe imaging is that the atomic arrangement at the defect can be seen.

12.2.2 *Grain Boundaries in Semiconductors.* An area of considerable importance is grain boundaries. The advantage of using 500 kV electron microscopy is clear from the work of Krivanek et al. on germanium,[78] illustrated in Figs. 3.41 and 3.42. Because of the increased point resolution with increasing voltage (see Fig. 1.5) point resolutions of approximately 2 Å become possible.

The boundary shown in Fig. 3.41 is a 39° tilt boundary observed end-on along the [011] direction at 500 kV under essentially kinematical conditions. The white dots in the image correspond to empty channels running through the crystal along the [011] direction, which are bounded by sixfold rings of germanium atoms. At the boundary 15 white dots are larger than the rest. These dots were interpreted as arising from larger channels, and a model in which they

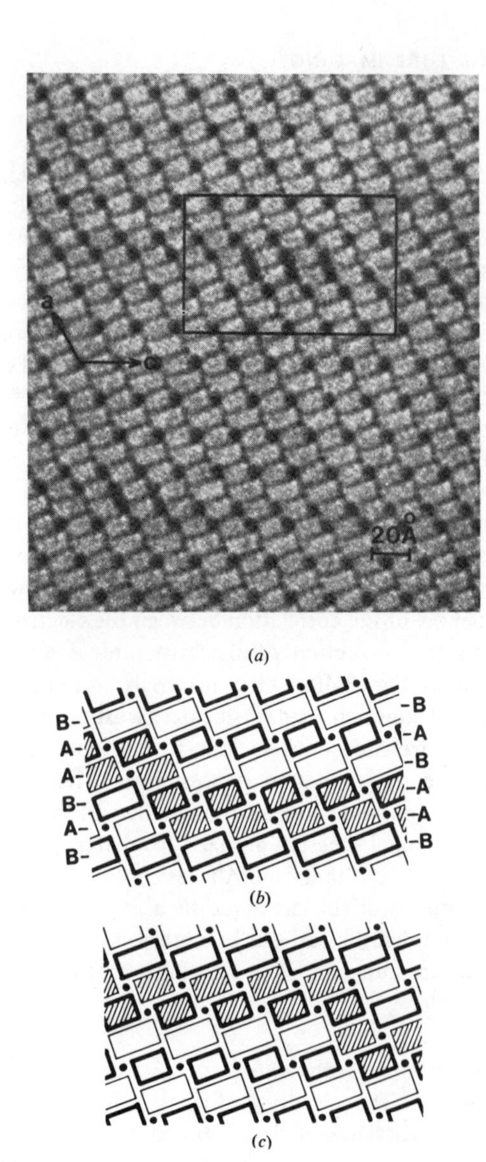

Fig. 3.40 (*a*) Structure image of H-Nb_2O_5 showing displacement defects. The usual black dots in the overlapping region of two defects (in the rectangular box) can be explained by a superposition of the two different arrangements of structural blocks shown in (*b*) and (*c*). Courtesy S. Iijima, *Acta Crystallographica*, **A29**, 18 (1973).

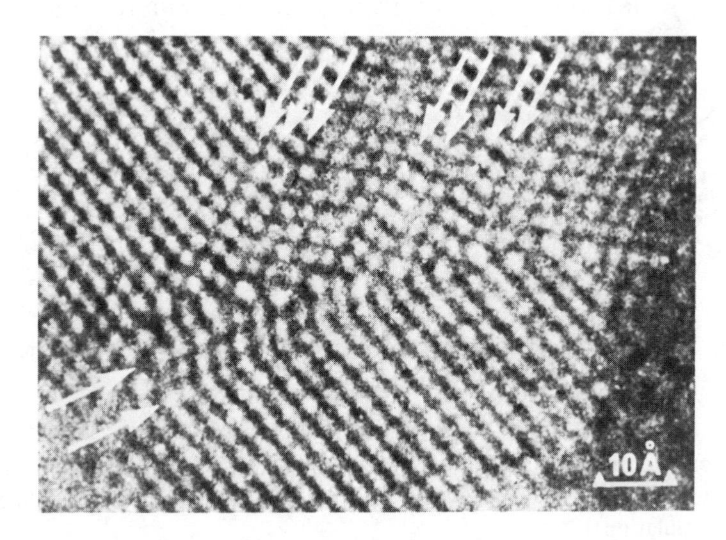

Fig. 3.41 Structure image of a grain boundary in germanium. Arrows mark the (111) twinning planes. Courtesy Taylor and Francis.[78]

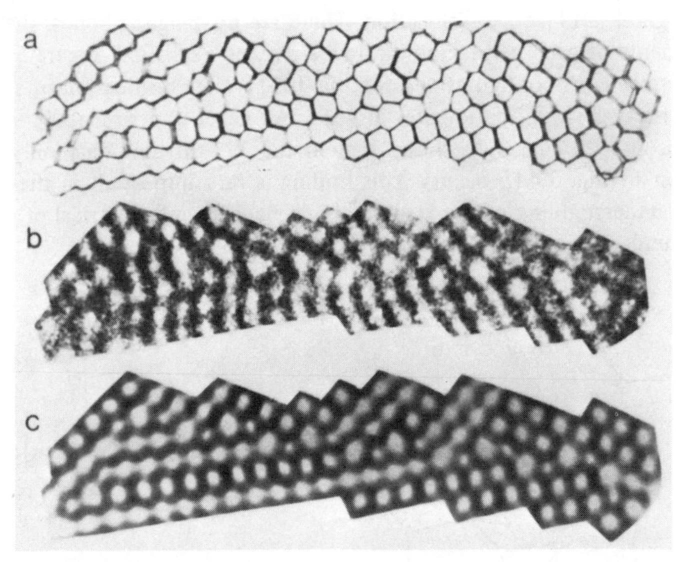

Fig. 3.42 Comparison of the model for the grain boundary (see Fig. 3.41) with the image: (*a*) the model, (*b*) the image (from Fig. 3.41), (*c*) blurred photograph of the model. Courtesy Taylor and Francis.[78]

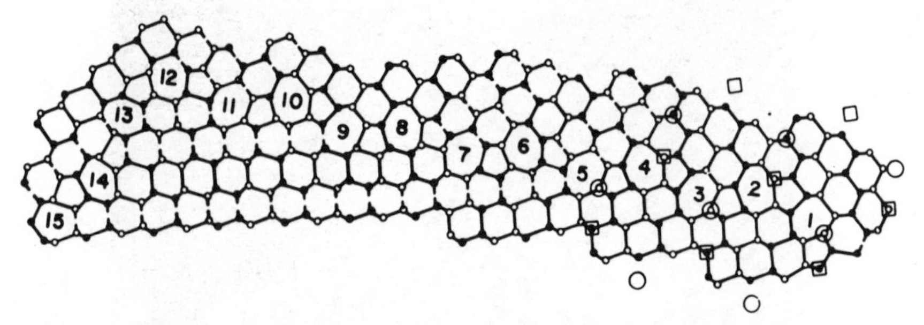

Fig. 3.43 Details of the model of the grain boundary. The seven-membered rings at the boundary are numbered. Courtesy Taylor and Francis.[78]

correspond to the channels of sevenfold rings shows excellent agreement with the original image [see Fig. 3.42*a* (model), Fig. 3.42*b* (experiment), and Fig. 3.42*c* (simulation)].

The details of the structure can be seen in Fig. 3.43. The tilt boundary consists of alternating pairs of fivefold and sevenfold rings, each pair of rings accommodating a $\frac{1}{2}$ [011] edge dislocation. There are no dangling bonds; the bonding requirements of each germanium atom are perfectly satisfied. Figure 3.44 shows two possible core configurations for the [011] edge dislocation in a diamond cubic lattice, derived by Hornstra in 1958 but never experimentally verified.[79] Comparison with Fig. 3.43 shows that in the 39° tilt boundary only the configuration in Fig. 3.44*b* occurs. This finding is very important in the quest for detailed understanding of the structural, mechanical, and electrical properties of grain boundaries and other internal surfaces.

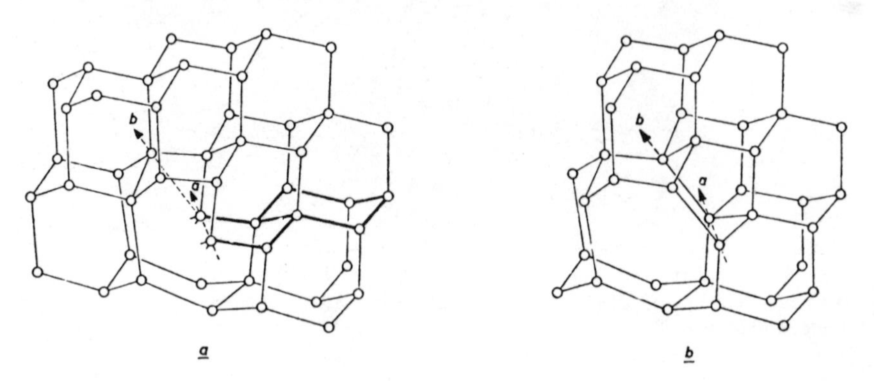

Fig. 3.44 Two models for the edge dislocation in germanium. Only type (*b*) was observed in the grain boundary study of Figs. 3.41–3.43. Courtesy Pergamon Press.[79]

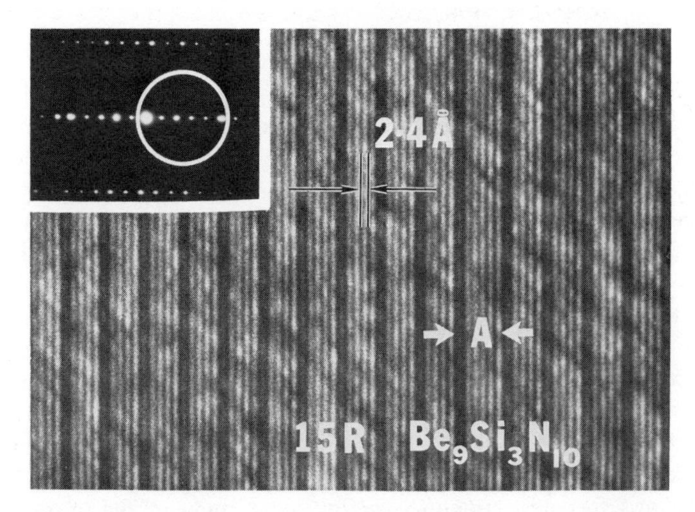

Fig. 3.45 Lattice fringe image of the $Be_9Si_3N_{10}$ showing predominantly the fivefold repeat of the 15*R* polytype. The inset diffraction pattern indicates the orientation and the objective aperture used, the optic axis being at the center of the aperture. A deviation to a six-layer structure is indicated at *A*. Courtesy *Science*.[80]

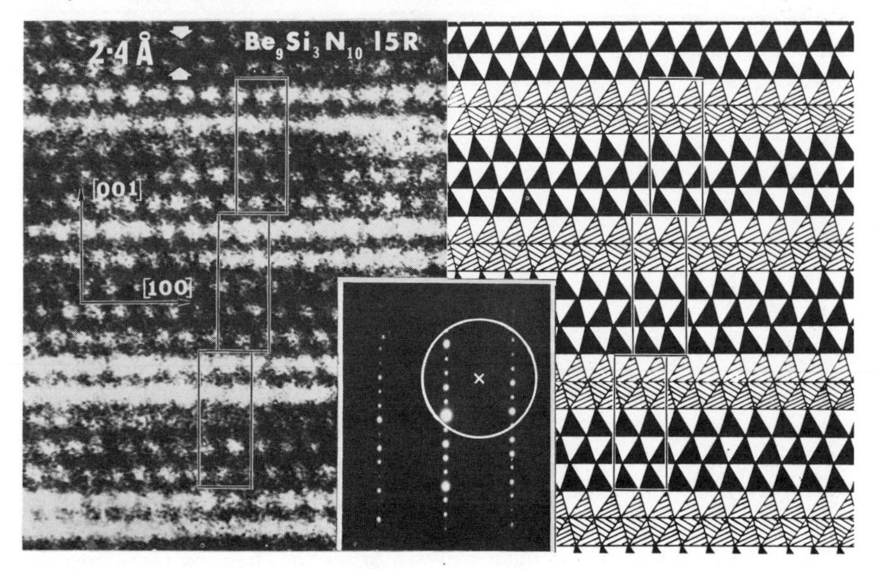

Fig. 3.46 (*a*) Two-dimensional structure image of the 15*R* polytype of Fig. 3.45 taken under the diffraction conditions indicated in the inset diffraction pattern. (*b*) Model of the structure for comparison. Courtesy *Science*.[80]

12.2.3 *Analysis of Polytypes.* A typical example of studies of polytypic materials is provided by the Be-Si-N system.[80] From X-ray diffraction studies and selected area electron diffraction patterns the material in question was found to be the 15*R* polytype of formula $Be_9Si_3N_{10}$. Direct lattice fringe imaging of the close-packed planes (using tilted illumination) shows that (ignoring, for the moment, region *A*) each block consists of five close-packed layers (Fig. 3.45). The change in fringe contrast every fifth layer suggests that a composition and/or stacking change occurs every fifth close-packed layer. These observations are consistent with the X-ray-determined structure, in which a M_2X layer with cubic stacking occurs every fifth layer in the 15*R* structure. The 15*R* model was further confirmed by forming a two-dimensional structural image, which is shown in Fig. 3.46, together with a model of the structure. The periodicities associated with the 15*R* structure are seen over most of the imaged area in two dimensions. In both cases the selected area electron diffraction pattern (inset) shows the ideal structure, careful direct study of the images being required to detect deviations. For example, the block denoted as *A* in Fig. 3.45 consists of six layers rather than the ideal five. This fault would be consistent with a local composition change (e.g., in anion-cation ratio). The significance of the structure image is that it shows the 15*R* structure to be made up of a faulted three-layer block, each block consisting of layers of cubic and hexagonal stacking in the ratio 2 : 3, and each block being displaced by $\frac{1}{3}$ [010].

13 Compositional Analysis by Lattice Imaging

13.1 Introduction

Compositional analysis using spectroscopy was described in Chapter 1 (see Section 5.4). Indirect methods of chemical analysis can also be carried out, such as accurate lattice parameter measurements, either from the diffraction pattern of the lattice image or by critical voltage measurements, which depend on changes in atomic scattering factors due to composition. These techniques are summarized in Table 3.6.

Thus an additional advantage of high resolution electron imaging is the opportunity it provides for localized compositional analysis. In contrast to the spectroscopic (X-ray or energy loss microanalysis) or microdiffraction techniques, this method relies exclusively on the imaging mode and is, in principle, capable of spatial resolution at the atomic plane level.[82]

The technique is based on the application of Vegard's law (assuming a linear relationship between lattice parameter and solute content). Measurements are made of the lattice image fringe spacing variations that are produced by changes

Table 3.6 Chemical Analysis Methods by Electron Microscopy

Method	Spatial (Å)	Spectroscopic Resolution (atomic fraction)	Lattice Parameter (%)	References in Book
STEM X-ray	~50[a]	10^{-5} ($Z > 11$)	–	Ch. 1, §5.4
STEM energy loss spectroscopy	~50–5000[b]	10^{-2}	–	Ch. 1, §5.4
Lattice imaging	~2	–	~0.5 (real space)	Ch. 3, §13.1
Microdiffraction	~20	–	~1 (real space)	Ch. 1, §5.3
Converging beam	~50	–	~0.1 (Kikuchi)	Ch. 1, §4.9
Critical voltage	~100	10^{-2}	–	Ch. 5, §6

[a] Although smaller spot sizes are possible, the signal-to-noise ratio determines the optimum.
[b] Modern developments suggest that 50 Å will be possible (limited by source brightness and signal detection).

in interatomic spacing due to compositional variation in the specimen. When a measurement standard is present in the lattice image (e.g., a region of single-phase or pure material with a well-established lattice parameter), it is possible to measure composition profiles on the same lattice image to within less than 1% error. Such measurements can be made from a scaling of direct photographic enlargements, from peak-to-peak distances on microdensitometer traces across the negative, or from optical diffractograms of the negative.[82]

Optical diffraction (from simulated atomic arrangements) has been used in the past to facilitate the interpretation of X-ray diffraction results[83] and is employed at present to complement high resolution electron micrography,[71] particularly of biological materials, but so far it has not been utilized very much for analyzing metallurgical or ceramic alloys. The optical system may be calibrated using pure gold 200 lattice images, for which the *d*-spacing and Bragg angle are well known.

A perfect crystal lattice image, formed by combining the transmitted beam and one diffracted electron beam, consists of a series of fringes parallel, and equal in spacing, to the relevant diffracting planes. By using these periodic fringe images as an optical amplitude grating, a Fraunhofer optical diffraction pattern with uniformly spaced diffracted intensities[84] is readily obtained. During a phase transformation (or near a lattice defect) distortions or imperfections in the crystal lattice may give rise to observable electron or X-ray diffraction effects. (A well-known example is the phenomenon of satellite spots around fundamental lattice reflections, arising from the periodic redistribution of atomic species during spinodal decomposition.) Thus, if lattice imaging is a reliable technique for studying fine-scale phenomena, the optical diffraction pattern from the image should also reproduce the essential features characteristic of the conventional electron diffraction pattern of the specimen, for the reflections that were used in the original image formation process. The main advantage is that diffraction is obtained from a very small area (approx. 10 Å in diameter), a feature not possible with electron diffraction because of limitations of spherical aberration (Chapter 1, Section 4.1). Applications of this simple experiment are illustrated below.

13.2 Some Applications

13.2.1 *Spinodal Decomposition in Au-Ni.* During spinodal decomposition solute segregation occurs by means of wavelike composition modulations in particular crystallographic directions, commonly ⟨100⟩ in alloys (for example Fig. 2.19). As the interplanar spacing is a function of composition, a corresponding *d*-spacing variation is also achieved. In the lattice image this is manifested by fringe spacing modulations, as recently demonstrated for Au-Ni.[85]

The micrograph shown in Fig. 3.47 was taken from a Au-Ni alloy aged for 21

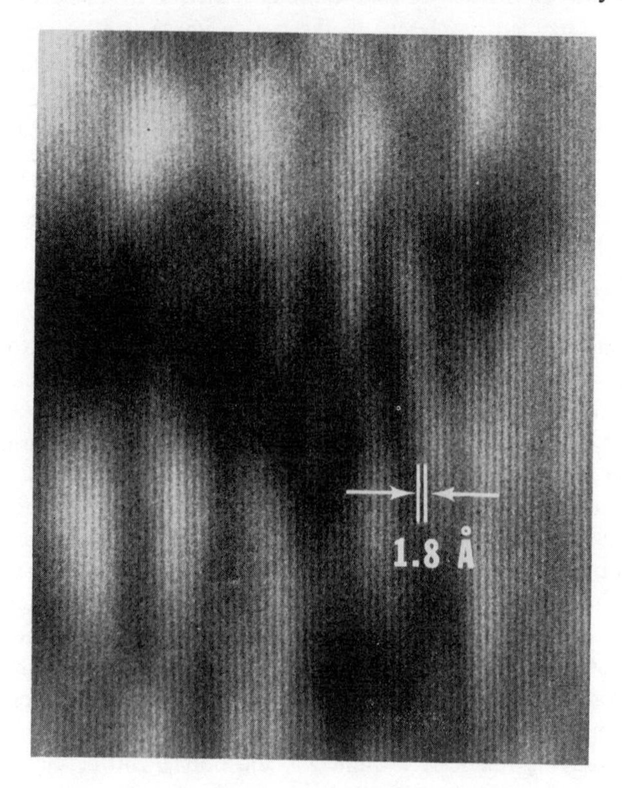

Fig. 3.47 Lattice image (positive print of the original plate) of spinodally decomposed Au-Ni. Courtesy Pergamon Press; *Acta Metallurgica.* [85]

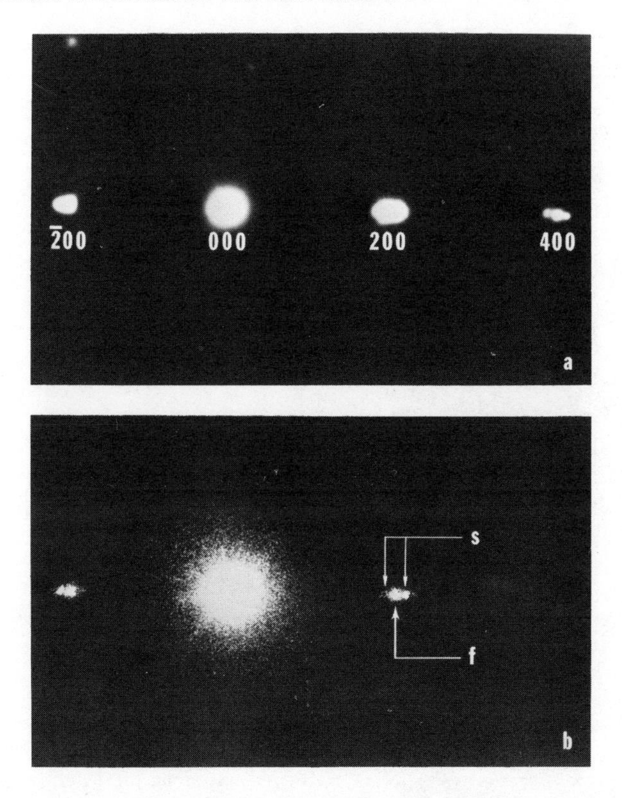

Fig. 3.48 Comparison of the electron diffraction pattern (*a*) from the decomposed Au-Ni specimen with the optical diffraction pattern (*b*) from the lattice image shown in Fig. 3.47. Satellites (*s*) about the fundamental (200) reflection (*f*) are arrowed. Courtesy Pergamon Press; *Acta Metallurgica.*[85]

hours at 150°C. It displays both the fine lattice fringe periodicity of the (200) planes and the coarse variation in image contrast that is typical of a modulated microstructure imaged under two-beam conditions. Satellites are clearly visible about the (400) spot in the electron diffraction pattern (Fig. 3.48*a*), which records the tilted illumination imaging condition used to form this lattice image. The corresponding optical diffraction pattern (Fig. 3.48*b*) identically reproduces all essential diffraction effects, namely, the intense high angle and weak low angle satellites symmetrically positioned along ⟨100⟩ reciprocal lattice directions about (*h*00) fundamental lattice reflections. The different intensities of the two satellites are due to the asymmetrical (nickel-rich) alloy composition.

Direct measurements from micrographs similar to and including Fig. 3.47 yield an average composition wavelength λ of 31 ± 5 Å. Complementary analysis of satellite spacings reveals that the optical diffraction patterns are once again in

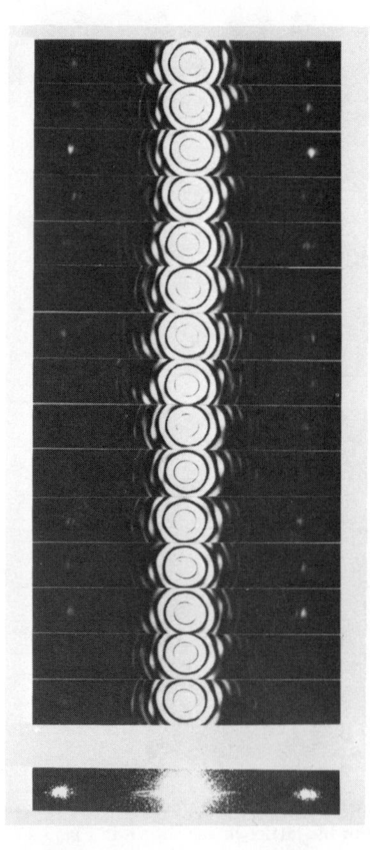

Fig. 3.49 Series of optical microdiffractograms from a lattice image of spinodally decomposed Au-Ni, taken at 10 Å intervals on the image, along the [100] direction. Courtesy R. Gronsky.

excellent quantitative agreement. As determined from the electron diffraction patterns, $\lambda = 29 \pm 3$ Å, while from the optical diffraction patterns $\lambda = 29 \pm 2$ Å. A series of microdiffraction optical patterns corresponding to a selected area of 20 Å, that is, within each region of composition modulating, is shown in Fig. 3.49. A smoothed plot of fringe spacing, as measured from the microdensitometer analysis against distance normal to a fringe, is shown in Fig. 3.50; each point $n(n = 1, 2, 3, \ldots)$ represents the average of raw data points n to $(n + 4)$. This figure gives direct evidence that the origin of satellites in the optical diffraction patterns is due to a periodic variation in fringe spacing. It can be further established that satellites are absent about (000) in the optical patterns by taking a series of photographs, using short exposure times, to determine the intensity profile near the intense transmitted beam. Such a series indicates that any contribution from a variation in fringe intensity (e.g., see Fig. 3.47) is negligible, and that the satellites therefore arise from the lattice spacing modulation. The periodicity of the modulation from this analysis, averaged over 12 wavelengths, is 29 ± 8 Å, also in excellent agreement with the diffraction data.

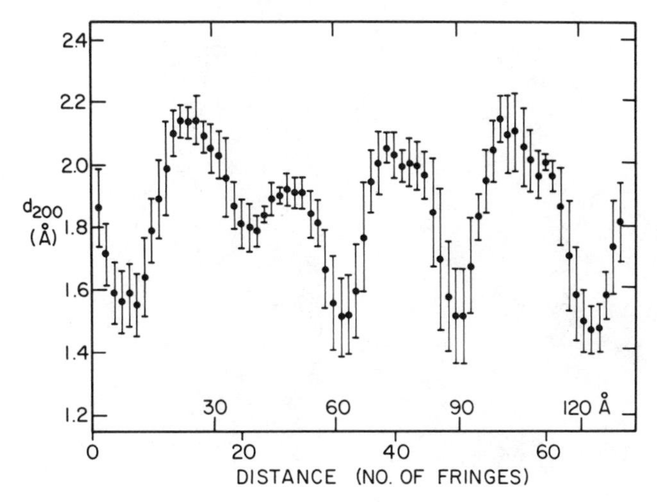

Fig. 3.50 Smoothed plot of fringe spacing with distance for Au-Ni as determined from a microdensitometer trace from the lattice image. Courtesy Pergamon Press; *Acta Metallurgica.*[85]

This type of plot may be utilized to establish the local composition profiles produced during a phase transformation, once the relationship between the interplanar spacing and the alloy composition is known. In the present case a continuous composition variation, characteristic of a spinodal reaction, is clearly illustrated. Similar results have been obtained for Cu-Ni-Cr.[58] In some regions, however, the modulation amplitude is greater than the maximum possible from the equilibrium phase compositions (Fig. 3.50), for reasons that are currently being further investigated.

13.2.2 *Grain Boundaries.* The ability to detect highly localized compositional variations is very desirable for experimental studies of grain boundaries. In current analyses of grain boundary precipitation reactions[87] by lattice imaging, fringe spacing measurements have given clear indications of composition profiles in the grain boundary vicinity with high precision. These results have been useful in identifying the reaction mechanisms involved and the particular role of grain boundaries in the precipitation processes.

Figure 3.51 is an example of a matrix (111) fringe lattice image across a grain boundary precipitate in an Al-9.5 at. % Zn alloy aged for 30 min at 180°C. The boxed region in Fig. 3.51*a* is shown enlarged in Fig. 3.51*b*, indicating the region on which compositional analysis was performed. Fringe spacings measured within both matrix (top) and precipitate (bottom) areas, at increasing distances from the grain boundary, are shown in Fig. 3.52. Each point in Fig. 3.52 indicates the average spacing of 10 fringes, with a representative error bar shown on the first datum point.

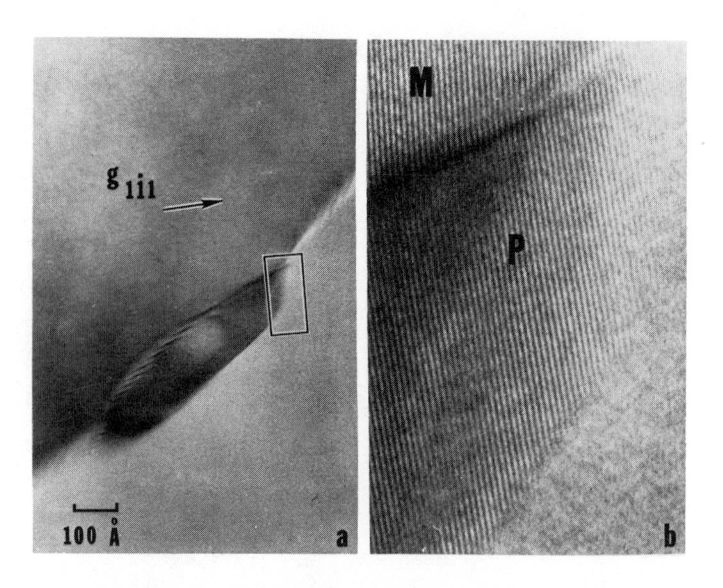

Fig. 3.51 (*a*) Lattice image of a grain boundary precipitate in an Al-9.5 at. % Zn alloy. (*b*) Enlargement of boxed region in (*a*), exhibiting 111 matrix and 0002 precipitate fringes. Courtesy Claitors Pub.[87]

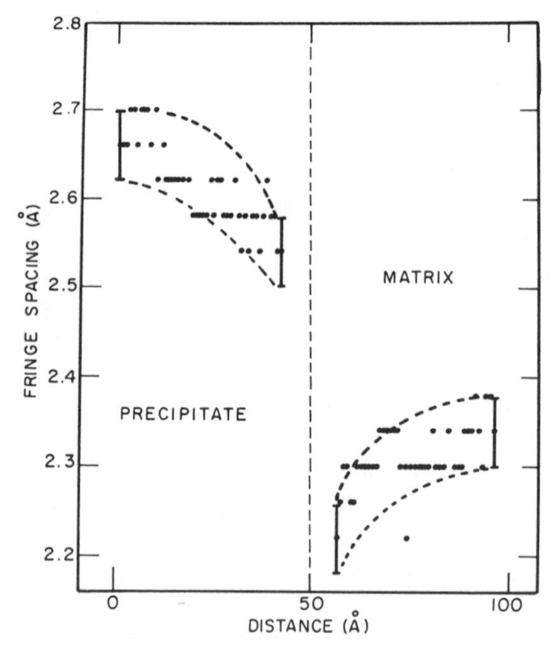

Fig. 3.52 Plot of fringe spacing as a function of distance from the grain boundary in Fig. 3.51 (vertical dashed line). Each point represents 10 measurements. The data clearly show a gradient in lattice parameter and hence composition. Courtesy Claitors Pub.[87]

This plot clearly indicates a decreasing fringe spacing as the boundary (dotted line) is approached from either side. It suggests that a solute gradient exists within both the matrix and the precipitate, and that concentration changes rapidly over a distance of only 50 Å. This method is clearly advantageous when resolutions better than those available by spectroscopic microanalysis are needed.

13.2.3 *Martensitic Interfaces.* Martensitic reactions are of great significance technologically, for example, in steels and shape-memory alloys. Although much has been written about the phenomenology of the transformation, little is known in detail about martensite nucleation or growth, the structure of the interfaces, or the factors that control the nature of the inhomogeneous shear (slip or twinning). Clearly much can be learned by applying lattice imaging, and here an example is shown of the potential of this technique for studies of complex dual-phase ferritic martensitic steels developed for improved strength/weight applications.[88] These steels have excellent strength and formability, and information is needed on the nature of the ferrite-martensite interface and also on the composition (i.e., solute distribution) of these phases. Lattice imaging is especially difficult because of the astigmatism corrections needed as a result of the ferromagnetism in the ferrite or martensitic phase steels.[89]

Figure 3.53 shows the ferrite (α)-martensite boundary in a transformed specimen of an Fe-2% Si-0.1% C alloy after quenching from the two-phase ($\alpha + \gamma$) field at 850°C. The image was produced by combining the transmitted beam with the closely spaced 101 reflections of both the bcc α-phase and the bct martensite, under tilted illumination conditions. It is seen that the tetragonality

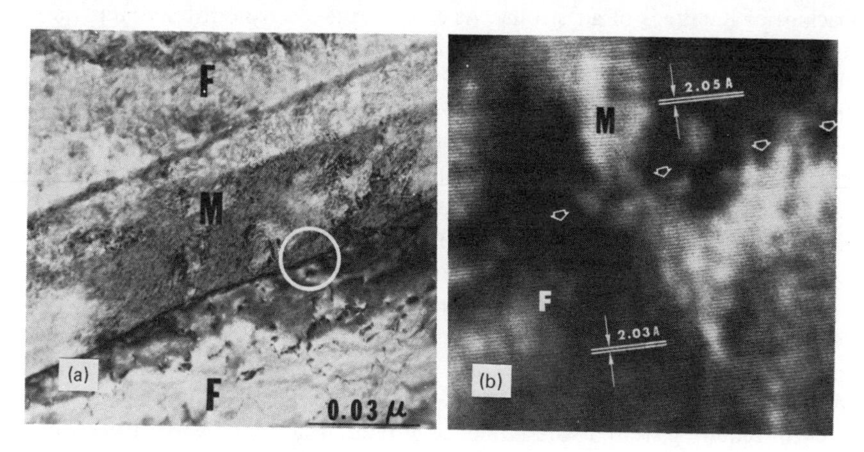

Fig. 3.53 Conventional bright field (*a*) and lattice image (*b*) of a α-martensite interface in a 2% Si DFM steel. The lattice image (*b*) was taken from the area encircled in (*a*). Martensite tetragonality creates the larger d_{101}-spacing in the martensite region (*M*). *F* is ferrite. The arrows indicate the interface. Courtesy Claitors Pub.[89]

of the martensite has induced a slightly larger interplanar spacing, which is reflected in the measured variation of fringe spacing recorded on the micrograph. From this result the carbon content of the martensite phase was estimated to be approximately 0.6%. Thus lattice imaging provides chemical information here which *cannot* be obtained by other methods.

13.3 Limits of Applicability

Although the use of optical diffraction in conjunction with lattice fringe imaging is a powerful method for identifying image periodicities, there are two inherent limitations to its application. First, it can only provide information about spatial periodicities in the fringe image and gives no indication of the crystallography of the stacking layers. Second, the total information in the optical diffraction pattern does not exceed that present in the electron diffraction pattern, although the former can often be more easily interpreted and analyzed. For example, the combination of lattice imaging and optical microdiffraction permits identification of stacking sequences that are not otherwise easily recognized. One example is in the analysis of complex polytypic structures.

In the example cited in Section 12.2.3 of beryllium silicon nitride (Fig. 3.45) the periodicity of the $15R$ stacking is regular over distances comparable to the resolution of selected area diffraction (apart from localized faults, as at A); hence the $15R$ polytype spacing is clearly resolved in the SAD pattern (insert in Fig. 3.45). However, more complex cases can arise, for example, in minerals and certain silicon carbides[64,81] in which the frequencies of faulting occur over spacings of hundreds of angstroms. As a result the electron diffraction patterns are streaked and difficult to interpret. In such cases the constituent periodicities cannot be identified from electron SAD patterns. In addition, if the volume fraction of a stacking sequence is small, its contribution to the total intensity in the diffraction pattern may well be too small for it to be recognized. After the optical diffraction patterns have been calibrated, the stacking sequences giving rise to the diffracted beams can be identified from the ratio of their positions in the diffraction pattern. In this manner stacking sequences in silicon carbide having periodicities of 8, 24, 36, and 48 composite Si-C layers have been recognized.[64]

As far as compositional analysis is concerned, accuracy depends upon the relationship between solute content and lattice parameter.[82] For large dependencies (as in interstitial solid solutions such as Fe-C) compositional differences of approximately 3% can be detected at spatial resolutions of about 20 Å. One problem requiring study is the influence of interfacial strains when lattice fringe spacings at coherent, strained interfaces are analyzed, so that compositional effects are not confused with interfacial elastic strain effects. In spite of these limitations it is clear that high resolution microscopy can provide very detailed structural and configurational information and, with care, solid solution compositional information at resolutions far better than are obtainable with any current spectroscopic method (Table 3.6).

Exercises

3.1 The kinematical intensity I_g for a diffracted beam from perfect crystal is given by eq. 2.14:

$$\frac{I_g}{I_0} = \left(\frac{F}{V_c}\right)^2 \frac{\sin^2 \pi ts}{(\pi s)^2}.$$

Define the terms involved, and describe the conditions under which the equation is valid.

3.2 A crystal is set so that a systematic row of reflections only is excited, and the Kikuchi band and lines appear as in Fig. 3.54. Obtain formulas for the value of s for the beams $-\mathbf{g}$ and $\mathbf{g}$. Comment on the usefulness of these conditions for obtaining weak beam images of dislocations in copper, using 100 keV electrons with g = 111, 200, and 220, respectively.

3.3 How would you distinguish between contrast from the following pairs?

(*a*) Moiré fringes and wedge thickness fringes.

(*b*) Extinction contours of high order and dislocations.

3.4 In studying silicon, you observe dislocations that lie in (111). List the reflections you should use (and hence the required foil orientation) to determine uniquely their Burger's vectors.

3.5 You observe a stacking fault on $(1\bar{1}\bar{1})$ in a fcc foil in orientation (and specimen surface normal) [012] (upward) with $\mathbf{g}$ = 200 operating. Sketch the geometry of this situation and the bright field images you would expect if the fault were (*a*) intrinsic, (*b*) extrinsic.

If two parallel, closely spaced faults were present, what would you expect for (*c*) both faults intrinsic and (*d*) one fault intrinsic and the other extrinsic?

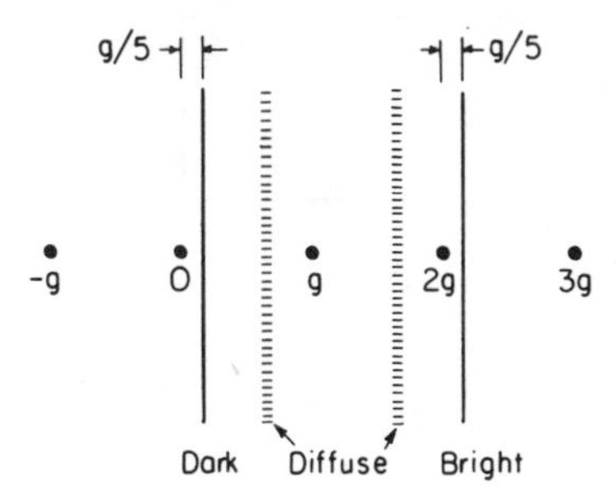

Fig. 3.54 Kikuchi line diagram for Exercise 3.2.

3.6 In a fcc crystal a dislocation with $b = \frac{1}{2}[110]$ is invisible when imaged using reflection g. If this dislocation splits into the two partials $\frac{1}{6}[121] + \frac{1}{6}[21\bar{1}]$ on either side of a stacking fault on $(1\bar{1}1)$, which of the following are possible?

(*a*) Both partials and fault visible.

(*b*) Both partials and fault invisible.

(*c*) One partial and fault visible.

(*d*) One partial and fault invisible.

(*e*) Both partials visible, fault invisible.

(*f*) Both partials invisible, fault visible.

3.7 In a fcc alloy long-range order is induced. What are the conditions required for observing (*a*) superlattice dislocations and (*b*) antiphase domain boundaries? How can you distinguish between APBs and stacking fault interfaces?

3.8 In Fig. 3.55 row 1 shows the strain contrast image (dark field near the reflecting position) for several types of defect. Sketch in rows 2 and 3 the expected appearance of the image, and in row 4 the displacement field vectors characterizing the defect.

3.9 Sketch a representative image that would be obtained with the aperture configurations (*a*), (*b*) and (*c*) in Fig. 3.56, the optic axis being always at the center of the aperture. Describe how each image might be influenced by changes in focus.

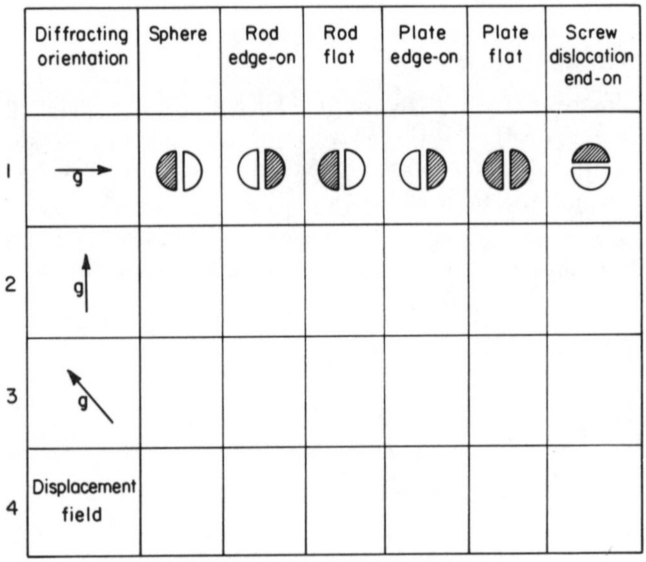

	Diffracting orientation	Sphere	Rod edge-on	Rod flat	Plate edge-on	Plate flat	Screw dislocation end-on
1	g						
2	g						
3	g						
4	Displacement field						

Fig. 3.55 Image contrast features for Exercise 3.8.

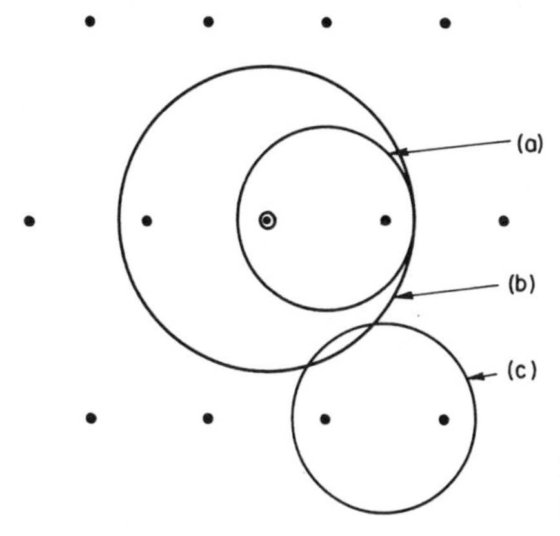

Fig. 3.56 Diffraction pattern and superimposed objective apertures for Exercise 3.9.

3.10 Assume that the best available microscope in your laboratory has an objective lens with $f = 3$ mm and $C_s = 1.8$ mm, and operates at 125 kV. Describe a high resolution experiment that will reveal the actual size and shape of radiation-induced defects in an aluminum alloy ($a = 4.09$ Å). The defects are known to produce rel-rods in $\{111\}$ directions in reciprocal space.

References

1. Cockayne, D. J. H., *Z. Naturforsch.*, **27**, 452 (1972).
2. Bell, W. L. and Thomas, G., *Proceedings of the 27th Conference of the Electron Microscopy Society of America*, Claitors, Baton Rouge, LA, 1969, p. 156.
3. Goringe, M. J., Hewat, E. A., Humphreys, C. J., and Thomas, G., *Proceedings of the 5th European Electron Microscopy Congress*, Institute of Physics, London, 1972, p. 538.
4. *Electron Microscopy and Structure of Materials* (Ed. G. Thomas), University of California Press, Berkeley, 1972, Chapters 1–3.
5. *Modern Diffraction and Imaging Techniques in Materials Science* (Eds. S. Amelinckx et al.), North Holland, Amsterdam, 1970. 2nd Ed., (rev.) 1978.
6. *Electron Microscopy in Material Science* (Ed. U. Valdrè), Academic, New York, 1972.
7. Siems, R., Delavignette, P., and Amelinckx, S., *Phys. Stat. Sol.*, **2** (1), 421 (1662). See also ref. 2, p. 158.
8. Bell, W. L. and Thomas, G., *Phys. Stat. Sol.*, **12**, 843 (1965).
9. Von Heimendahl, M., *Micron*, **4**, 111 (1973).
10. Okamoto, P. R., Levine, E., and Thomas, G., *J. Appl. Phys.*, 38, 289 (1967).
11. Okamoto, P. R. and Thomas, G., *Phys. Stat. Sol.*, **25**, 81 (1968).
12. Hirsch, P. B., et al., *Electron Microscopy of Thin Crystals*, Butterworths, London, 1965.
13. Levine, E., Bell, W. L., and Thomas, G., *J. Appl. Phys.*, **37**, 2141 (1966).
14. Okamoto, P. R. and Thomas, G., *Acta Met.*, **15**, 1325 (1967).

15. Bell, W. L. and Thomas, G., *Phil. Mag.*, **13,** 395 (1966).
16. Head, A. K., *Austr. J. Phys.*, **20,** 557 (1967).
17. Clarebrough, L. M., *Austr. J. Phys.*, **24,** 79 (1971).
18. Bell, W. L., Roser, W. R., and Thomas, G., *Acta Met.*, **12,** 1247 (1964).
19. Thomas, G. and Bell, W. L., *Lattice Defects and Their Interactions* (Ed. R. Hasiguti), Gordon and Breach, New York, 1967, p. 477.
20. Whelan, M. J. and Hirsch, P. B., *Phil. Mag.*, **2** 1121 (1957).
21. Hashimoto, H., Howie, A., and Whelan, M. J., *Proc. Roy. Soc.*, **A269,** 80 (1962). See also Gevers, R., Art, A., and Amelinckx, S., *Phys. Stat. Sol.*, **3,** 1563 (1963).
22. Goringe, M. J. and Valdrè, U., *Proc. Roy. Soc.*, **A295,** 192 (1966). See also Gevers, R., Delavignette, P., Blank, H., van Landuyt, J., and Amelinckx, S., *Phys. Stat. Sol.*, **4,** 383 (1964); **5,** 595 (1964); and Villagrana, R. and Thomas, G., *ibid.*, **9,** 499 (1965).
23. Van der Biest, O. and Thomas, G., *Acta. Cryst.*, **A31,** 70 (1975).
24. Serneels, R., Snykers, M., Delavignette, P., Gevers, R. and Amelinckx, S., *Phys. Stat. Sol. (b)*, **58,** 277 (1973).
25. Amelinckx, S., ref. 5, p. 257.
26. *Surface Science*, **31,** 1972 (Publication of papers given at an International Conference on Interfaces, IBM, New York, 1971).
27. Marcinkowski, M. J., *Electron Microscopy and Structure of Materials*, (Ed. G. Thomas), University of California Press, Berkeley, 1972, p. 382.
28. Clark, W. A. T., Pond, R. C., Smith, D. A., and Goringe, M. J., *Developments in Electron Microscopy and Analysis* (Ed. J. A. Venables), Academic Press, New York, 1976, pp. 433 and 453.
29. Thomas, G., ref. 5, p. 131.
30. Fillingham, P. J., Leamy, H. J., and Tanner, L. E., ref. 4, p. 163.
31. Phillips, V. A. and Tanner, L. E., *Acta Met.*, **21,** 441 (1973).
32. Phillips, V. A., *Acta Met.*, **21,** 219 (1973).
33. Koo, J. Y., private communication.
34. Rao, B. V. N., Ph.D. thesis, University of California, Berkeley, LBL Report #7361.
35. Rao, B. V. N., Miller, R. W., and Thomas, G., Proc. 16th. Intl. Heat-Treatment Conference, The Metals Society, London, p. 75, 1976.
36. Huang, D. and Thomas, G., *Met. Trans.*, **2,** 1587 (1971).
37. Huang, D. and Thomas, G., *Met. Trans.*, **8A,** 1661 (1977).
38. Wilkens, M., ref. 5, p. 233.
39. Bell, W. L., Proceedings of the 7th International Congress of Electron Microscopy, Grenoble, 1970 (Société Francaise de Microscopie Electronique, Paris, 1970), p. 337.
40. Chen, L-J. and Thomas, G., *Phys. Stat. Sol.*, **25,** 193 (1974).
41. Osiecki, R., de Jonghe, L. C., Bell, W. L., and Thomas, G., unpublished.
42. Grundy, P. J., *Contemporary Phys.*, **18,** 47 (1977).
43. Jakubovics, J. P., *Phil. Mag.*, **13,** 85 (1966).
44. de Jonghe, L. C., Ph.D. thesis, University of California, Berkeley (Lawrence Radiation Laboratory Report UCRL–20369), 1970.
45. Menter, J. W., *Proc. Roy. Soc.*, **A236,** 119 (1956).
46. Scherzer, O., *J. Appl. Phys.*, **20,** 20 (1949).

47. Komoda, T., *Japan J. Appl. Phys.*, **5**, 603 (1966).
48. Parsons, J. R. and Hoelke, C. W., *J. Appl. Phys.*, **40**, 866 (1969).
49. Phillips, V. A. and Hugo, J. A., *Acta Met.*, **18**, 123 (1970).
50. Parsons, J. R. and Hoelke, C. W., *Phil. Mag.*, **22**, 1071 (1970).
51. Cockayne, D. J. H., Parsons, J. R., and Hoelke, C. W., *Phil. Mag.*, **24**, 139 (1971).
52. Phillips, V. A. and Wagner, R., *J. Appl. Phys.*, **44**, 4252 (1973).
53. Phillips, V. A., *Acta Met.*, **20**, 1143 (1972).
54. Parsons, J. R., Rainville, M., and Hoelke, C. W., *Phil. Mag.*, **21**, 1105 (1970).
55. Howe, L. M. and Rainville, M., *Rad. Eff.*, **16**, 203 (1972).
56. Phillips, V. A., *Acta Met.*, **23**, 751 (1975).
57. Chang, A. L. J., Sass, S. L. and Krakow, W., *Acta Cryst.*, **A33**, 672 (1976).
58. Kuan, T. S. and Sass, S. L., Report no 2803, Materials Science Center, Cornell University, Ithaca, NY.
59. Sinclair, R. and Thomas, G., *J. Appl. Cryst.*, **8**, 206 (1975).
60. Dutkiewicz, J. and Thomas, G., *Met. Trans.*, **6A**, 1919 (1975).
61. Sinclair, R., Schneider, K., and Thomas, G., *Acta Met.*, **23**, 873 (1975).
62. Dutkiewicz, J. and Thomas, G., *Thin Sol. Films*, **32**, 329 (1976).
63. Sinclair, R. and Dutkiewicz, J., *Acta Met.*, **25**, 235 (1977).
64. Clarke, D. R. and Thomas, G., Proc. 6th European Congress on Electron Microscopy, Jerusalem, 1976, p. 564.
65. Okamoto, P. R. and Thomas, G., *Acta Met.*, **19**, 825 (1971).
66. Clarke, D. R. and Thomas, G., *J. Am. Ceram. Soc.*, **60**, 461 (1977); **61**, 114 (1978).
67. Heidenreich, R. D., Hess, W. M., and Ban, L. L., *J. Appl. Cryst.*, **1**, 1 (1968).
68. Wu, C. K., Sinclair, R., and Thomas, G., *Met. Trans.*, **9A**, 381 (1978).
69. Rudee, M. L. and Howie, A., *Phil. Mag.*, **25**, 1001 (1972).
70. Herd, S. R. and Chaudhari, P., *Phys. Stat. Sol.*, **A26**, 627 (1974).
71. Krivanek, O. L., Gaskell, P. H., and Howie, A., *Nature,* **262**, 454 (1976); see also Krivanek, O. L. and Howie, A., *J. Appl. Cryst.*, **8**, 213 (1975).
72. Jenkins, G. M., Kawamura, K., and Ban, L. L., *Proc. Roy. Soc.*, **A327**, 501 (1972).
73. Cowley, J. M. and Iijima, S., *Z. Naturforsch.*, **27a**, 445 (1972).
74. Allpress, J. G. and Sanders, J. V., *J. Appl. Cryst.*, **6**, 165 (1973).
75. Skarnulis, A. J., Iijima, S., and Cowley, J. M., *Acta Cryst.*, **A32**, 799 (1976).
76. Bursill, L. A. and Wilson, A. R., *Acta Cryst.*, **A33**, 672 (1977).
77. Iijima, S. and Allpress, J. G., *Acta Cryst.*, **A30**, 22 (1974).
78. Krivanek, O. L., Isoda, S., and Kobayashi, K., *Phil. Mag.*, **36**, 931 (1977).
79. Hornstra, J., *J. Phys. Chem. Sol.*, **5**, 129 (1958).
80. Shaw, T. M., LBL–6930, M.S. thesis, University of California, Berkeley, 1977. Also: Shaw, T. M., and Thomas, G., Science *202*, 625 (1978).
81. Yessik, M., Shinozaki, S., and Sato, H., *Acta Cryst.*, **A31**, 764 (1975).
82. Sinclair, R. and Thomas, G., *Met. Trans.*, **9A**, 373 (1978).
83. Taylor, C. A., Hinde, R. M., and Lipson, H., *Acta Cryst.*, **4**, 261 (1951).
84. Gronsky, R., Sinclair, R., and Thomas, G., Proceedings 34th Annual Meeting Electron Microscopy Society of America, Claitors, Baton Rouge, LA, p. 494, 1976.

85. Sinclair, R., Gronsky, R., and Thomas, G., *Acta. Met.*, **24**, 789 (1976).
86. Van der Biest, O., *Acta Cryst.*, **A33**, 618 (1977).
87. Gronsky, R. and Thomas, G., Proceedings, 35th. Annual Meeting Electron Microscopy Society of America, Claitors, Baton Rouge, LA, p. 116, 1977.
88. Koo, J. Y. and Thomas, G., *Met. Trans.*, **8A**, 525 (1977).
89. Koo, J. Y. and Thomas, G., Proceedings 35th Annual Meeting Electron Microscopy Society of America, Claitors, Baton Rouge, LA, p. 118, 1977.

FOUR

THEORY OF DIFFRACTION CONTRAST

1 Introduction

In the preceding chapters a large number of examples were presented of deductions that may be made from the geometry of the diffraction pattern and from observation of how the contrast in the images of defects changes with the exact diffracting conditions. The purpose of this chapter and the following one is to present the theoretical background of image contrast and to indicate how any particular problem may, if not covered by more general "rules," be handled by way of a model contrast calculation. In the present chapter the theory is developed in a formal way to obtain all the equations required for the calculation of image contrast; in Chapter 5 methods of using these equations are discussed by means of typical examples. It is thus possible for the present chapter to be omitted at a first reading, to be returned to after a "feel" for the subject has been engendered by the examples of Chapter 5. For a fuller exposition of the basis of the theory the reader is referred to Hirsch et al.[1]

The problem to be solved concerns the interaction of high energy electrons (100 keV and greater) with the potential variations inside a solid specimen (perhaps a few volts or tens of volts). In general, this problem involves a discussion of Schrödinger's equation for a general potential distribution, for which the solution is intractable. However, for the special case of crystalline materials the potential distribution is periodic, which makes it possible to find a solution reasonably easily, as will be seen in Section 5. This solution, which is similar in many respects to that obtained in the "nearly free electron" treatment of conduction electrons found in a number of textbooks (e.g., ref. 11), involves a number of approximations. The physical justification for making these approximations can be more easily understood if a simpler view is first taken of the specimen potential. Furthermore, the simplified theory developed in the next section allows discussion of several approximations to the full solution, namely, phase grating, kinematical, and two-beam dynamical, and shows the relationships between them.

2 A Simplified General Theory

2.1 Neglect of Back Reflection

As a first approximation assume that a specimen of thickness t may be defined as a structureless region of attractive electrostatic potential V. Electrons of energy E_0 impinging normally on the entrance surface of such a specimen are transmitted and reflected according to the standard "quantum-mechanical barrier" rules. With the quantities as defined in Fig. 4.1, electrons traveling in the z-direction are considered to have a vacuum wave vector χ and a local wave vector inside the specimen χ_l, where

$$\chi = \sqrt{\frac{2mE_0}{h^2}} \quad \text{and} \quad \chi_l = \sqrt{\frac{2m(E_0 + eV)}{h^2}}, \tag{4.1}$$

an incident wave of unit amplitude, a reflected wave of amplitude A, and a transmitted wave of amplitude B. The magnitudes of the wave vectors are in units of reciprocal lengths, depending on accelerating potential; some values are given in Table 1.2. When the wave functions and their spatial derivatives at the entrance surface ($z = 0$) are matched, the reflected amplitude (ignoring any effects from the exit surface) is given by

$$A = \frac{\chi - \chi_l}{\chi + \chi_l} \simeq \frac{-eV}{4E_0}. \tag{4.2}$$

For $E_0 \sim 100$ keV and V a few volts eq. 4.2 indicates that the reflected amplitude is of order 10^{-5}. Any reflection at the exit surface will be similarly small. Thus in the following sections backreflection can be and will be ignored.

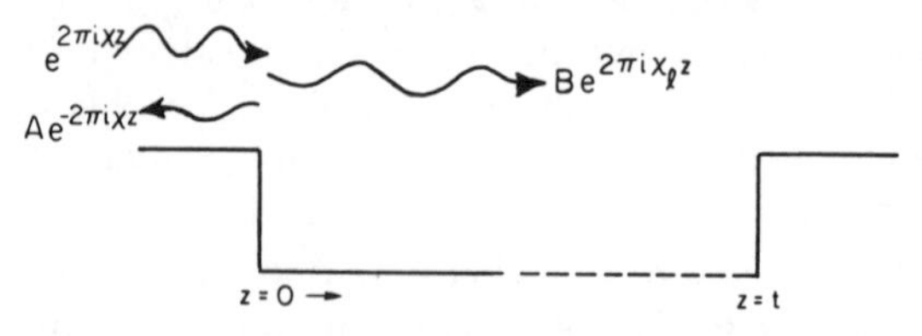

Fig. 4.1 Schematic representation of an electron wave entering a specimen, suffering some reflection of amplitude A.

For the energies used in transmission electron microscopy (50 keV and above) backreflection from the specimen surfaces is negligible even for large angular deviations from normal incidence; only at grazing incident angles does it become appreciable (and potentially useful in reflection microscopy). For lower energy electrons, however, such as those used in low energy electron diffraction (LEED), the reflection coefficients may become appreciable, consistently with the back-reflection mode of LEED.[2] It should also be noted that in the scanning electron microscope primary backreflected electrons, although small in number, play a significant role in image formation.[3]

2.2 The Phase Grating Approximation

Comparing the expressions for χ and χ_l in eqs. 4.1 shows that

$$\chi_l \simeq \chi\left(1 + \frac{eV}{2E_0}\right) = \chi + \frac{m_0 eV}{h^2 \chi}, \tag{4.3}$$

that is, the magnitude of the wave vector χ is increased to the new local value χ_l. In optical terminology *refraction* takes place. When a plane wave (of amplitude ϕ_0), $\psi_i = \phi_0 \exp(2\pi i \boldsymbol{\chi} \cdot \mathbf{r})$, is incident normally on a thin slab specimen of thickness dz, an extra phase shift will occur (with respect to the comparable path through a vacuum) because of this change in wave vector. The exit wave is thus

$$\psi_e = \phi_0 \exp(2\pi i \boldsymbol{\chi} \cdot \mathbf{r}') \exp\left[\frac{2\pi i m_0 e V(\mathbf{r})\, dz}{h^2 \chi}\right], \tag{4.4}$$

where $\mathbf{r}' = \mathbf{r} + dz$.

Integration of eq. 4.4 through a specimen of thickness t yields the *phase grating integral:*

$$\psi_e(x, y, t) = \phi_0 \exp(2\pi i \boldsymbol{\chi} \cdot \mathbf{r}') \exp\left[\frac{2\pi i m_0 e}{h^2 \chi} \int_0^t V(x, y, z)\, dz\right]. \tag{4.5}$$

Integration in this way is legitimate only if the resultant phase shift term (the argument of the second exponential) is small. To obtain an estimate define "small" as less than 1 rad and assume that the specimen is structureless once more. The maximum thickness for which eq. 4.5 is valid is then t_{max}, where

$$t_{max} = \frac{h^2 \chi}{2\pi m_0 e V}. \tag{4.6}$$

When reasonable values of $V \simeq 5$ V and $\chi = 27 \times 10^{10}$ m^{-1} (100 keV electrons) are substituted, $t_{max} \simeq 2 \times 10^{-8}$ m (20 nm, 200 Å). For specimens thinner than this, direct integration of the potential distribution gives the phase modulation of the exit wave. If the potential $V(x, y, z)$ is periodic, the result is a periodic phase-modulated wave–hence the term "phase grating," used to describe this approach. Many of the lattice images discussed in Chapter 3, Sections 12 and 13, may be described theoretically by means of such phase gratings. However, the approximation is not restricted to periodic structures such as crystals; it is equally applicable to noncrystalline materials.[4]

2.3 The Kinematical Approximation

The potential experienced by the electrons in a crystalline solid is shown schematically in Fig. 4.2*a* for a layer of atoms in the x-y plane, the potential contour lines shown indicating the potential wells at the atom cores. If the slab crystal is of thickness $t = \Delta z$ (= one atomic layer), then $V(\mathbf{r})$ varies for the different electron trajectories of Fig. 4.2*a* as shown in Fig. 4.2*b*. In principle the phase integration (eq. 4.5) should be performed for each trajectory. However, since the phase shift for a single atom layer is sufficiently small (Δz is very much less than t_{max} in eq. 4.6), $V(x, y, z)$ may be replaced by a projected value $\overline{V}_z(x, y)$ which varies only in the x–y plane, as shown in Fig. 4.2*c*, such that the

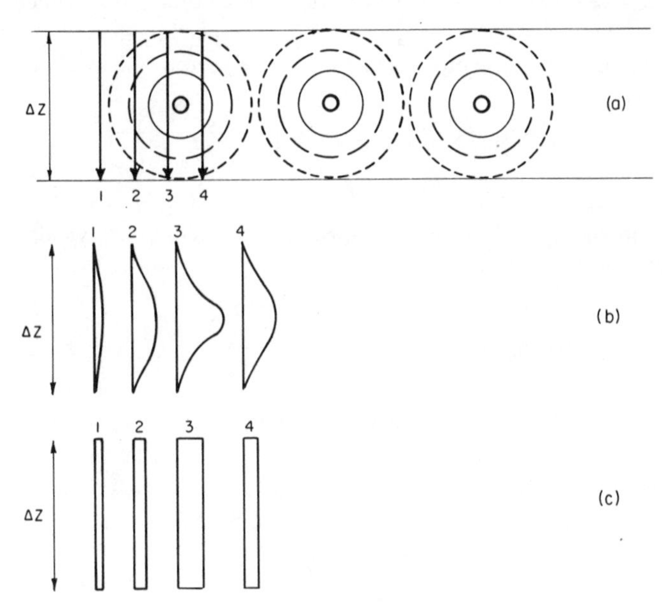

Fig. 4.2 The potential experienced by an electron traversing a crystal slice of thickness Δz equal to any atomic layer. (*a*) the actual potential contours around the atoms (deepening well toward the atom centers). (*b*) The potential variation along the representative trajectories indicated in (*a*). (*c*) The average potentials of the same "area" as in (*b*) assumed in the theoretical development.

phase shift (area under the curve) is the same as in Fig. 4.2*b*. In terms of the equations this is equivalent to rewriting eq. 4.5 as

$$\psi_e(x, y, \Delta z) = \phi_0 \exp(2\pi i \boldsymbol{\chi} \cdot \mathbf{r}') \exp\left[\frac{2\pi i m_0 e}{h^2 \chi} \overline{V}_z(x, y)\, \Delta z\right]$$

or, since Δz is small,

$$\psi_e(x, y, \Delta z) = \phi_0 \exp(2\pi i \boldsymbol{\chi} \cdot \mathbf{r}') \left[1 + \frac{2\pi i m_0 e}{h^2 \chi} \overline{V}_z(x, y)\, \Delta z\right]. \quad (4.7)$$

It is now convenient to write the projected periodic potential of a perfect crystal, $\overline{V}_z(x,y)$, as a Fourier series:

$$\overline{V}_z(x, y) = \sum_{\mathbf{g}} (V_{\mathbf{g}} + iV'_{\mathbf{g}}) \exp(2\pi i \mathbf{g} \cdot \mathbf{r}), \quad (4.8)$$

where $\mathbf{g}$ is a reciprocal lattice vector in the (real space) x–y plane. For generality the potential is made complex to introduce the possibility of electrons being absorbed (removed from the electron microscope image in some way), since from eq. 4.3 it may be seen that a complex potential V leads to a complex wave vector χ and hence to attenuation of the wave. It is reasonable that absorption should be structure dependent and hence expressible as a Fourier series in the same way as the real potential so far considered. The principal absorption mechanisms, such as thermal diffuse scattering and single-electron excitations, are associated with the atoms, and expressions for $V'_{\mathbf{g}}$ may be calculated,[5,6] although another important mechanism, namely, plasmon excitation, is not similarly periodic and does not lead so simply to expressions for $V'_{\mathbf{g}}$.[7] The values so derived, however, all serve to justify a further necessary approximation, namely, that the imaginary Fourier component $V'_{\mathbf{g}}$ of the potential is small compared with the real component $V_{\mathbf{g}}$.

If now the Fourier series potential of eq. 4.8 is substituted into eq. 4.7, another expression for the exit wave follows:

$$\psi_e(x, y, \Delta z) = \phi_0 \exp(2\pi i \boldsymbol{\chi} \cdot \mathbf{r}') + \phi_0 \left(\frac{2\pi i m_0 e\, \Delta z}{h^2 \chi}\right) \sum_{\mathbf{g}} (V_{\mathbf{g}} + iV'_{\mathbf{g}}) \cdot \exp[2\pi i (\boldsymbol{\chi} + \mathbf{g}) \cdot \mathbf{r}]. \quad (4.9)$$

Equation 4.9 shows that, in addition to the incident wave of wave vector $\boldsymbol{\chi}$, the exit wave consists of a series of waves of vectors $\boldsymbol{\chi} + \mathbf{g}$ and amplitude $\Delta\phi_{\mathbf{g}}$ traveling in directions $\boldsymbol{\chi}_{\mathbf{g}}$ (see Fig. 4.3 for the familiar Ewald sphere construction), where

$$\Delta\phi_{\mathbf{g}} \exp(2\pi i \boldsymbol{\chi}_{\mathbf{g}} \cdot \mathbf{r}) = \phi_0 \left(\frac{2\pi i m_0 e}{h^2 \chi}\right) \Delta z (V_{\mathbf{g}} + iV'_{\mathbf{g}}) \exp[2\pi i (\boldsymbol{\chi} + \mathbf{g}) \cdot \mathbf{r}], \quad (4.10)$$

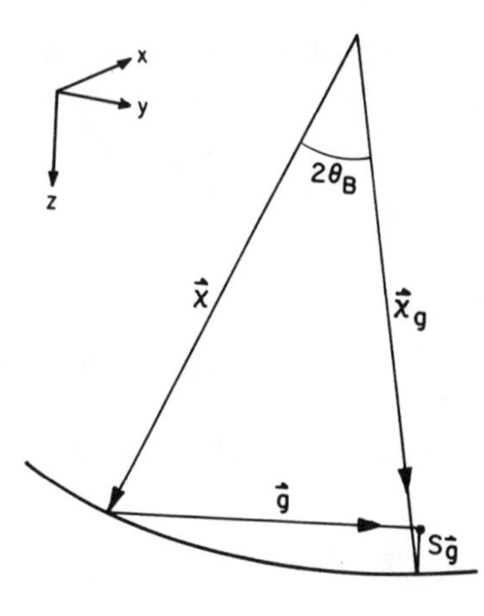

Fig. 4.3 The Ewald sphere construction, relating incident wave of wave vector χ, diffracted wave χ, reflecting vector **g**, and deviation parameter $\mathbf{s_g}$.

that is, diffracted waves have been produced. The "scattering" mentioned in Chapter 1, Section 2, thus follows automatically from the potential distribution in the specimen. Inspection of Fig. 4.3 relates the vectors $\boldsymbol{\chi}$, $\boldsymbol{\chi}_g$, **g** and the deviation parameter $\mathbf{s}_g$, already discussed in Chapter 2 (Fig. 2.6 and Section 3.1), by

$$\boldsymbol{\chi}_g - \boldsymbol{\chi} = \mathbf{g} + \mathbf{s}_g. \tag{4.11}$$

If only the z-component of $\mathbf{s}_g$, s_g, is considered eqs. 4.10 and 4.11 (rewritten for convenience in differential form) combine to yield

$$\frac{d\phi_g}{dz} = \pi\left(\frac{i}{\xi_g} - \frac{1}{\xi_g'}\right)\phi_0 \exp(-2\pi i s_g z), \tag{4.12}$$

where quantities with dimensions of length, ξ_g* (extinction distance) and ξ_g' (absorption length), have been defined by

$$\xi_g = \frac{h^2\chi}{2m_0 eV_g}, \qquad \xi_g' = \frac{h^2\chi}{2m_0 eV_g'}. \tag{4.13}$$

Equation 4.9 also predicts a change $d\phi_0$ in the amplitude of the incident wave ϕ_0, because of absorption and refraction, where

$$\frac{d\phi_0}{dz} = \pi\left(\frac{i}{\xi_0} - \frac{1}{\xi_0'}\right)\phi_0. \tag{4.14}$$

*The values of V_g and ξ_g are usually calculated for atoms at rest (i.e., at absolute zero) and the effect of temperature introduced, as in the X-ray case, by means of Debye-Waller factors $\xi_g \to \xi_g \exp(M_g)$, where $M_g = 2\pi^2 u_p^2 g^2$, u_p^2 being the mean-square atomic displacements perpendicular to the crystal planes under consideration. For the isotropic situation $M_g \propto g^2$, so that higher order reflected waves are affected more than lower order ones, all extinction distances in general increasing with rising temperature.

If absorption is neglected (i.e., set $\xi_g' \to \infty$), integration of eq. 4.12 gives the *kinematical integral for perfect crystals*, ϕ_0 being constant (apart from the phase change introduced by the refraction term V_0):

$$\phi_g = \phi_0 \left(\frac{\pi i}{\xi_g}\right) \int_0^t \exp(-2\pi i s_g z)\, dz = \phi_0 \left(\frac{\pi i}{\xi_g}\right) \exp(-\pi i t s_g) \frac{\sin \pi s_g t}{\pi s_g},$$

and a dark field intensity I_g, where

$$\frac{I_g}{I_0} = \frac{\pi^2 \sin^2 \pi s_g t}{\xi_g^2 (\pi s_g)^2}. \tag{4.15}$$

When eq. 4.15 is compared with eq. 2.14, which was derived under essentially the same conditions, and with eq. 4.13, the following relationships between atomic scattering factors F_g, Fourier components V_g of the lattice potential, and extinction distances ξ_g emerge:

$$\xi_g = \frac{h^2 \chi}{2 m_0 V_g} = \frac{\pi V_c}{F_g}, \tag{4.16}$$

where V_c is the volume of the unit cell.

Imperfect crystals may be treated by considering displacements of the crystal by $\mathbf{R}$,* so that the potential at any point $\mathbf{r}$ in the imperfect crystal is the same as that at point $\mathbf{r} - \mathbf{R}(\mathbf{r})$ in the perfect crystal, that is, eq. 4.8 is replaced by

$$V(\mathbf{r}) = \sum_g (V_g + iV_g') \exp[2\pi i \mathbf{g} \cdot (\mathbf{r} - \mathbf{R})]. \tag{4.17}$$

When the argument is followed through as before, eq. 4.12 becomes

$$\frac{d\phi_g}{dz} = \pi \left(\frac{i}{\xi_g} - \frac{1}{\xi_g'}\right) \phi_0 \exp[-2\pi i (s_g z + \mathbf{g} \cdot \mathbf{R})], \tag{4.18}$$

eq. 4.14 being unaltered.

On the assumption that ϕ_0 is constant as before, and with absorption again neglected, integration of eq. 4.18 yields

$$\phi_g = \frac{\phi_0 \pi i}{\xi_g} \int_0^t \exp[-2\pi i (s_g z + \mathbf{g} \cdot \mathbf{R})]\, dz, \tag{4.19}$$

which is the *kinematical integral for imperfect crystals* previously developed in eq. 3.4. The kinematical integral of eq. 4.19 was considerably used in the early

*These displacements may be either physical displacements such as are produced by dislocations, and so on, or effective displacements produced by electric or magnetic fields, for example.

development of the subject of contrast theory and has recently returned to prominence, having relevance in "weak beam" situations (see Chapter 5, Section 7.1).

2.4 The Two-Beam Dynamical Approximation

To develop the two-beam dynamical theory, consider only two beams, the incident beam of amplitude ϕ_0 and the principal diffracted beam of amplitude ϕ_g. Obviously, if ϕ_0 may be scattered into ϕ_g, and itself decay according to eqs. 4.18 and 4.14, respectively, then ϕ_g (considered now as an incident wave in that direction) may be similarly scattered back to ϕ_0, and itself decay. However, in the latter case the change $\boldsymbol{\chi}_g \rightarrow \boldsymbol{\chi}$ is the reverse of $\boldsymbol{\chi} \rightarrow \boldsymbol{\chi}_g$, and thus the sign of the phase factor $(s_g z + \mathbf{g} \cdot \mathbf{R})$ must be reversed.* Hence the (coupled equations become

$$\left.\begin{aligned} \frac{d\phi_0}{dz} &= \pi\left(\frac{i}{\xi_0} - \frac{1}{\xi_0'}\right)\phi_0 + \pi\left(\frac{i}{\xi_g} - \frac{1}{\xi_g'}\right)\phi_g \exp\left[2\pi i(s_g z + \mathbf{g} \cdot \mathbf{R})\right], \\ \frac{d\phi_g}{dz} &= \pi\left(\frac{i}{\xi_g} - \frac{1}{\xi_g'}\right)\phi_0 \exp\left[-2\pi i(s_g z + \mathbf{g} \cdot \mathbf{R})\right] + \pi\left(\frac{i}{\xi_0} - \frac{1}{\xi_0'}\right)\phi_g. \end{aligned}\right\} \quad (4.20)$$

The problem of calculating the expected bright field and dark field images of a defect in a crystal is, then, that of integration (numerical, in general) of eqs. 4.20 through the specimen of thickness t, from the known starting conditions at the top of the crystal, $\phi_0 = (1, 0)$ and $\phi_g = (0, 0)$.† To assist in this procedure certain simplifying substitutions may be made to remove unwanted terms. As a result of transforming $\phi_0 \rightarrow \phi_0 \exp(\pi i z/\xi_0)$ and $\phi_g \rightarrow \phi_g \exp(\pi i z/\xi_0 - 2\pi i s_g z - 2\pi i \mathbf{g} \cdot \mathbf{R})$ eqs. 4.20 become

$$\left.\begin{aligned} \frac{d\phi_0}{dz} &= -\frac{\pi}{\xi_0'}\phi_0 + \pi\left(\frac{i}{\xi_g} - \frac{1}{\xi_g'}\right)\phi_g, \\ \frac{d\phi_g}{dz} &= \pi\left(\frac{i}{\xi_g} - \frac{1}{\xi_g'}\right)\phi_0 + \pi\left[2i(s_g + \beta_g') - \frac{1}{\xi_0'}\right]\phi_g, \end{aligned}\right\} \quad (4.21)$$

where

$$\beta_g' = \mathbf{g} \cdot \frac{d\mathbf{R}}{dz}. \quad (4.22)$$

*This statement strictly applies only to the situation where the scattering object is centrosymmetric. The situation for the noncentrosymmetric crystal is discussed in Section 6.

†The quantities ϕ_0 and ϕ_g are written as (real, imaginary) to emphasize their complex nature. At other points in the argument phase and amplitude are used to the same end.

In eqs. 4.21 and 4.22 it is apparent that atomic displacements affect the beam amplitudes through β'_g, a factor that locally modifies the deviation parameter s_g—the lattice planes are bent near a defect.

In principle all that is required to calculate the expected image under two-beam conditions (the many-beam situation is similar in form, but more complicated) is numerical integration of eqs. 4.20 or 4.21 through the specimen, the choice depending, for example, on knowledge of expressions for the strain fields in terms of displacements **R** or strains $d\mathbf{R}/dz$. However, as will be seen in Section 3 and in Chapter 5, Sections 3 and 5, unnecessary parts of the integration (e.g., perfect crystals) can be avoided by the use of explicit solutions, scattering matrices, or the Bloch wave formulation.

Equations 4.20 to 4.22 are not quite the whole story though; they imply that the only variations that are important are z variations. This is tenable for a perfect crystal, but for a deformed crystal amounts to making the *column approximation*, a step that requires further investigation.

2.5 The Column Approximation

To establish the problem consider Fig. 4.4, after Takagi.[8] In the two-beam case the beams emitted at B, traveling in directions $\boldsymbol{\chi}$(incident) and $\boldsymbol{\chi} + \mathbf{g} + \mathbf{s}_g$ (diffracted), can have been produced only by multiple reflection from atoms in the triangular region $AA'B$. The angle ABA' is almost exactly $2\theta_B$, where θ_B is the Bragg angle for the reflection concerned. Since, it will be recalled, for the wavelengths used in electron microscopy (see Table 2.2; e.g., for 100 keV electrons $\lambda = 3.7 \times 10^{-12}$ m) a typical low index reflection has $\theta_B \simeq 0.01$ rad (= 0.5°), AA' (= $2\theta_B t$, where t is the specimen thickness, typically 10^{-7} m) is only 2×10^{-9} m (20 Å). [It should also be noted that the greater thickness of specimen penetrated by higher energy electrons (e.g., 3 to 5 times greater for 1 MeV)

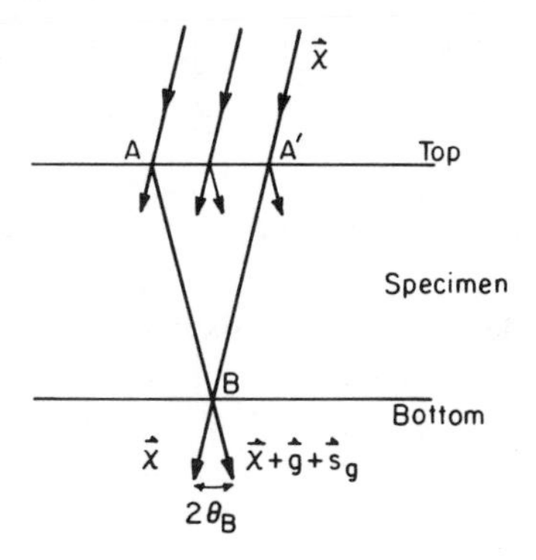

Fig. 4.4 An electron microscope specimen, illustrating how only information in the triangle $AA'B$ may be present in the electron beams emitted at the lower surface at B.

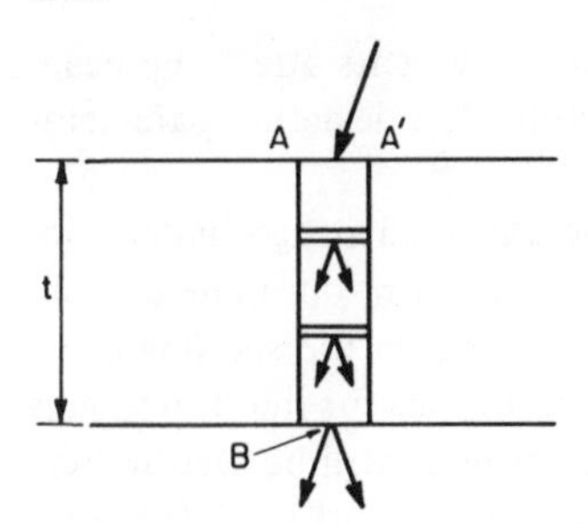

Fig. 4.5 Illustrating the column approximation.

is compensated for by a similar factor of reduction in the wavelength and hence the Bragg angle.] This produces one of the bases of the column approximation—if the atomic displacements do not vary appreciably *across* the column of width AA' (i.e., in the x–y plane), from top to bottom of the crystal (see Fig. 4.5), the result at B is independent of what happens outside the column and hence is the same as that calculated for a crystal of infinite lateral extent with displacements $\mathbf{R}$ a function of z only.

The column approximation is common to most of the calculations that have been performed to date, whether in two-beam, many-beam, or Bloch wave formulations (see below), as well as in the original kinematical formulation, where its justification was made in terms of the wave optics of Fresnel zones (see, e.g., Hirsch et al.,[1] Chapter 7). Howie and Basinski[9] have discussed the validity of the column approximation in some detail and conclude that the approximation is actually *more accurate* than discussed above; the effective divergence angle $2\theta_D$ is generally significantly less than $2\theta_B$ (e.g., for typical dislocations $\theta_D \sim 10^{-4}$ but is larger for defects with a shorter ranging strain field). One cautionary note is needed in regard to this small value of the column width: it depends on the physical crystal being much greater in extent than the column width, that is, the crystal must be composed of a number of adjacent columns differing only slightly from each other. For diffraction to occur at all without large shape-factor effects (see Chapter 2) the specimen must be a large number of interplanar spacings d in width.

3 Two-Beam Analytical Solution for Perfect Crystals

As noted above, eqs. 4.21 have within them, in the two-beam approximation, all the information necessary to solve the problem of contrast from a particular defect. All that is needed is some experience of which ratios of ξ_g'/ξ_g, and so on, to use, a good estimate of β_g' (eq. 4.22), and some computational skills. Certain simple problems have explicit solutions, so these may be discussed first before developing the theory more fully.

For a perfect crystal $\beta_g' = 0$, so eqs. 4.21 may be solved analytically for both ϕ_0 and ϕ_g. Using trial solutions of complex wave vector $\gamma = (\gamma_{re} + i\gamma_{im})$ of the form $\exp[2\pi i(\gamma_{re} + i\gamma_{im})z]$, after some algebra, and assuming that $\gamma_{re} \gg \gamma_{im}$ (i.e., small absorption), yields two solutions with

$$\gamma_{re}^{(1)(2)} = \frac{w \pm \sqrt{1 + w^2}}{2\xi_g}, \tag{4.23)*$$

$$\gamma_{im}^{(1)(2)} = \frac{1}{2\xi_0'} \pm \frac{1}{2\xi_g'\sqrt{1 + w^2}}, \tag{4.24}$$

where

$$w = s\xi_g, \tag{4.25}$$

and the superscripts 1, 2 refer to the branches of the dispersion surface (Fig. 4.10) see Sec. 4.2.

The general solution for ϕ_0 and ϕ_g must be linear combinations of the form

$$\phi_0 = D_0^{(1)} \exp(2\pi i\gamma^{(1)}z) + D_0^{(2)} \exp(2\pi i\gamma^{(2)}z), \tag{4.26}$$

$$\phi_g = D_g^{(1)} \exp(2\pi i\gamma^{(1)}z) + D_g^{(2)} \exp(2\pi i\gamma^{(2)}z), \tag{4.27}$$

where the D's are, for the moment, arbitrary constants. Neglecting absorption in eqs. 4.21 and substituting for ϕ_0 and ϕ_g yields the following relationship between the D's:

$$\frac{D_g^{(1)}}{D_0^{(1)}} = 2\gamma^{(1)}\xi_g = w + \sqrt{1 + w^2}, \tag{4.28}$$

$$\frac{D_g^{(2)}}{D_0^{(2)}} = 2\gamma^{(2)}\xi_g = w - \sqrt{1 + w^2}. \tag{4.29}$$

Strictly, absorption would make the coefficients D complex, but since absorption is small the imaginary part is usually ignored without introducing appreciable error.

At the top of the crystal ($z = 0$) only the incident beam has significant intensity, so that the matching conditions are

$$\left.\begin{aligned} \phi_0(0) &= 1 = D_0^{(1)} + D_0^{(2)}, \\ \phi_g(0) &= 0 = D_g^{(1)} + D_g^{(2)}. \end{aligned}\right\} \tag{4.30}$$

*Here and subsequently an ordered numbering scheme has been adopted for the solutions in descending value of γ_{re}, in agreement with more recent discussions of many-beam theory (see, e.g., ref. 10). Unfortunately this convention is the opposite of that adopted originally in two-beam theory (see, e.g., ref. 1).

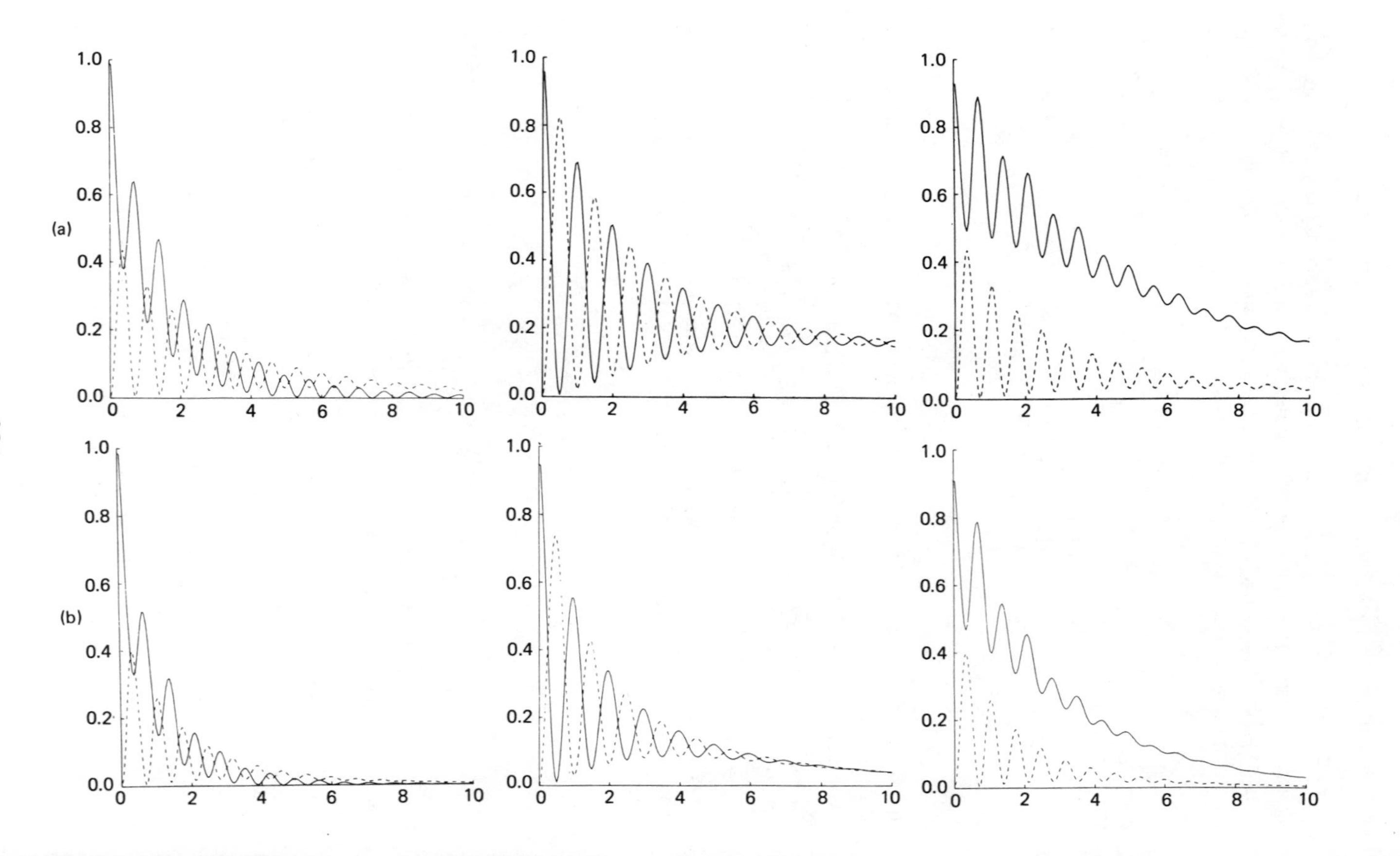

(a)
(b)
1.0
0.8
0.6
0.4
0.2
0.0
0
2
4
6
8
10

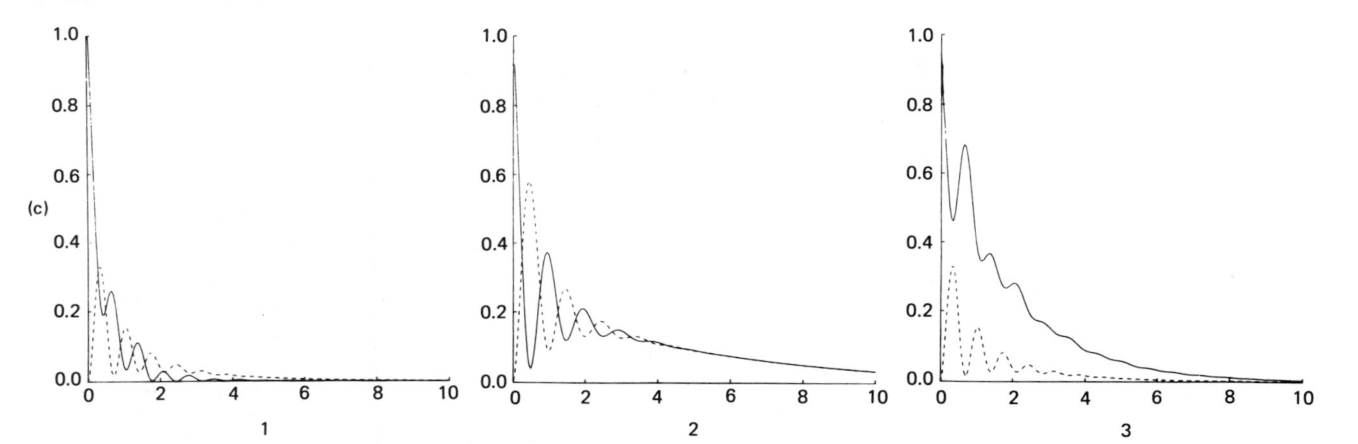

Fig. 4.6 Typical two-beam bright field (solid line) and dark field (dashed) thickness fringe profiles, for these different deviations from the reflecting positions: (1) $w = -1.0$, (2) $w = 0$, and (3) $w = +1.0$. The plots are all for $\mathbf{g} = 111$ for the three typical materials (*a*) aluminum, (*b*) copper, and (*c*) gold as defined in Table 4.1, and in each case the thickness is in units of the extinction distance.

After making the convenient substitution eqs. 4.28 to 4.30 yield

$$w = \cot \beta \tag{4.31}$$

$$\left.\begin{aligned} D_0^{(1)} &= \sin^2 \frac{\beta}{2}, \qquad D_0^{(2)} = \cos^2 \frac{\beta}{2}, \\ D_g^{(1)} &= \sin \frac{\beta}{2} \cos \frac{\beta}{2}, \qquad D_g^{(2)} = -\sin \frac{\beta}{2} \cos \frac{\beta}{2}. \end{aligned}\right\} \tag{4.32}$$

Hence the solutions for the beams at depth z in a slice of perfect crystal are found by substitution into eqs. 4.26 and 4.27 for the D's and γ's. The equations become

$$\phi_0(z) = \left[\sin^2 \frac{\beta}{2} \exp(iXz) + \cos^2 \frac{\beta}{2} \exp(-iXz)\right] \exp\left(\frac{-\pi z}{\xi_0'}\right), \tag{4.33}$$

$$\phi_g(z) = [\exp(iXz) - \exp(-iXz)] \sin \frac{\beta}{2} \cos \frac{\beta}{2} \exp\left(\frac{-\pi z}{\xi_0'}\right), \tag{4.34}$$

where

$$X = \frac{\pi\sqrt{1 + w^2}}{\xi_g} + \frac{\pi i}{\xi_g' \sqrt{1 + w^2}}, \tag{4.35}$$

and the common phase factor $\exp(\pi i w z/\xi_g)$ has been removed.

These expressions may be used to calculate bright field ($= |\phi_0|^2$) and dark field ($= |\phi_g|^2$) intensity profiles as a function of thickness z at constant orientation (thickness fringes), or of orientation w at constant thickness (rocking curves), or of both. Typical curves are shown in Figs. 4.6 to 4.9, where theoretical plots and experimental micrographs may be compared. Note that the thickness fringes (Figs. 4.6 and 4.7) have a periodicity (effective extinction distance) given by the real part of X (eq. 4.35) and a decreasing amplitude (the electrons are absorbed) given by the imaginary part. The fading of fringes with increasing thickness may also be seen in the thicker parts of the "bend contour" crystal shown in Fig. 4.9, as well as in the calculated "rocking curves" of Fig. 4.8. For the calculations presented in Figs. 4.6 and 4.8 values of the parameters ξ_0', ξ_g, and ξ_g' have been used that correspond to typical light (aluminum), medium (copper), and heavy (gold) materials. For reference now and later the values used for a number of reflections in these materials are assembled in Table 4.1.

In the present context the following points should be noted.

1. The greater the atomic weight the shorter the extinction distance, and hence the thinner is the specimen at, say, the fifth thickness fringe (compare Figs. 4.6*b*, *e*, and *h*, and note that the plots are in units of the extinction distance).

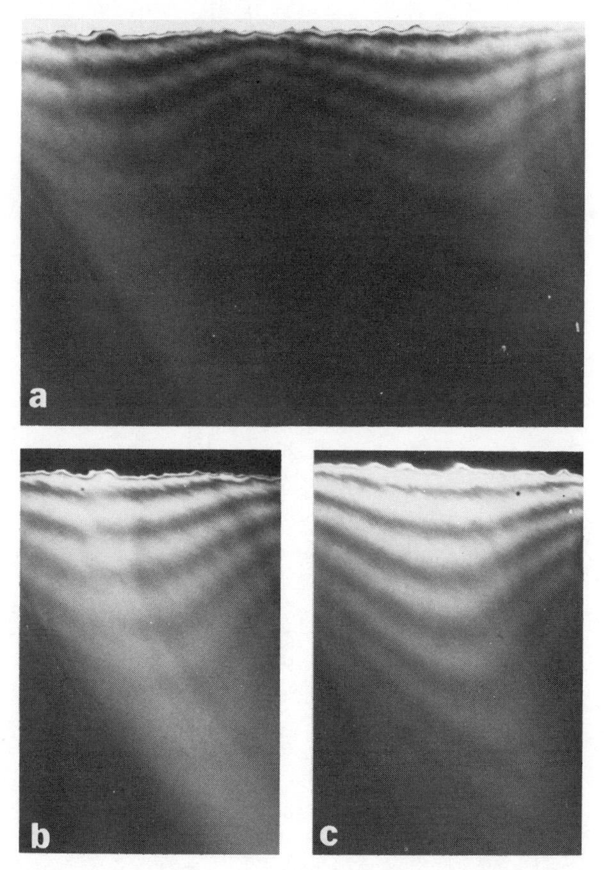

Fig. 4.7 Micrographs of a tapered specimen showing thickness fringes to compare with the profiles of Fig. 4.6: (*a*) bright field; (*b*) dark field, corresponding to the left-hand contour; (*c*) right-hand dark field.

2. The greater the atomic weight the shorter the absorption lengths relative to the extinction distances, and hence the smaller is the penetration (compare Figs. 4.6*b*, *e*, and *h*); because the extinction distances are smaller for the heavier elements, the penetration in absolute terms is, of course, even less.
3. In bright field penetration is greater for positive deviations from the reflecting position and smaller for negative deviations; in dark field either deviation leads to lower penetration (compare Figs. 4.6*b* 1 to 3, for example). Both effects are greater when the absorption is larger (compare Figs. 4.6*a*, *b*, and *c* with *g*, *h*, *i*.) These effects may perhaps be seen even more clearly in the rocking curves of Fig. 4.8.

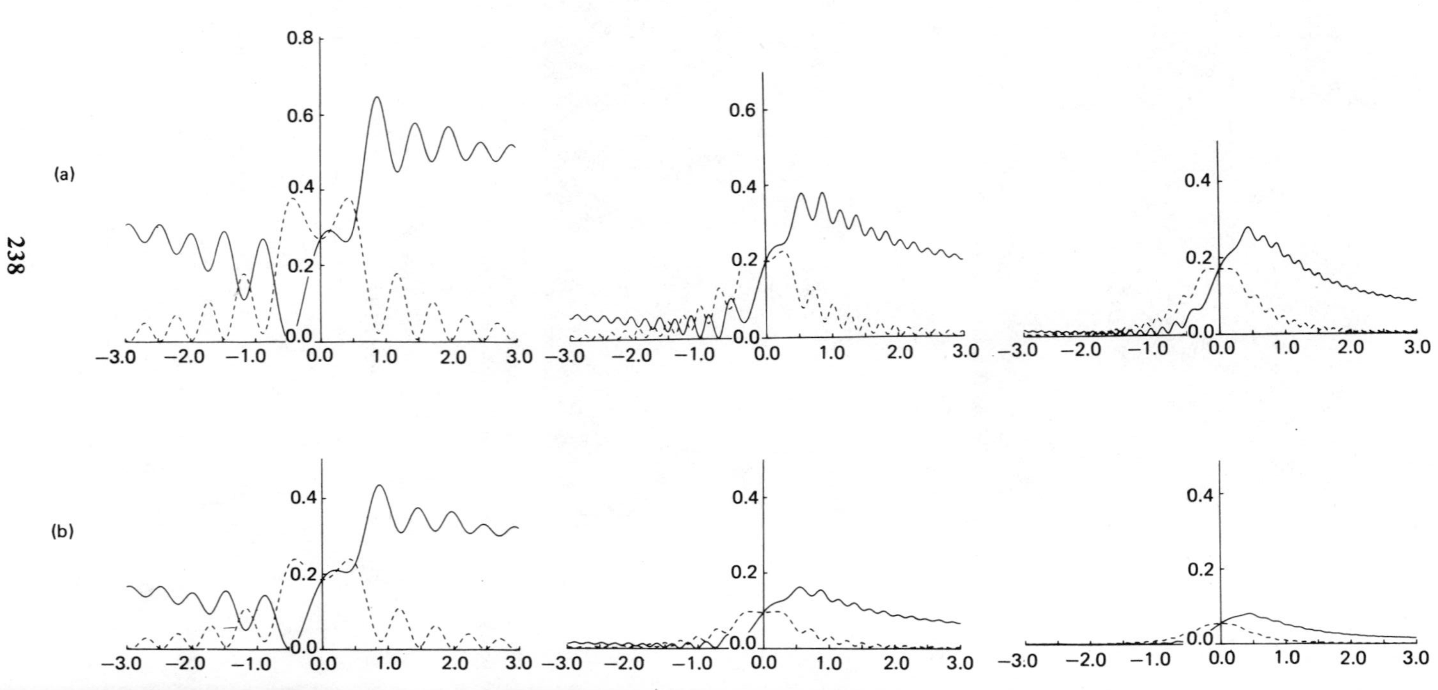

(a)
0.8
0.6
0.4
0.2
0.0
−3.0 −2.0 −1.0 0.0 1.0 2.0 3.0
0.6
0.4
0.2
0.0
−3.0 −2.0 −1.0 0.0 1.0 2.0 3.0
0.4
0.2
0.0
−3.0 −2.0 −1.0 0.0 1.0 2.0 3.0
(b)
0.4
0.2
0.0
−3.0 −2.0 −1.0 0.0 1.0 2.0 3.0
0.4
0.2
0.0
−3.0 −2.0 −1.0 0.0 1.0 2.0 3.0
0.4
0.2
0.0
−3.0 −2.0 −1.0 0.0 1.0 2.0 3.0

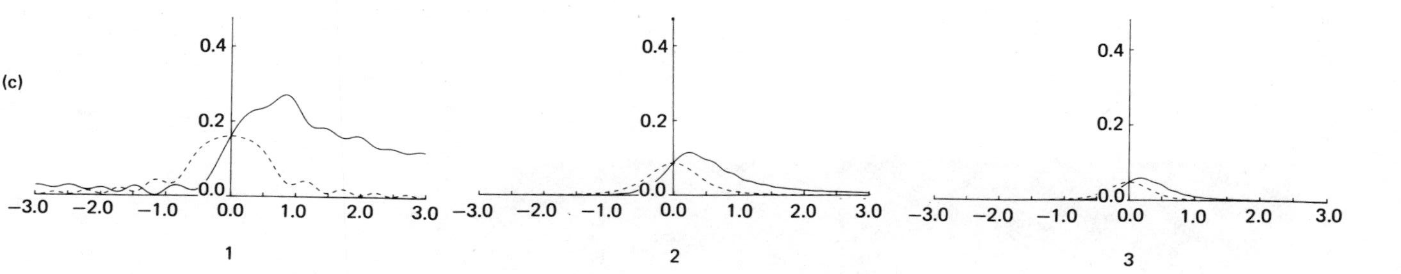

Fig. 4.8 Calculated two-beam "rocking curves" for various crystals of uniform thickness. Bright field (solid line) and dark field (dashed) curves. Note the asymmetry of the bright field curves and symmetry of the dark field. As in Fig. 4.6, the plots are for the three typical materials defined by Table 4.1 with $\mathbf{g} = 111$: (*a*) aluminum, (*b*) copper, and (*c*) gold, at three different thicknesses t/ξ_g of (1) 2.25, (2) 5.25, and (3) 8.25.

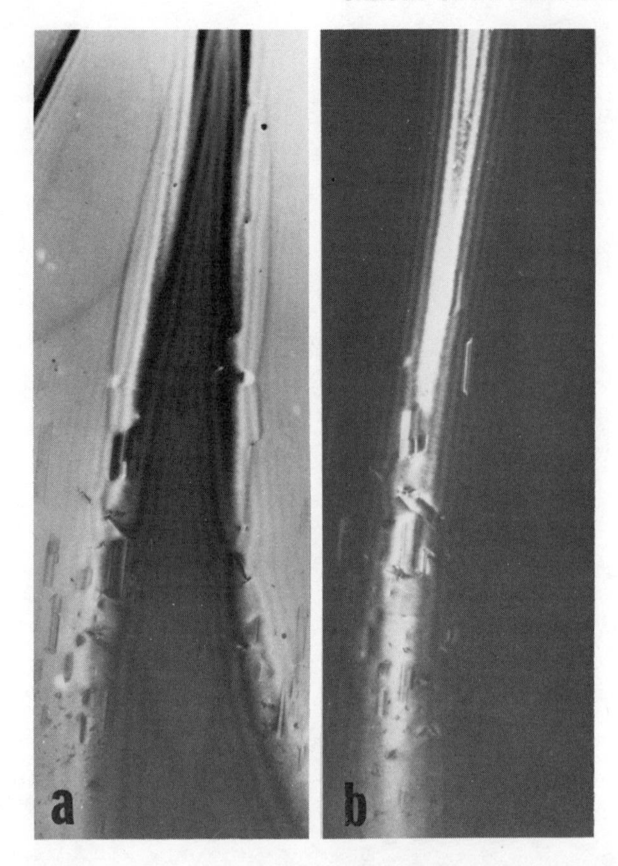

Fig 4.9 Micrographs of a bent specimen showing bend contours to compare with the rocking curves of Fig. 4.8: (*a*) bright field, (*b*) dark field, corresponding to the left-hand part of (*a*).

4 Further Development of the Theory

4.1 Beams and Bloch Waves

The total wave function $\psi(\mathbf{r})$ (eq. 4.9) in the crystal is

$$\psi(\mathbf{r}) = \phi_0 \exp(2\pi i \boldsymbol{\chi} \cdot \mathbf{r}) + \phi_g \exp(2\pi i \boldsymbol{\chi}_g \cdot \mathbf{r}),$$

where ϕ_0 and ϕ_g are given by eqs. 4.26 and 4.27, respectively, and $\boldsymbol{\chi}_g = \boldsymbol{\chi} + \mathbf{g}$. Setting

$$D_0^{(j)} = \epsilon^{(j)} C_0^{(j)}, D_g^{(j)} = \epsilon^{(j)} C_g^{(j)}, j = 1, 2 \tag{4.36}$$

Table 4.1 Extinction Distances and Absorption Lengths (nm) for 100 keV Electrons Used in Calculations for Typical Light, Medium, and Heavy Elements[a]

	Light (Aluminum)		Medium (Copper)		Heavy (Gold)	
g	ξ_g	ξ_g'	ξ_g	ξ_g'	ξ_g	ξ_g'
000	(20.0)	886	(15.0)	293	(11.4)	89.8
111	57.3	1000	30.2	410	18.3	107
200	68.3	1100	33.6	430	20.1	112
220	107	1570	45.1	500	26.5	121
222	140	1820	55.6	530	32.5	130
400	168	1930	65.9	599	38.2	132
333	230	2500	92.7	742	53.1	148
444	362	3930	149	851	78.6	175
555	526	5720	223	1210	110	229
666	705	7660	303	1640	145	290

[a]Values based on calculations of Doyle and Turner,[18] Humphreys and Hirsch,[19] and Radi.[20]

and rearranging slightly to group the terms in $\mathbf{k}^{(j)}$ where $\mathbf{k}^{(j)} = \boldsymbol{\chi} + \gamma^{(j)}, j = 1, 2$, and splitting into component vectors in the x–y plane, $\mathbf{k}_{xy}^{(j)}$, and z-direction, $\mathbf{k}_z^{(j)}$, yields

$$\begin{aligned}\psi(\mathbf{r}) = \epsilon^{(1)}[C_0^{(1)} \exp(2\pi i \mathbf{k}_{xy}^{(1)} \cdot \mathbf{r}_{xy}) + C_g^{(1)} \exp(2\pi i(\mathbf{k}_{xy}^{(1)} + \mathbf{g}) \cdot \mathbf{r}_{xy})]\\ \cdot \exp(2\pi i k_z^{(1)} z) + \epsilon^{(2)}[C_0^{(2)} \exp(2\pi i \mathbf{k}_{xy}^{(2)} \cdot \mathbf{r}_{xy}) + C_g^{(2)}\\ \cdot \exp(2\pi i(\mathbf{k}_{xy}^{(2)} + \mathbf{g}) \cdot \mathbf{r}_{xy})] \cdot \exp(2\pi i k_z^{(2)} z).\end{aligned} \tag{4.37}$$

It may now be seen that the square-bracketed terms are of the form of *Bloch waves** in the x–y plane, that is, they are solutions for a lattice periodic in that plane. For convenience "excitation" terms $\epsilon^{(j)}$ have been extracted, allowing normalization of the Bloch wave components C_0, C_g by

$$(C_0^{(j)})^2 + (C_g^{(j)})^2 = 1, \quad j = 1, 2, \tag{4.38}$$

and also "propagation" terms, $\exp(2\pi i k_z^{(j)} z)$, which describe the changing phase and amplitude relationships between the Bloch waves as they pass through the crystal (increasing z).

*For a full explanation of the term "Bloch waves" or "Bloch functions" see, for example, ref. 11. For present purposes it is sufficient to appreciate that they are periodic functions in the x–y plane describing the electron distribution in that plane.

The full definition of the Bloch wave amplitudes is completed by using eqs. 4.32, 4.36 and 4.38 to obtain

$$\left.\begin{aligned} \epsilon^{(1)} &= C_0^{(1)} = -C_g^{(2)} = \sin\frac{\beta}{2}, \\ \epsilon^{(2)} &= C_0^{(2)} = C_g^{(1)} = \cos\frac{\beta}{2}. \end{aligned}\right\} \tag{4.39}$$

In a strict sense, of course, Bloch waves should not propagate–they are standing or almost standing waves–and the term "Bloch wave" is being used here to describe the electron densities in the x-y plane; the propagation is in the z-direction.

Thus in a more concise notation

$$\psi(\mathbf{r}) = \sum_{j=1,2} b^{(j)}\epsilon^{(j)} \exp(2\pi i\gamma^{(j)}z), \tag{4.40}$$

where the Bloch waves $b^{(j)}$ are the square-bracketed terms in eq. 4.37, and the phase factor $\exp(2\pi i\chi_z z)$, which is common to both, has been dropped.

In terms of the "beam" notation used earlier (see eqs. 4.26 and 4.27)

$$\phi_0(z) = \sum_j \epsilon^{(j)}C_0^{(j)} \exp(2\pi i\gamma^{(j)}z), \tag{4.41}$$

$$\phi_g(z) = \sum_j \epsilon^{(j)}C_g^{(j)} \exp(2\pi i\gamma^{(j)}z). \tag{4.42}$$

4.2 Dispersion Surfaces

The dispersion surface for electrons of fixed total energy is the locus of points describing the real parts of the wave vectors of allowed Bloch waves, that is, $\mathbf{k}^{(j)}$, and is made up of a number of branches (one for each value of j). Such a pair of branches is shown in Fig. 4.10, where the incident wave vector $\boldsymbol{\chi}$ is drawn for a particular orientation defined by the deviation parameter s_g, with its value $\mathbf{k}$ after correction for the refraction produced by the mean inner potential (V_0 in eq. 4.17), the values of $\gamma^{(1)}$ and $\gamma^{(2)}$, and the two beams making up the Bloch wave (i) of wave vectors $\mathbf{k}^{(i)}$ and $\mathbf{k}^{(i)} + \mathbf{g}$ ($i = 1, 2$). The dispersion surface branches are traced out by varying the incident orientation; $\gamma^{(j)}$ are functions of s (eq. 4.23).* From eqs. 4.23 and 4.35 the effective extinction distance ξ^{eff} (the

*Note the similarity between the dispersion surface diagrams (e.g., Fig. 4.10, and even more notably Fig. 4.13) with standard "nearly free electron" $E(k)$ curves (see, e.g., ref. 11) for conduction electrons in metals. The dispersion surface is, of course, in contrast to the $E(k)$ curve, for fixed *total* energy, but changes of position on any branch correspond to

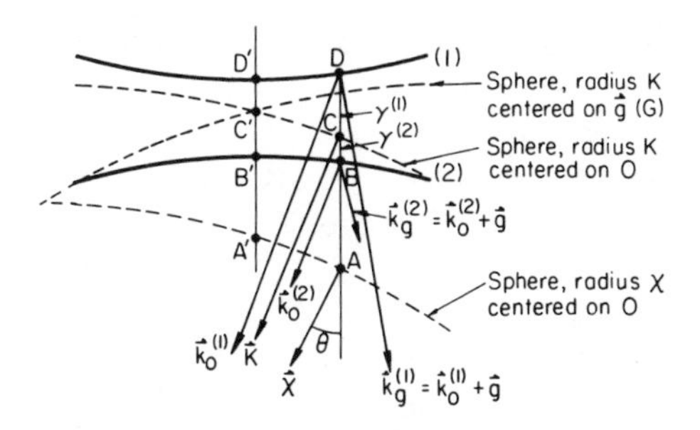

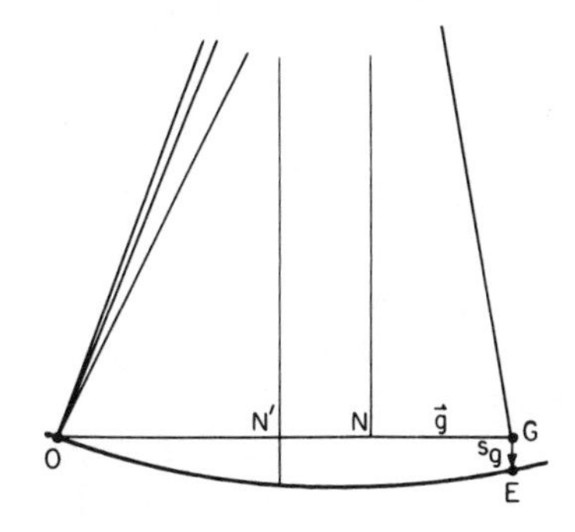

Fig. 4.10 Relationship between the Ewald sphere constructions and the dispersion surfaces for incident electrons deviated from the reflecting position (unprimed letters). The corresponding positions for incidence at the exact reflecting position are indicated by primed letters on the Brillouin zone boundary $D'N'$. Note that the horizontal and vertical scales are exaggerated with respect to each other (see Exercise 1.6.).

thickness fringe spacing Δz) is seen to be equal to the reciprocal of the separation of the dispersion surface branches:

$$\Delta\gamma = \frac{\sqrt{1+w^2}}{\xi_g} = \frac{1}{\Delta z} = \frac{1}{\xi^{\text{eff}}}. \tag{4.43}$$

This definition of the effective extinction distance is general; it is the "extinction" or "interference" distance for the beating between any pair of Bloch waves. Whether the thickness fringes corresponding to this beating are visible depends, of course, on the relative excitations and absorptions of the pair, and of any others that may be significantly excited (see Sections 4.4 and 5 on many-beam theory).

Also illustrated in Fig. 4.10 is one of the important aspects of wave matching—for near-normal incidence there can be no immediate change in the x–y momentum after entry, so the points A, B, C, and D are linked by a vertical line. (The x–y momentum is, of course, in time changed by Bragg reflection—by the build-up in the amplitude of the diffracted wave with increasing thickness.) This is an illustration of the more general principle that at an interface the excitations of the Bloch waves change in general, but not in a random way. The forms of the

---

changes in the way this total energy is distributed between potential energy and "forward" (z-direction) and "sideways" kinetic energies, thus allowing the curves to be similar in appearance.

Bloch waves are always related by points lying on lines drawn perpendicular to the boundary (see, e.g., ref. 12). Since $\chi \gg \gamma$, the excitations and forms of the Bloch waves given by eq. 4.39 are not appreciably in error unless the angle of inclination of the boundary to the beam is nearly 90°, that is, nearly grazing rather than the normal incidence assumed in the discussion so far.

4.3 Bloch Waves and Absorption

Earlier in the discussion of beams it was shown that the two solutions for the imaginary part $\gamma_{\text{im}}^{(j)}$ of the propagation factors $\gamma_{\text{re}}^{(j)} + i\gamma_{\text{im}}^{(j)}$ were different (eq. 4.24). In the present context this means that the two Bloch waves set up in the crystal are absorbed differently, and since $\gamma_{\text{im}}^{(1)} > \gamma_{\text{im}}^{(2)}$ the Bloch wave with the larger k_z is absorbed more strongly than the other. Thus in the two-beam approximation one may speak of the well-transmitted Bloch wave (2) and the heavily absorbed Bloch wave (1) (see Fig. 4.10).

Applying eqs. 4.39 for excitations and 4.24 for absorptions, it may now be seen why the bright field bend contours of Figs. 4.8 and 4.9 are asymmetric and the dark field symmetric. At the exact reflecting position (Fig. 4.11, $w = 0$, $\beta = \pi/2$) both waves are equally excited; for w negative, however, the more heavily absorbed Bloch wave is more strongly excited than the well-transmitted wave, that is, less amplitude is available at the bottom of the crystal in total, while the reverse situation applies when w is positive. Now the transmitted beam amplitude C for bright field $C_0^{(2)}$ is of the same form, and therefore there is more emission in the bright field beam when more of the well-transmitted Bloch wave is present, but the converse is true for the dark field beam $C_g^{(2)}$—the product $\epsilon^{(2)} * C_g^{(2)}$ of ("excitation" × "emission") is symmetrical. Hence bright and dark field bend contours in crystals that *are thick enough to absorb one of the waves* must be asymmetrical and symmetrical, respectively, as seen in the examples of Fig. 4.9 and the calculated rocking curves of Fig. 4.8. Similarly, thickness fringes occur because the Bloch waves are interfering, and this can happen only if two are present! Therefore thickness fringes should not occur in thicker crystals, as observed (see Fig. 4.7).

Examination of eqs. 4.37 and 4.39 and inspection of the dispersion surface construction of Fig. 4.10 shows that, at the exact reflecting position,

$$\mathbf{k}_{xy}^{(1)} = \mathbf{k}_{xy}^{(2)} = \frac{-\mathbf{g}}{2}$$

(the left-hand equality being true in general for incidence normal to the crystal surface), that is, the Bloch waves $b^{(j)}$ of eq. 4.40 become

$$b^{(1)} = \sin\frac{\beta}{2}\exp(-\pi i\mathbf{g}\cdot\mathbf{r}) + \cos\frac{\beta}{2}\exp(\pi i\mathbf{g}\cdot\mathbf{r})$$

$$= \sqrt{2}\cos\pi\mathbf{g}\cdot\mathbf{r}$$

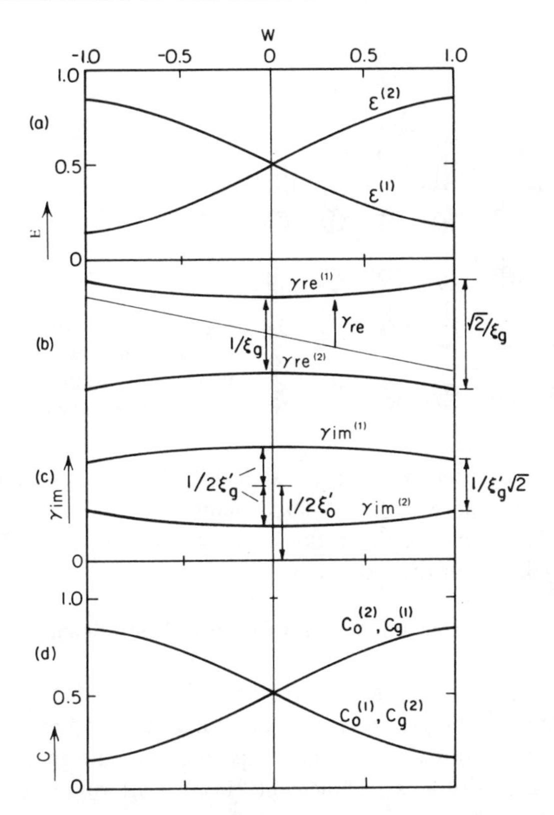

Fig. 4.11 Schematic variations with orientation of the parameters of the Bloch waves: (*a*) excitation factors ($\epsilon^{(i)}$), (*b*) propagation factors $\gamma_{\mathrm{re}}^{(i)}$ (measured with respect to the "free electron" sphere—the sloping line here), (*c*) absorption factors $\gamma_{\mathrm{im}}^{(i)}$, (*d*) "beam" amplitudes $C_0^{(i)}$, $C_g^{(i)}$ comprising the Bloch waves. In (*b*) γ_{re} is measured from the sloping line as indicated by the arrow.

and

$$b^{(2)} = i\sqrt{2}\,\sin \pi \mathbf{g} \cdot \mathbf{r}.$$

When drawn out (Fig. 4.12), it may be seen that the electron densities in the two cases are both of the $\cos^2$ type, but the density of $b^{(1)}$ is peaked at the atom cores, while that of $b^{(2)}$ is peaked between them. It is then clear physically why the two waves should have different propagation vectors: the two solutions must correspond to the same *total energy* (the incident electron beam's energy), but type 1 peaks in regions of large *negative* potential energy near the atoms and hence from Schrödinger's equation must possess greater kinetic energy than the mean value. The converse is true for the type 2 wave; it has lower kinetic energy (and hence a smaller γ).

This model is also consistent with what would be expected for absorption;

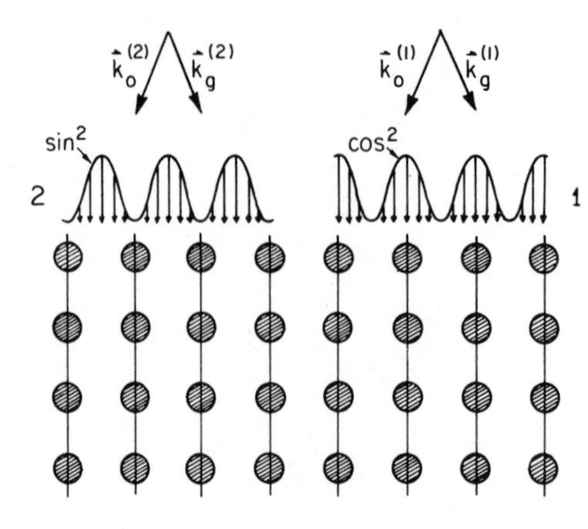

Fig. 4.12 Schematic diagram of the electron distributions in the two types of Bloch wave, together with the atoms on the reflecting planes.

possible absorption processes such as excitation of X-rays and thermal diffuse scattering are associated with the atoms, schematically indicated by the shaded areas in Fig. 4.12. Therefore absorption is more likely for the wave that has its electron density higher at the atom sites.

4.4 Extension of Formulation to Many Beams

In Section 2.4 only two beams were considered: the incident beam and one diffracted beam. In principle there is no difficulty in extending the formulation to any number of beams. For example, eqs. 4.20 may be generalized by considering each beam in turn as an "incident" beam scattering into all the others. In this way a set of equations follows, of the form

$$\frac{d\phi_k}{dz} = \sum_{j=1}^{n} \pi \left(\frac{i}{\xi_{\Delta g}} - \frac{1}{\xi'_{\Delta g}} \right) \phi_j \exp\left[-2\pi i (\Delta s_g z + \Delta \mathbf{g} \cdot \mathbf{R})\right], \quad k = 1, 2, \ldots, \tag{4.44}$$

where

$$\Delta \mathbf{g} = \mathbf{g}_k - \mathbf{g}_j \quad \text{and} \quad \Delta s_g = s_{gk} - s_{gj}.$$

Equations 4.21 may be similarly generalized. However, many-beam situations are in many ways more easily dealt with in the Bloch wave formulation, which will be developed in the next section.

5 Development of the Bloch Wave Formulation

As noted earlier, the incident electrons must obey Schrödinger's equation (neglecting spin effects, which are expected to be small):

$$\nabla^2 \psi(\mathbf{r}) + \left(\frac{8\pi^2 m}{h^2} \right) [E + eV(\mathbf{r})]\, \psi(\mathbf{r}) = 0. \tag{4.45}$$

The attractive potential in the crystal is of the form

$$V(\mathbf{r}) = \frac{h^2}{2me} \sum_g U_g \exp(2\pi i \mathbf{g} \cdot \mathbf{r}) = \sum_g V_g \exp(2\pi i \mathbf{g} \cdot \mathbf{r}), \tag{4.46}$$

which is an extension of the real part of eq. 4.8 to allow relativistic corrections, included here through

$$\left. \begin{aligned} m &= \frac{m_0}{\sqrt{1 - v^2/c^2}} = m_0 \left(1 + \frac{E_0}{m_0 c^2}\right) \\ E &= \frac{E_0(1 + E_0/2m_0c^2)}{1 + E_0/m_0c^2}, \end{aligned} \right\} \tag{4.47}$$

where m_0 and E_0 are the rest mass and incident electron energy, respectively. With these last two corrections the free space wave vector $\boldsymbol{\chi}$ becomes

$$\chi = \frac{\sqrt{2m_0E_0(1 + E_0/2m_0c^2)}}{h},$$

rather than as used before (eq. 4.2).

Also, for the moment it is assumed that the potential $V(\mathbf{r})$ is real, that is, $V(\mathbf{r}) = V^*(\mathbf{r})$ and $U_g = U^*_{-g}$. Furthermore, if the crystal is, for the moment,* taken to be centrosymmetric, then $V(\mathbf{r}) = V(-\mathbf{r})$, that is, $U_g = U_{-g}$, or

$$U_g = U_{-g} = U^*_g. \tag{4.48}$$

A solution exists in the form of a Bloch wave:

$$\psi(\mathbf{r}) = b(\mathbf{k}, \mathbf{r}) = \sum_g C_g(\mathbf{k}) \exp[2\pi i(\mathbf{k} + \mathbf{g}) \cdot \mathbf{r}] \tag{4.49}$$

that is, a many-beam Bloch wave solution.

Substituting for $\psi(\mathbf{r})$ from eq. 4.49 and for $V(\mathbf{r})$ from eq. 4.46 into the Schrödinger equation 4.45 and considering the coefficients of terms of the form $\exp[2\pi i(\mathbf{k} + \mathbf{g}) \cdot \mathbf{r})$ yields a set of N equations:

$$[K^2 - (\mathbf{k} + \mathbf{g})^2]\, C_g(\mathbf{k}) + \sum_h{}' U_h C_{g-h}(\mathbf{k}) = 0 \tag{4.50}$$

where, in the sum over h, the prime indicates that $h = 0$ is omitted. This term (equivalent to refraction by the mean inner potential V_0) is included in K by

$$K^2 = \frac{2mE}{h^2} + U_0. \tag{4.51}$$

*The removal of this restriction to include noncentrosymmetric crystals is discussed in Section 6.

5.1 Two-beam Approximation (Zero Absorption)

The two-beam approximation may now be made by selecting only the two equations **g** = 0, **g** from the set of eqs. 4.50, giving the pair of equations

$$(K^2 - k^2)\, C_0(\mathbf{k}) + U_{-g} C_g(\mathbf{k}) = 0, \tag{4.52}$$

$$U_g C_0(\mathbf{k}) + [K^2 - (\mathbf{k} + \mathbf{g})^2]\; C_g(\mathbf{k}) = 0. \tag{4.53}$$

This has a nontrivial solution for the C's if the determinant of the coefficients vanishes, that is, if

$$(K^2 - k^2)\,[K^2 - (\mathbf{k} + \mathbf{g})^2] = U_g U_{-g}.$$

Writing $\mathbf{K} = k_{xy}\hat{\mathbf{u}} + k_z\hat{\mathbf{w}}$, and so on (where $\hat{\mathbf{u}}$ and $\hat{\mathbf{w}}$ are unit vectors in the $x - y$ plane and z direction, respectively), remembering that $g = g_{xy} = g$, and considering the locus of pairs of points with the same x-y momentum (the dispersion surface), that is, $\mathbf{K}_{xy} = \mathbf{k}_{xy}$, one obtains from the last equation

$$(K_z^2 - k_z^2)\;[(K_z^2 - k_z^2) - g(2k_{xy} + g)] = U_g U_{-g}. \tag{4.54}$$

Now in the high energy limit applicable to electron microscopy the z-components are large compared both with x-y components and with differences, so that $2K_z$ may be extracted from each difference of squares. Also, for a centrosymmetric crystal, $U_g = U_{-g}$, reducing eq. 4.54 to

$$(K_z - k_z)\left[K_z - k_z - \frac{g(2k_{xy} + g)}{2K_z}\right] = \frac{U_g^2}{4K_z^2}. \tag{4.55}$$

Inspection of these equations allows certain properties of the dispersion surface to be deduced immediately.

1. For an incident beam exactly at the reflecting position (see Fig. 4.10) $K_{xy} = -g/2$, that is, the incident beam vector **K** originates on the Brillouin zone boundary (see, e.g., Kittel[11] for a fuller discussion of terminology and, incidentally, a derivation of the properties of conduction electrons in metals and semiconductors which parallels the present discussion). Equation 4.55 then reduces to

$$(K_z - k_z)^2 = \frac{U_g^2}{4K_z^2},$$

and, since $K_z = K\cos\theta_\mathrm{B}$, the two solutions for k_z are

$$k_z = K_z \pm \frac{U_g}{2K\cos\theta_\mathrm{B}} \tag{4.56}$$

that is, two solutions separated by Δk_z, where

$$\Delta k_z = \frac{U_g}{K \cos \theta_B}.$$

Remembering that the extinction distance is equal to the reciprocal of the separation of the dispersion surface branches (which is Δk_z here),

$$\xi_g = \frac{K \cos \theta_B}{U_g} \simeq \frac{K}{U_g}. \tag{4.57}$$

2. For U_g small compared with K^2 (the high energy approximation) the asymptotic solutions for incident beams far from the reflecting positions are two circles, of radius K, centered on O and **g** (Fig. 4.10).
3. Near the Brillouin zone boundary the two solutions for k_z describe hyperbolas, the separation between the branches varying as $\sqrt{1 + w^2}$. This may be seen by first relating the deviation parameter s_g to the values of k_{xy} and g: Applying the intersecting chord theorem to Fig. 4.10 (neglecting small differences between K, s, and their z-components, and noting that ON = $-k_{xy}$) gives

$$s = \frac{-g(2k_{xy} + g)}{2K_z}. \tag{4.58}$$

Substitution into eq. 4.55 yields the result

$$k_z^{(1)(2)} = K_z + \frac{s \pm \sqrt{s^2 + U_g^2/K_z^2}}{2},$$

or, from eqs. 4.57 and 4.25,

$$k_z^{(1)(2)} = K_z + \frac{s \pm \sqrt{w^2 + 1/\xi_g}}{2} \tag{4.59}$$

where again (1), (2) refer to the two branches (Eq. 4.10). The separation between the branches is then

$$\Delta k = k_z^{(1)} - k_z^{(2)} = \frac{\sqrt{1 + w^2}}{\xi_g},$$

consistently with the earlier definitions of extinction distance and thickness fringe spacing (eqs. 4.35 and 4.34).

Having sketched the dispersion surface and seen its identity with that developed earlier, it is now necessary to confirm that the coefficients C are identical in the two formulations. When eqs. 4.52, 4.58, and 4.57 are compared, it may be seen that

$$\frac{C_g^{\substack{(1)\\(2)}}}{C_0^{\substack{(1)\\(2)}}} = w \pm \sqrt{1 + w^2}.$$

Inspection of eqs. 4.28, 4.29 and 4.36 is then sufficient to show that the problem is the same as that discussed previously (absorption and many-beam effects will be reintroduced in the next sections).

5.2 Many-Beam Solution (Zero Absorption)

It is appropriate now to return to the point in the development of the solution of Schrödinger's equation before the two-beam approximation was made. Equation 4.50 may be developed to produce the n-beam analogue of eq. 4.55:

$$\left[(K_z - k_z) - \frac{g(2k_{xy} + g)}{2K_z}\right] C_g + \sum_h{}' \frac{U_h C_{g-h}}{2K_z} = 0. \tag{4.60}$$

Note that in this equation $K_z - k_z = -\gamma$ and $-g(2k_{xy} + g)/2K_z = s_g$ (eq. 4.58), that is,

$$(s_g - \gamma)\, C_g + \sum_h{}' \frac{U_h C_{g-h}}{2K \cos \theta_g} = 0,$$

with $s_g = 0$ for $g = 0$. This is a standard eigenvalue equation* of the form

$$\mathsf{A}\mathbf{C}^{(i)} = \gamma^{(i)}\mathbf{C}^{(i)}, \qquad i = 1, 2, \ldots, n, \tag{4.61}$$

where the eigenvector $\mathbf{C}^{(i)}$ is a column vector whose components are the wave amplitudes of the ith Bloch wave of eigenvalue (propagation vector) $\gamma^{(i)}$, and the matrix A has the form

$$A_{00} = 0, \qquad A_{gg} = s_g, \qquad A_{gh} = \frac{U_{g-h}}{2K \cos \theta_g} \tag{4.62}$$

or, in the case of the third factor,

$$A_{gh} = \tfrac{1}{2}\xi_{g-h} \tag{4.63}$$

*For a description of the mathematics of eigenequations see, for example, ref. 13.

by analogy with two-beam extinction distances and separation of the branches of the dispersion surface.

For a centrosymmetric crystal $U_{g-h} = U_{h-g}$, so $\tilde{\mathsf{A}} = \mathsf{A}$. Also, since U_{g-h} is real, A_{gh} is real; therefore $\mathsf{A}^* = \mathsf{A}$, that is, A is a real, symmetric matrix and therefore its eigenvalues $\gamma^{(i)}$ are real and eigenvectors $\mathbf{C}^{(i)}$ are also all real and orthogonal.

5.3 Inclusion of Absorption

Formally absorption may be incorporated by reintroducing the imaginary part of U_g. Thus eq. 4.60 becomes

$$[K^2 - (\mathrm{k} + \mathrm{g})^2 + iU_0'] \; C_g(\mathrm{k}) + \sum_h{}' \, (U_h + iU_h') \, C_{g-h}(\mathrm{k}) = 0.$$

Since the imaginary parts are small, an eigenequation of the form of eq. 4.61 follows immediately, eq. 4.62 being extended to

$$A_{00} = \frac{i}{2\xi_0'}, \qquad A_{gg} = s_g + \frac{i}{2\xi_0'}, \qquad A_{gh} = \frac{1}{2\xi_{g-h}} + \frac{i}{2\xi_{g-h}'}, \tag{4.64}$$

where the extinction and absorption lengths are all of the form

$$\xi = \frac{K \cos \theta}{U}.$$

The matrix A is obviously no longer real, although it is still symmetric for the centrosymmetric crystal. This means that the eigenvectors $\mathbf{C}^{(i)}$ must be complex, as also must be the eigenvalues $\gamma^{(i)}$. Although standard computer routines exist to find the eigenvectors and eigenvalues of the complex matrix A, conventionally, and sufficiently accurately for most purposes, it is convenient to do what has effectively been done before in the two-beam case (Sections 2.4 and 5.1), namely, assume that the C's are real and find the imaginary part of γ by introducing iU_g' as a perturbation, that is, split A into its real (A_{re}) and imaginary ($i\mathsf{A}_{\mathrm{im}}$) parts and solve

$$\mathsf{A}_{\mathrm{re}} \mathbf{C}^{(i)} = \gamma_{\mathrm{re}}^{(i)} \mathbf{C}^{(i)}, \qquad i = 1, 2, \ldots, n, \tag{4.65}$$

for $\gamma_{\mathrm{re}}^{(i)}, \mathbf{C}^{(i)}$, and

$$\gamma_{\mathrm{im}}^{(i)} = \tilde{\mathbf{C}}^{(i)} \mathsf{A}_{\mathrm{im}} \mathbf{C}^{(i)}, \qquad i = 1, 2, \ldots, n, \tag{4.66}$$

for $\gamma_{\text{im}}^{(i)}$, where the right-hand side of eq. 4.66 is immediately seen to be the standard $\langle\psi|\mathcal{H}'|\psi\rangle$ operation of first-order perturbation theory[14], where $\mathcal{H}'$ is the Hamiltonian.

The excitation of the Bloch waves may be written, concentrating on the C's, as

$$\boldsymbol{\phi} = \mathsf{C}\,\boldsymbol{\epsilon} \tag{4.67}$$

where C is a matrix with the ith *column* made up of the elements $\mathbf{C}_g^{(i)}$ and $\boldsymbol{\epsilon}$ is a column vector whose elements are the excitations of the Bloch waves on the various branches (defined by $\gamma_{\text{re}}^{(i)}$) of the dispersion surface taken in the same order. Now the Bloch waves are orthogonal (the C's being eigenfunctions of the real symmetric matrix A_{re}). This means that $\mathsf{C}^{-1} = \tilde{\mathsf{C}}$, and hence when eq. 4.67 is premultiplied by C^{-1} the excitations are given by

$$\boldsymbol{\epsilon} = \mathsf{C}^{-1}\boldsymbol{\phi} = \tilde{\mathsf{C}}\boldsymbol{\phi}.$$

At the top of the crystal $\boldsymbol{\phi} = (1, 0, 0, \ldots, 0)$, that is, the values of the $\epsilon^{(i)}$ are the first column of $\tilde{\mathsf{C}}$ and hence the first row of C—always $C_0^{(i)}$. Thus the general result (of which eq. 4.39 was a particular example) is

$$\epsilon^{(i)} = C_0^{(i)}. \tag{4.68}$$

5.4 Many-Beam Dispersion Surface

The n-beam dispersion surface, like the two-beam surface, is the locus of points of equal total energies drawn for the real part of the propagation vector, γ_{re}. Figure 4.13 shows a section along a systematic row of reflections, where the branches are numbered in descending order of γ ($\gamma = 0$ is between 1 and 2 and is measured positively upward), that is, in order of increasing kinetic energy. Physically the higher numbered branches have smaller values of γ (corresponding to lower momentum in the z-direction) because they have larger components from higher order reflections (which have larger x-y momentum). As implied by use of the term "surface," the general solution is three dimensional, but it is convenient (and for a crystal set so that the dominant feature in its diffraction pattern is a single, systematic row of reflections, realistic) to restrict the diagrams to a two-dimensional section. Another point which is illustrated in Fig. 4.13 is that the branches, being properties of Bloch waves, must be periodic in reciprocal space. Hence, after the γ values (both real—shown here—and imaginary) in the first Brillouin zone have been found, they are seen to be repeated in all other zones. What changes from zone to zone is the *excitation* of the waves; near the first Brillouin zone boundary (near the first-order Bragg reflecting position), $\mathbf{k}_{xy} = -\mathbf{g}/2$, branches 1 and 2 will be most strongly excited. Near $\mathbf{k}_{xy} = -\mathbf{g}$ the the most strongly excited branches will be 2 and 3; near $\mathbf{k}_{xy} = -3\mathbf{g}/2$, 3 and 4;

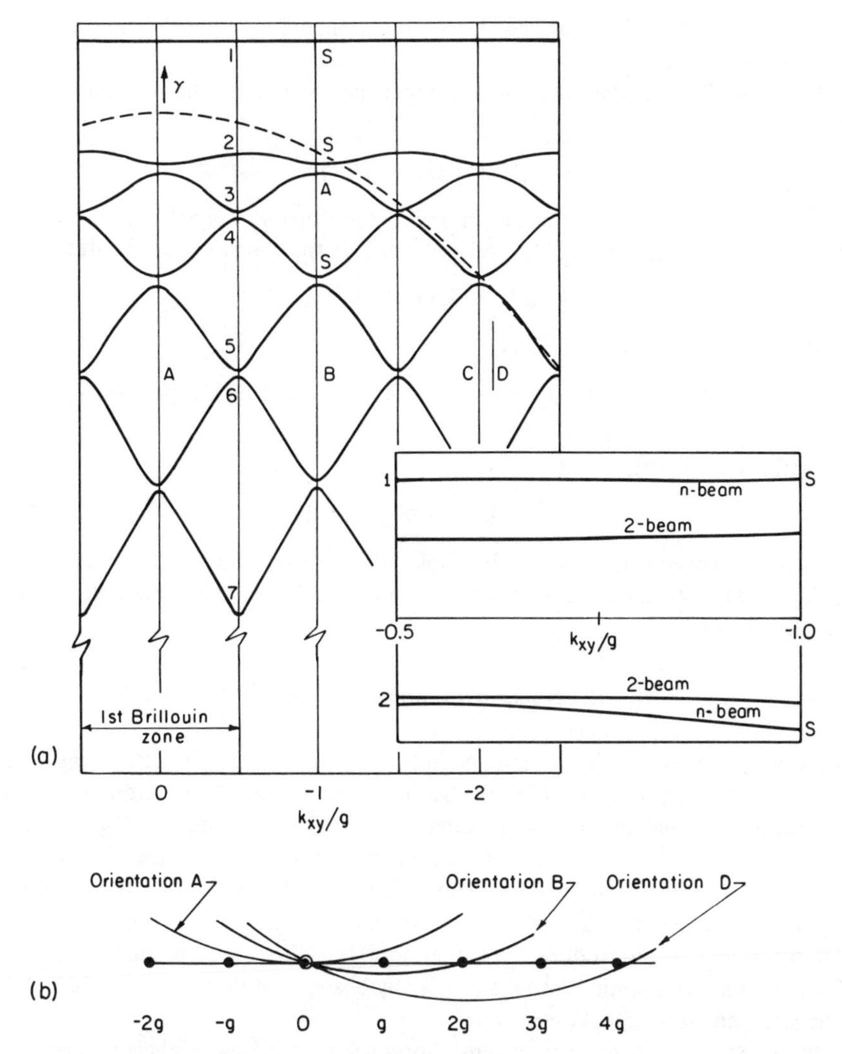

Fig. 4.13 (*a*) Many-beam dispersion surfaces for a systematic row of reflections with (inset) an indication of how the two-beam situation is usually modified by the influence of other reflections. (*b*) The Ewald sphere intersecting the systematic row at two particular orientations *A* and *B* in (*a*).

and so on, that is, the branches closest to the modified (for refraction, by U_0) free electron sphere.

Formally Bloch's theorem for the repetitive nature of the solution of eq. 4.61 may be written as

$$\gamma^{(i)}(\mathbf{k} + \mathbf{G}) = \gamma^{(i)}(\mathbf{k}),$$

where $\mathbf{G}$ is a reciprocal lattice vector and $\gamma^{(i)}$ is either $\gamma^{(i)}_{\mathrm{re}}$ or $\gamma^{(i)}_{\mathrm{im}}$. Similarly, the Bloch wave coefficients $C^{(i)}_g$ are related since, if the Bloch waves are identical,

$$b'(\mathbf{k} + \mathbf{G}, \mathbf{r}) = \mathbf{b}\,(\mathbf{k}, \mathbf{r}),$$

and from eq. 4.49 it then follows that

$$\sum_g C'_g\,(\mathbf{k} + \mathbf{G}) \exp\,[2\pi i\,(\mathbf{k} + \mathbf{G} + \mathbf{g}) \cdot \mathbf{r}] = \sum_g C_g(\mathbf{k}) \exp\,[2\pi i(\mathbf{k} + \mathbf{g}) \cdot \mathbf{r}] .$$

Equating coefficients yields

$$C'_{\mathbf{g}}\,(\mathbf{k} + \mathbf{G}) = C_{\mathbf{g}+\mathbf{G}}\,(\mathbf{k}),$$

that is, the coefficients are merely displaced in the series by $\mathbf{G}$. The excitations being maximized near the free electron sphere constitutes an example of this displacement.

6 Noncentrosymmetric Crystals

For simplicity it has been assumed until now that the diffracting crystal is centrosymmetric, that is, that in Section 2.4 the scattering from $0 \rightarrow \mathbf{g}$ is identical with that in the reverse direction, $\mathbf{g} \rightarrow 0$, or, in Section 5, $U_g = U_{-g}$. This is, of course, not true in general, but does cover a vast range of materials commonly encountered in engineering. For an understanding of the contrast features in materials that are not centrosymmetric, however, the complicating factors must be reintroduced. Furthermore the differences produced by this lack of symmetry appear only under many-beam conditions, making their discussion even more complicated.

In terms of the "scattered beams" approach of the Howie-Whelan equations (eq. 4.21, generalized to many beams as eq. 4.44), the lack of symmetry is admitted by noting that $1/\xi_{\Delta g} \neq 1/\xi_{-\Delta g}$. Both factors have the same magnitude but are complex quantities differing in phase. In terms of the eigenvalue equation approach of eq. 4.61 the matrix A becomes complex, even in the absence of absorption, but under this last restriction remains Hermitean. For "real" situations, however, absorption must of course be included, making the matrix non-Hermitean. The equations must then be solved by either integration of the more complicated Howie-Whelan equations or diagonalization of the complex

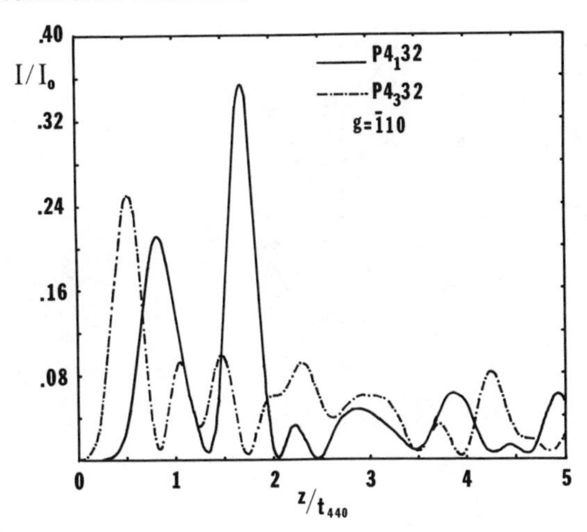

Fig. 4.14 Thickness fringe profiles calculated for the symmetric {332} orientation for lithium ferrite ($LiFe_5O_8$) for two noncentrosymmetric enantiomorphs $P4_132$ and $P4_332$. Both profiles are for dark field ($\mathbf{g} = \bar{1}10$), the thickness scale is in terms of the extinction distance for $\mathbf{g} = 440$, $\xi_{440} = 90.6$ nm, and electron energy is 650 kV. Absorption is included by setting $\xi_0/\xi'_0 = 0.07$ and $\xi_g/\xi'_g = 0.05$ for all $\mathbf{g}$. The bright field profiles are identical for the two structures. Courtesy *Acta Crystallographica.*[15]

matrix, both of which, in the many-beam situation, are undertaken only with the aid of a computer. An example of a set of calculations[15] integrating the differential equations for two enantiomorphic, noncentrosymmetric forms of a lithium ferrite is shown in Fig. 4.14, which may be compared with the micrograph of Fig. 4.15. The differences between areas 1 and 2 in the dark field micrograph arise because the two enantiomorphs produce different dark field intensities even though they have the same thickness (see Fig. 4.14). The difference between them is produced solely by the fact that in one set of domains the $+\mathbf{g}$ reflection is operating, whereas in the other set $-\mathbf{g}$ operates, and these reflections are nonequivalent.

The example cited here bears out some of the general principles of contrast for noncentrosymmetric crystals, which may be summarized as follows:

1. No differences may be detected under two-beam conditions.
2. Similarly, no differences appear under many-beam systematic conditions.
3. Differences appear only under many-beam nonsystematic conditions and in certain dark field images.
4. The bright field image is invariant under all conditions.

These conditions are discussed in detail in ref. 16.

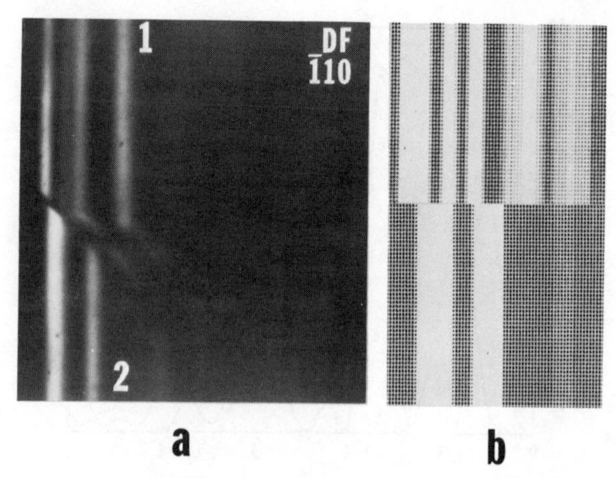

Fig. 4.15 (*a*) Dark field ($\mathbf{g} = \bar{1}\bar{1}0$) micrograph of a wedge-shaped crystal of lithium ferrite. The domain labeled 1 has the $P4_332$ structure; that labeled 2, has the $P4_132$ structure. (*b*) Simulated micrograph of (*a*). The boundary has not been simulated. $\xi_0/\xi_0' = 0.07$, $\xi_g/\xi_g' = 0.05$ for all $\mathbf{g}$. Symmetric [332] orientation. The corresponding thickness-fringe profiles are shown in Fig. 4.14. Courtesy *Acta Crystallographica.*[15]

7 Other Formulations

The wave mechanical treatments of electron diffraction are based on the early work of Bethe[21] which is itself based on Ewald's dynamical theory for X-rays developed ten years earlier. Many beam effects were described in 1962 by Sturkey[22] using the scattering matrix formulation. Nowadays the Cowley-Moodie multislice method (see e.g., ref. 17) is achieving prominence. This method follows the line explored here only to the phase grating level (Section 2.2). A full discussion of the limitations imposed by some of the approximations has not been made, such as the column approximation (Section 2.5), the relaxation of which has been fully covered by Howie and Basinski.[9] However, the limited number of comparisons made here, and the many more contained in the next chapter, show that the theory outlined in this chapter makes possible the extraction of a great deal of accurate information from electron micrographs of crystalline materials.

Exercises

4.1 A thin crystal of thickness t (in the z-direction, t small) has a potential of the form

$$V(\mathbf{r}) = V_0 + V_1 \cos 2\pi\mathbf{g} \cdot \mathbf{r}$$

where g lies in the x–y plane. Describe the beams emitted from the lower surface when a high energy electron beam traveling in the z-direction is incident perpendicularly on the upper surface.

4.2 What are the effects of higher voltage operation on the range of validity of the answers to Exercises 3.1 and 4.1?

4.3 A crystal of thickness t has a phase grating potential given by

$$V_0 + V_1 \cos(2\pi g_x x) + V_1 \cos(2\pi g_y y) + V_2 \cos[2\pi(g_x x + g_y y)],$$

where $V_1 >> V_2$. Calculate the intensity (on the phase grating approximation) of the two diffracted beams having $\mathbf{g} = (g_x, 0, 0)$ and $\mathbf{g} = (g_x, g_y, 0)$, respectively.

4.4 A hypothetical absorption process gives rise to a uniform probability of absorption in a cube of side αd centered on each atom, where d is the spacing of the Bragg planes and $0 < \alpha < 1$. What is the ratio of the absorption distance of the well-transmitted Bloch wave to that of the strongly absorbed wave at the exact reflecting orientation (two-beam)? What happens to the ratio as α tends to (*a*) zero and (*b*) unity?

4.5 A high energy electron beam is incident on a fcc crystal near $[1\bar{1}1]$, such that both 220 and $20\bar{2}$ reflections are exactly satisfied. Derive expressions for the Bloch waves set up in the crystal, including the electron distributions around the atoms. Compare the results with those of the two-beam case (i.e., only one 220 reflection satisfied).

4.6 A crystal of silicon is prepared in the form of a wedge with its upper surface parallel to (100) and its lower surface parallel to (210). Electrons of energy 100 keV enter through the upper surface and travel in the (001) plane perpendicular to the edge of the wedge, falling on the (022) planes at the exact Bragg angle. If the value of ξ_g for this reflection is 76 nm, calculate the angular splitting of the reflected beam due to refraction as it leaves through the lower surface. Comment on the relationship between the splitting and the thickness fringes observed in the dark field image.

4.7 Discuss the relationship between dispersion surfaces as used in electron diffraction theory and the energy-momentum diagrams utilized to describe the behavior of conduction electrons in metals.

References

1. Hirsch, P. B., Howie, A., Nicholson, R. B., Pashley, D. W., and Whelan, M. J., *Electron Microscopy of Thin Crystals*, Butterworths, London, 1965.
2. Pendry, J. B., *Low Energy Electron Diffraction* (LEED): *The Theory and Its Application to Determination of Surface Structure*, Academic, London and New York, 1974.
3. Thornton, P. R., *Scanning Electron Microscopy*, Chapman and Hall, London, 1968.

4. Howie, A., in *Electron Microscopy in Material Science* (Ed. U. Valdrè), Academic, New York, 1971.
5. Hall, C. R. and Hirsch, P. B., *Proc. R. Soc.*, **A286,** 158 (1965).
6. Whelan, M. J., *J. Appl. Phys.*, **36,** 2099, 2103 (1965).
7. Cundy, S. L., Metherell, A. J. F., and Whelan, J., *Phil Mag.*, **15,** 623 (1967).
8. Takagi, S., *Acta Crystallogr.*, **15,** 1311 (1962).
9. Howie, A. and Basinski, Z. S., *Phil. Mag.*, **17,** 1039 (1968).
10. Humphreys, C. J. and Fisher, R. M., *Acta Crystallogr.*, **A27,** 42 (1971).
11. Kittel, C., *Introduction to Solid State Physics*, 5th ed., John Wiley, New York, 1975.
12. Whelan, M. J. and Hirsch, P. B., *Phil. Mag.*, **2,** 1121, 1302 (1957).
13. Matthews, P. T., *Introduction to Quantum Mechanics*, McGraw-Hill, New York, 1968.
14. Dicke, R. H. and Wittke, J. P., *Introduction to Quantum Mechanics*, Addison-Wesley, Reading, MA, 1960, Chapter 14.
15. Van der Biest, O. and Thomas, G., *Acta Crystallogr.*, **A33,** 618 (1977).
16. Serneels, R., Snykers, M., Delavignette, P., Gevers, R., and Amelinckx, S., *Phys. Status Solidi*, **B58,** 277 (1973).
17. Cowley, J. M., *Prog. Mater. Sci.*, **13,** 269 (1967). See also J. M. Cowley *Diffraction Physics* N. Holland/Elsevier Amsterdam, 1975.
18. Doyle, A. P. and Turner, P. S., *Acta Crystallogr.*, **A24,** 390 (1968).
19. Humphreys, C. J. and Hirsch, P. B., *Phil. Mag.*, **18,** 115 (1968).
20. Radi, G., *Acta Crystallogr.*, **A26,** 41 (1970).
21. Bethe, H. A., *Ann. d. Physik,* **87,** 55 (1928).
22. Sturkey, L., *Proc. Phys. Soc.*, **80,** 321 (1962).

FIVE

APPLICATION OF DYNAMICAL THEORY TO THE CALCULATION OF IMAGE CONTRAST

1 Introduction

In Chapter 4 the dynamical theory of electron diffraction was developed in a formal way to produce the governing differential equations applicable to both perfect and imperfect crystals. These equations were solved analytically in the two-beam approximation for a perfect crystal. The purpose of the present chapter is to discuss how the equations may be handled in more general cases and how best to compare the resultant calculated contrast effects with observed micrographs.

Initially in this chapter the two-beam approximation is again assumed, although, as indicated in Chapter 4, this is a convenience rather than a necessity. In later sections this approximation is relaxed, allowing more accurate prediction of contrast effects than is possible with two-beam theory. Also, a number of effects that are essentially many-beam are then discussed.

2 Scattering Matrices

The first of the simplifying techniques to be discussed involves principally notation; as will be seen below, the solutions of the differential equations may be more conveniently handled in many cases by using a matrix notation, applicable to both the "beam" and "Bloch wave" approaches.

2.1 Scattering of Beams: Perfect Crystal

Equations 4.41 and 4.42 relating Bloch waves and beams in the two-beam approximation may be written in matrix form as

$$\begin{pmatrix} \phi_0(z) \\ \phi_g(z) \end{pmatrix} = \begin{pmatrix} C_0^{(1)} C_0^{(2)} \\ C_g^{(1)} C_g^{(2)} \end{pmatrix} \begin{pmatrix} \exp(2\pi i\gamma^{(1)}z) & 0 \\ 0 & \exp(2\pi i\gamma^{(2)}z) \end{pmatrix} \begin{pmatrix} \epsilon^{(1)} \\ \epsilon^{(2)} \end{pmatrix}. \quad (5.1)$$

The wave matching that defines the C's and ϵ's (eqs. 4.30) may be written as

$$\begin{pmatrix} C_0^{(1)} & C_0^{(2)} \\ C_g^{(1)} & C_g^{(2)} \end{pmatrix} \begin{pmatrix} \epsilon^{(1)} \\ \epsilon^{(2)} \end{pmatrix} = \begin{pmatrix} \phi_0(0) \\ \phi_g(0) \end{pmatrix}. \quad (5.2)$$

Denoting a matrix as C, premultiplying eq. 5.2 by C^{-1}, and substituting for ϵ in eq. 5.1 yields the complex matrix equation

$$\boldsymbol{\phi}(z) = \mathsf{C}\,\{\exp(2\pi i\gamma z)\}\,\mathsf{C}^{-1}\boldsymbol{\phi}(0), \quad (5.3)$$

where the brackets, { }, indicate a diagonal matrix. The matrix

$$\mathsf{P} = \mathsf{C}\,\{\exp(2\pi i\gamma z)\}\,\mathsf{C}^{-1}, \quad (5.4)$$

which relates the incident amplitudes $\boldsymbol{\phi}(0)$ and the resultant amplitudes at depth z, $\boldsymbol{\phi}(z)$, is termed the *scattering matrix*, and in this case is 2×2 complex. Note (from eqs. 4.39) the simplifying result that $\mathsf{C}^{-1} = \widetilde{\mathsf{C}}$, where $\widetilde{\mathsf{C}}$ is the transpose of C.

As a trivial example it may be confirmed (after some algebra) that, if a single slab of perfect crystal is split into two slabs by an imaginary fault plane (which does not affect the beams in any way; see Fig. 5.1), then

$$\boldsymbol{\phi}(t) = \boldsymbol{\phi}(t_1 + t_2) = \mathsf{P}(t_2)\,\mathsf{P}(t_1)\boldsymbol{\phi}(0), \quad (5.5)$$

that is, thicker segments of perfect crystal may be discussed by multiplying together the scattering matrices for thinner crystal (see also ref. 22, Chap. 4).

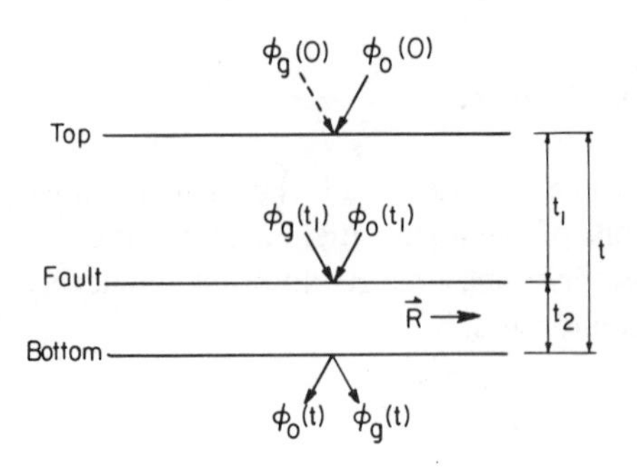

Fig. 5.1 Schematic diagram showing the parameters used to describe a stacking fault parallel to the specimen surface. For an inclined fault t_1 and t_2 vary so that $t_1 + t_2 = t$.

2.2 Scattering of Beams: Planar Faults

The simplest type of imperfect crystal is one containing a planar displacement fault (e.g., a stacking fault) lying perpendicular to the beams (Fig. 5.1), that is, a plane separating two slabs of perfect crystal with a relative displacement **R**. The beams just above the fault must be related to the incoming beams by a perfect crystal scattering matrix, that is,

$$\boldsymbol{\phi}(t_1) = \mathsf{P}(t_1)\boldsymbol{\phi}(0).$$

These beams are incident on another slab of perfect crystal, displaced by **R**. In the lower crystal the Bloch waves must be described by modified forms of eqs. 4.37 and 4.40;

$$b^{(j)}(\mathbf{k}^{(j)} \cdot \mathbf{r}) = \sum_g C_g^{(j)} \exp\,[2\pi i(\mathbf{k}^{(j)} + \mathbf{g}) \cdot (\mathbf{r} - \mathbf{R})]$$

$$= \exp\,(-2\pi i\mathbf{k}^{(j)} \cdot \mathbf{R}) \sum_g C_g^{(j)}$$

$$\cdot \exp\,(-2\pi i\mathbf{g} \cdot \mathbf{R}) \exp\,[2\pi i(\mathbf{k}^{(j)} + \mathbf{g}) \circ \mathbf{r}]. \qquad (5.6)$$

Performing the wave matching again (eqs. 4.41, 4.42, and 4.36) for the displaced crystal gives new matrices C' which are related to the previous matrices C by changes in the "diffracted" components C_g only, of the form

$$C_g' = C_g \exp\,(-i\alpha),$$

where $\alpha = 2\pi\mathbf{g} \cdot \mathbf{R}$.

Thus the solution for the faulted crystal becomes

$$\boldsymbol{\phi}(t) = \mathsf{P}'(t_2)\,\mathsf{P}(t_1)\boldsymbol{\phi}(0). \qquad (5.7)$$

It is more instructive, although apparently more laborious, to rewrite eq. 5.7 in terms of unmodified perfect crystal matrices P and fault matrices F. With F^+ defined as

$$\mathsf{F}^+ = \begin{pmatrix} 1 & 0 \\ 0 & \exp\,(i\alpha) \end{pmatrix} \qquad (5.8)$$

the effect of the two slabs of crystal and the interfacial fault is given by

$$\boldsymbol{\phi}(t) = \mathsf{F}^-\mathsf{P}(t_2)\,\mathsf{F}^+\mathsf{P}(t_1)\boldsymbol{\phi}(0), \qquad (5.9)$$

where F^- is similar to F^+ except that α is replaced by $-\alpha$. Equation 5.9 mirrors what is happening physically: the abrupt change at the interface scatters the beams into each other–matrix F^+–The slabs of perfect crystal behaving identi-

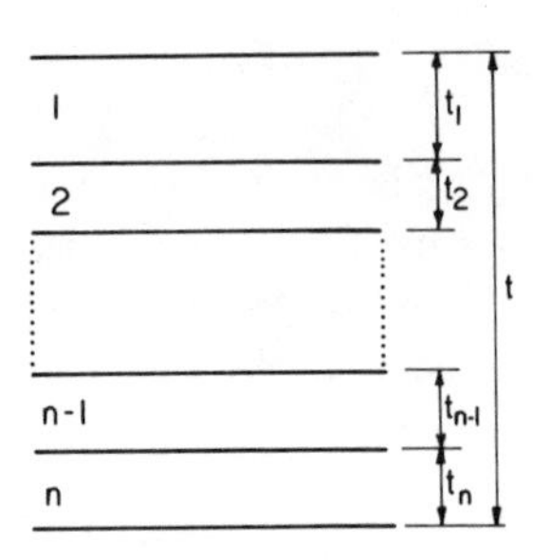

Fig. 5.2 Schematic diagram showing a number of faults parallel to the surface.

cally (except for differences in thickness)–matrices P. The relative phases of the emergent beams are modified, however, by the displacement–matrix F^-.

For a number of slabs of crystal (see Fig. 5.2) the result is

$$\boldsymbol{\phi}(t) = F_n^- P_n F_n^+ F_{n-1}^- P_{n-1} \cdots F_3^+ F_2^- P_2 F_2^+ P_1 \boldsymbol{\phi}(0),$$

where F_j is defined as in eq. 5.8, α_j being the phase change as measured with respect to the first slab. It is convenient to rewrite the product matrices $F_j^+ F_{j-1}^-$ as

$$F_j^+ F_{j-1}^- = F_{jk}, \qquad k = j - 1, \tag{5.10}$$

where F_{jk} is of the same form as eq. 5.8, α_{jk} being the phase *difference* between slabs j and $j - 1$. Thus (since $F_2^+ \equiv F_{21}$)

$$\boldsymbol{\phi}(t) = F_n^- P_n F_{n,n-1} \cdots F_{32} P_2 F_{21} P_1 \boldsymbol{\phi}(0); \tag{5.11}$$

and if the phase relationship between ϕ_0 and ϕ_g is unimportant (e.g., if only intensities are required), the matrix F_n^- may be omitted. Neglecting F_n^- in eq. 5.11 it may be seen that the effect of each fault is to introduce a phase change in the diffracted beam only. It should be noted that this equation applies to two situations of considerable interest: (*a*) stacking faults separating slabs of perfect crystal, all with the same orientation, the so-called α-faults, and (*b*) coherent

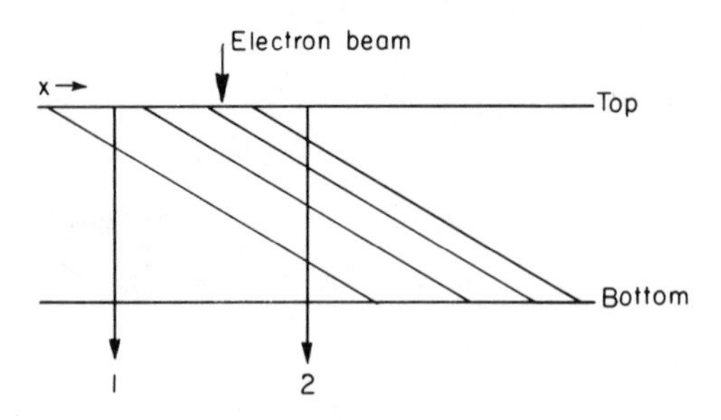

Fig. 5.3 Schematic representation of a number of inclined faults.

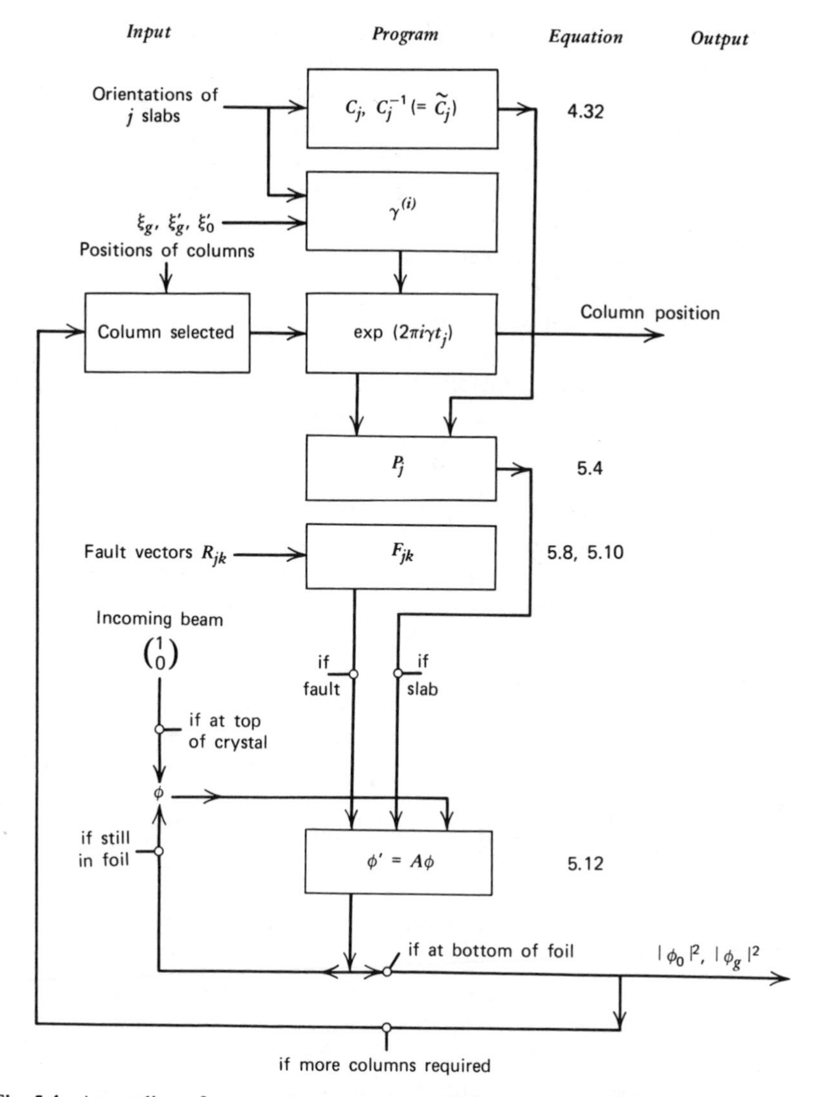

Fig. 5.4 An outline of a computer program to calculate two-beam intensities from columns passing through a number of slabs of perfect crystal separated by faults.

twin boundaries, where there is no phase change at the "fault" plane (i.e., F_{jk} = I, all j, where I is the identity matrix), but the slabs of perfect crystal are differently oriented (i.e., the values of C_j comprising P_j are different–eq. 5.4), the so-called δ-faults. Of course, any combination of (*a*) and (*b*) may also arise.

The situation that is usually encountered experimentally and hence the one for which calculated image intensities are required involves faults that are inclined in the specimen (Fig. 5.3), when the components of the matrices P_j depend on the position x of the column considered, as does the number of faults encountered. [Strictly, inclined faults modify both the form and the excitations of the Bloch waves because of the requirement to match by means of the lines perpendicular to the boundary. However, since Δk_z is small ($\sim 5 \times 10^7$ m^{-1}), the resulting changes in k_{xy} and hence in the effective value of ω, are negligible.] In general, in electronic computation a subroutine is set up to calculate

$$\boldsymbol{\phi}' = \mathsf{A}\boldsymbol{\phi}, \tag{5.12}$$

$\boldsymbol{\phi}$ and $\boldsymbol{\phi}'$ being 2×1 (complex) vectors describing the beams before and after the operation, respectively, and the 2×2 (complex) matrix A being alternately a perfect crystal scattering matrix P and a fault matrix F. The subroutine is used until the bottom of the crystal is reached, when $\boldsymbol{\phi}'$ is the required solution for the amplitudes of the beams. An outline diagram of the computational method is shown in Figure 5.4.

Of course in simple cases it is sometimes preferable to calculate expected images from algebraic expressions for ϕ_0 and ϕ_g (such as those obtained by multiplying out eqs 5.11), for example, in the case of a single stacking fault or coherent twin boundary or even a pair of such boundaries. However, for two or more boundaries a mechanical method is usually preferred. Discussion of the results of calculations is deferred until Section 3.

2.3 Scattering of Beams: General Strain Fields

The scattering matrices P (eq. 5.4) are, of course, descriptions of the integration of the basic differential equations (eqs 4.21) through a slab of crystal of thickness z and orientation defined by s_g, but zero strain field, $\beta_g' = 0$. However, since β_g' and s appear together as a sum, similar scattering matrices $\mathsf{P}(s + \beta_g', \Delta z)$ may be used to describe the changes in the amplitudes and phases of the beams on traversing a small thickness Δz of a distorted crystal (Δz small enough that β_g' does not vary significantly), that is,

$$\boldsymbol{\phi}(z + \Delta z) = \mathsf{P}(s + \beta_g', \Delta z)\boldsymbol{\phi}(z),$$

where β_g' is obviously in general a function of z, and P is defined by a suitably modified form of eq. 5.4. The integration of the differential equations may then be achieved by successive matrix multiplications. This is the basis of the technique which Thölén has used to calculate images and for which an improvement in speed of a factor of 10X is claimed[1] over the direct integration method.

2.4 Scattering of Bloch Waves: Perfect Crystal and Planar Faults

The principal advantage of the Bloch wave notation over the beam notation in terms of calculation is illustrated by the fact that Bloch waves, once set up at the top of the crystal, propagate unchanged through perfect crystal (apart from decaying according to the well-known exponential law through the imaginary component q of the propagation vector), while beams are continually scattering into each other. In the case of Bloch waves, then, the term "scattering" is not relevant to perfect crystals.

In the case of planar faults the problem is how to match the Bloch waves in the upper perfect crystal (see Fig. 5.5) to similar Bloch waves in the lower, displaced crystal. Obviously the total disturbances on the two sides of the boundary must match, that is,

$$\Psi_u = \Psi_l,$$

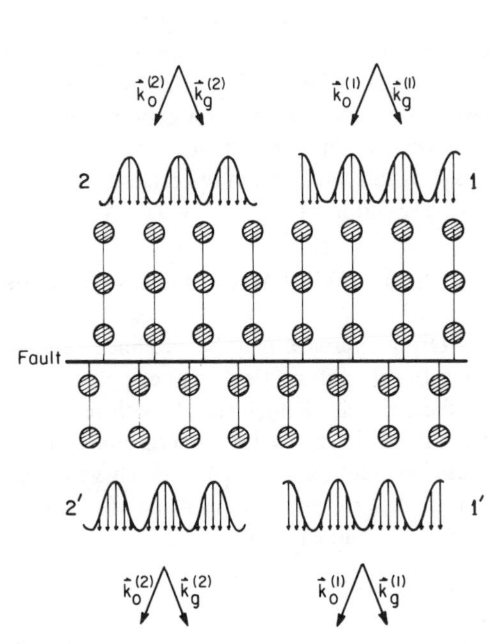

Fig. 5.5 The Bloch waves in the upper crystal must be "scattered" at the fault to match those to be set up in the lower crystal.

where Ψ_u in the upper crystal is given by eq. 4.37 and Ψ_l by a similar equation derived from eq. 5.6:

$$\sum_j \epsilon_1^{(j)} \sum_g C_g^{(j)} \exp\,[2\pi i(\mathbf{k}+\mathbf{g})\cdot\mathbf{r}]\,\exp\,(2\pi i\gamma^{(j)}t_1)$$

$$= \sum_j \epsilon_2^{(j)} \sum_g C_g^{(j)} \exp\,(-i\alpha_{12})\exp\,[2\pi i(\mathbf{k}+\mathbf{g})\cdot\mathbf{r}]\,\exp\,(2\pi i\gamma^{(j)}t_1), \quad (5.13)$$

where eq. 4.36 has been used in its form corrected for refraction, namely,

$$k^{(j)} = k + \gamma^{(j)}.$$

Rewriting eq. 5.13 in matrix form:

$$\mathsf{C}\,\{\exp\,(2\pi i\gamma t_1)\}\;\boldsymbol{\epsilon}_1 = \mathsf{C}'\,\{\exp\,(2\pi i\gamma t_1)\}\;\boldsymbol{\epsilon}_2$$

and premultiplying by $\{\exp\,(-2\pi i\gamma t_1)\}\,(\mathsf{C}')^{-1}$ yields

$$\boldsymbol{\epsilon}_2 = \{\exp\,(-2\pi i\gamma t_1)\}\,\mathsf{C}^{-1}\,\mathsf{F}_{12}\mathsf{C}\,\{\exp(2\pi i\gamma t_1)\}\,\boldsymbol{\epsilon}_1 \quad (5.14)$$

as the equation governing Bloch wave scattering by a planar fault, at a depth t_1, in the crystal. Computationally eq. 5.14 is more complicated than eq. 5.9, since the emergent beams must be reformed from the Bloch waves at the bottom of the crystal. However, the Bloch wave formulation may be preferable in the situation of continuously varying strain fields, which is discussed in the next section. It also makes possible a discussion of certain features of fault contrast in a qualitative fashion (see Section 7).

2.5 Scattering of Bloch Waves: General Strain Fields

The Bloch wave scattering approach may be extended to continuously varying strain fields by noting that the strain field can be approximated by an assembly of thin slices of perfect crystal of thickness dz with relative displacements between them (an alternative but equivalent description to the thin slices of different orientations used in Section 2.3). If the Bloch wave excitations at depth z at the top of the thin slice in question are $\boldsymbol{\epsilon}$ and after the fault at the bottom of the slice are $\boldsymbol{\epsilon} + d\boldsymbol{\epsilon}$, then from eq. 5.14

$$\boldsymbol{\epsilon} + d\boldsymbol{\epsilon} = \{\exp\,(-2\pi i\gamma z)\}\,\mathsf{C}^{-1}\,\mathsf{F}_{\text{slice}}\,\mathsf{C}\,\{\exp\,(2\pi i\gamma z)\}\,\boldsymbol{\epsilon}\,. \quad (5.15)$$

But the components α_g in $\mathsf{F}_{\text{slice}}$ (see eqs. 5.8 and 4.22) are small, being of the form

$$\alpha_g = 2\pi\,\mathbf{g}\cdot\frac{d\mathbf{R}}{dz}\,dz = 2\pi\,\beta_g'\,dz,$$

$d\mathbf{R}$ being the displacement of the bottom of the slice relative to its top. Thus $\mathsf{F}_{\text{slice}}$ may be expanded as

$$\mathsf{F}_{\text{slice}} = \mathsf{I} + 2\pi i \{\beta_g'\}, \tag{5.16}$$

where I is the identity matrix. Substituting from eq. 5.16 into 5.15 then yields

$$\frac{d\boldsymbol{\epsilon}}{dz} = 2\pi i \{\exp(-2\pi i\gamma z)\} \mathsf{C}^{-1} \{\beta_g'\} \mathsf{C} \{\exp(2\pi i\gamma z)\}\boldsymbol{\epsilon} \tag{5.17}$$

as the equation governing the scattering of Bloch waves by continuously varying strain fields. Equation 5.17 restates the earlier proposition that, apart from absorption, which is present here as the imaginary part of the propagation vectors γ, the Bloch waves pass through the crystal unchanged *unless the crystal is distorted.* In terms of computation this means that integration is trivial (nothing changes) unless the crystal is distorted. This is in contrast to integrations of the $\boldsymbol{\phi}$ formulation of eqs. 4.21, where the beams scatter into each other throughout the crystal, and hence time-consuming numerical integration is required even in regions that are essentially perfect.

On the other hand, the Bloch wave components and excitations must be calculated at the top of the crystal from the various matching equation and definitions used previously, that is, in the two-beam case by eqs. 4.39. The exit beams are reconstituted from the Bloch waves at the bottom of the crystal by an equation analogous to eq. 5.1 for perfect crystal:

$$\boldsymbol{\phi}(z) = \{Q(z)\}^{-} \mathsf{C} \{\exp(2\pi i\gamma z)\} \boldsymbol{\epsilon}(z), \tag{5.18}$$

where the factor $\{Q(z)\}^{-}$ is a diagonal matrix, defined by the displacement $\mathbf{R}(z)$ of the bottom of the crystal, of the same form as F_n^-, derived for the faulted crystal in eq. 5.11. However, for a number of two-beam calculations and certainly for many-beam calculations the time saved in the integration itself outweighs the extra steps necessary at the top and bottom. An outline of a typical computer program (for the many-beam case) for such an integration is shown in Fig. 5.24. Further consideration is given in Section 3.2.2 to yet other methods by which computations may be made more quickly.

3 Solutions for Typical Defects (Two-Beam)

In this section a number of defects are considered, some of which have been touched on in preceding chapters. Part of the discussion deals with deciding which formulation of the theory is best suited to the class of defect in question and whether particular methods of calculation or presentation of the results of calculation (be they accurate, numerical, or approximate) are more suitable than others. Here "suitability" is measured ultimately by whether unambiguous identification of the defect causing the contrast is possible.

3.1 Planar Defects: α- and δ-Fringes

Planar defects have already been discussed in terms of both scattering matrices for beams and the scattering of Bloch waves. With regard to ease of calculation the two techniques are comparable, entailing (obviously, since they are so closely related) similar numbers of operations. The information obtained from the micrographs is deduced initially from the symmetry properties of the image, which are determined by the nature of the fault, as discussed in Chapter 3, Section 8. The most common faults are stacking faults with $\alpha = \pm 2\pi/3$ (e.g., in fcc and hcp structures; see Tables 3.2 to 3.4), but faults with $\alpha = \pm\pi/4, \pm\pi/3, \pm\pi/2$, and π may also occur in certain ordered structures (see Chapter 3, Section 8.3). Abrupt orientation changes, without immediate phase changes, can occur at coherent twin boundaries and lead to pure δ-fringes. Fringes resulting from a combination of the two effects are obviously also possible; for example, at certain antiphase boundaries ($\alpha = \pi$) there are also changes in the deviation parameter s.

The geometry of the situation usually computed is that of the inclined fault running through the specimen foil from top to bottom (see, e.g., Figs. 3.13*a* and 5.3). Some results of calculations have already been presented, but for completeness all cases will be considered again here, reference being made where appropriate to the examples quoted previously.

3.1.1 *Stacking Faults:* $\alpha = \pm 2\pi/3$. The profiles of Figs. 3.13 and 5.6*c* and *d* and the experimental micrographs of Figs. 3.15 and 5.7 exhibit the following features characteristic of pictures taken near the reflecting position, in fairly thick absorbing crystals:

1. The bright field image is symmetrical with respect to the center of the specimen.
2. The dark field image is approximately antisymmetrical.
3. The bright and dark field images are similar when the fault is near the top surface and approximately complementary when the fault is near the bottom.
4. The sign of α (i.e., $\pm 2\pi/3$) may be determined from the sign of the external fringes; α is positive when the "alike" fringes in bright and dark field are bright, and negative if the reverse is true.
5. The fringes are spaced as would be thickness fringes in perfect crystal tapered at the thinner section only (i.e., more closely spaced as $|w|$ increases from its value of zero at the reflecting position), with complicated or blank regions near the middle.
6. The visibility of the fringes in the dark field image is symmetrical only at the exact reflecting position; for negative values of the deviation parameter the

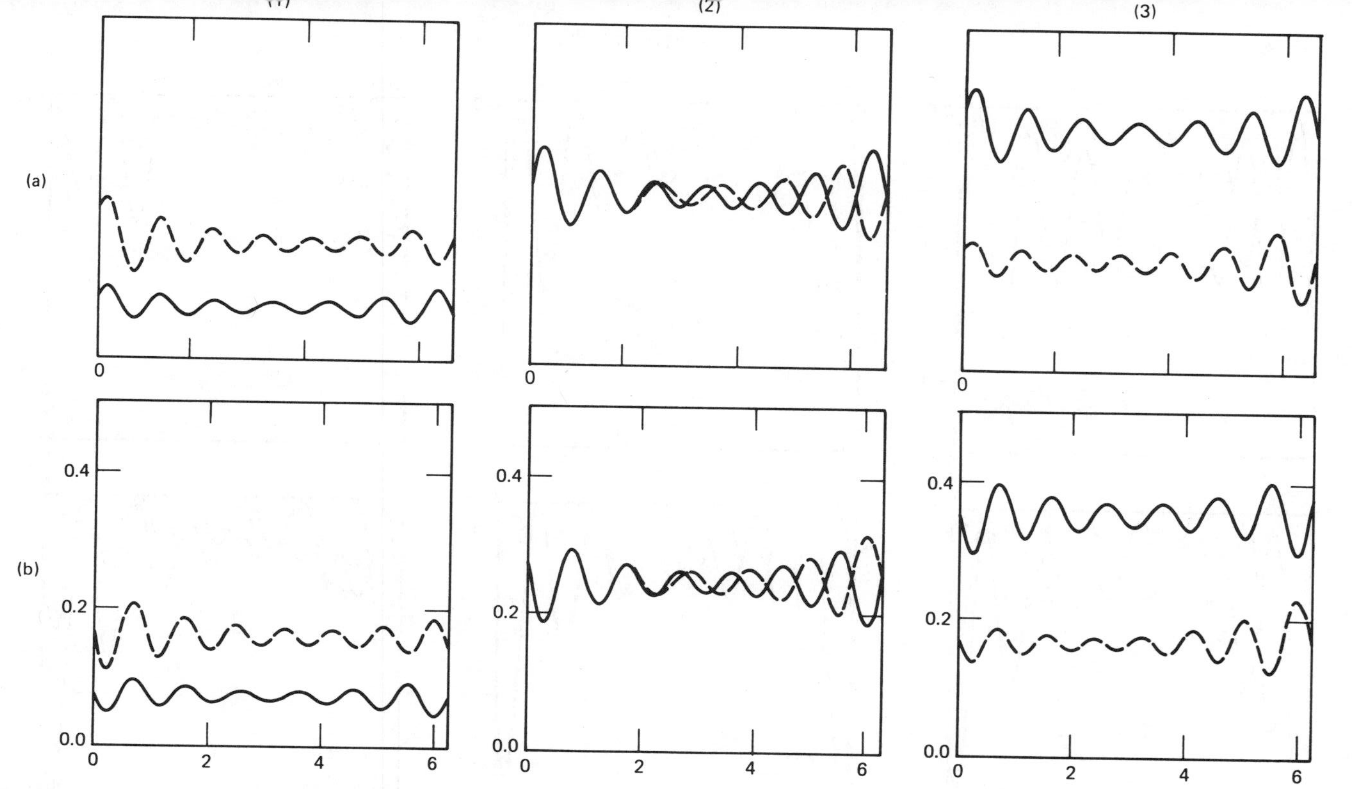

Fig. 5.6 Calculated fringe profiles for single inclined stacking faults imaged under various conditions for a typical medium atomic weight element ($\xi_g/\xi'_g = \xi_g/\xi'_0 = 0.1$) of reasonable thickness ($6.25\ \xi_g$). In all cases the fault is near the top (electron beam entrance surface) at the left, and the three sets of reflecting conditions are (1) $w = -0.4$, (2) $w = 0$, and (3) $w = +0.4$. The fault phase angles, $\alpha = 2\pi \mathbf{g} \cdot \mathbf{R}$ are (*a*) $\alpha = +0.1\pi$, (*b*) $\alpha = -0.1\pi$, (*c*) $\alpha = +2\pi/3$, (*d*) $\alpha = -2\pi/3$, and (*e*) $\alpha = \pi$. The profiles show both bright field (solid) and dark field (dashed) intensities as a function of the depth of the fault.

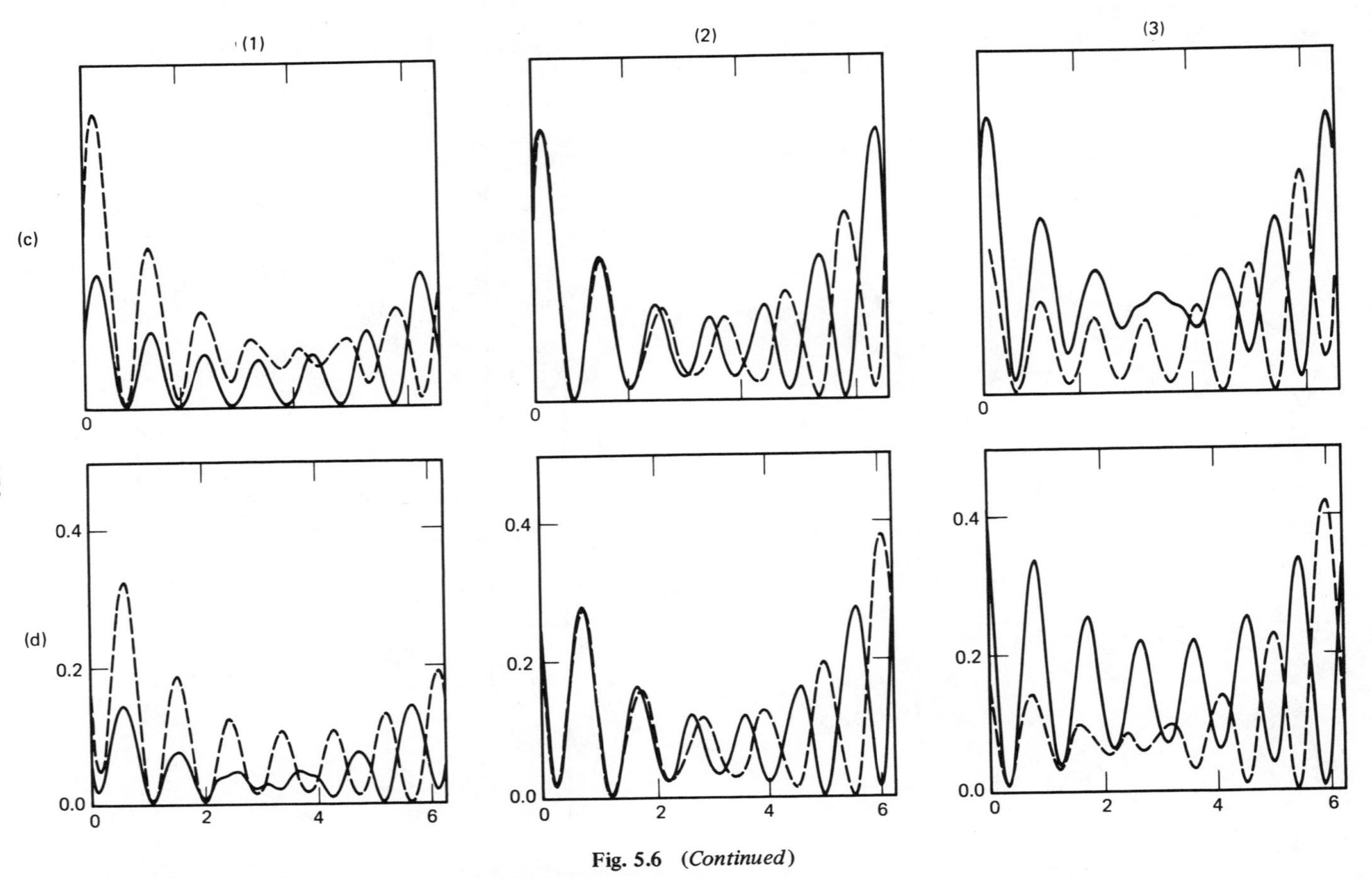

Fig. 5.6 (Continued)

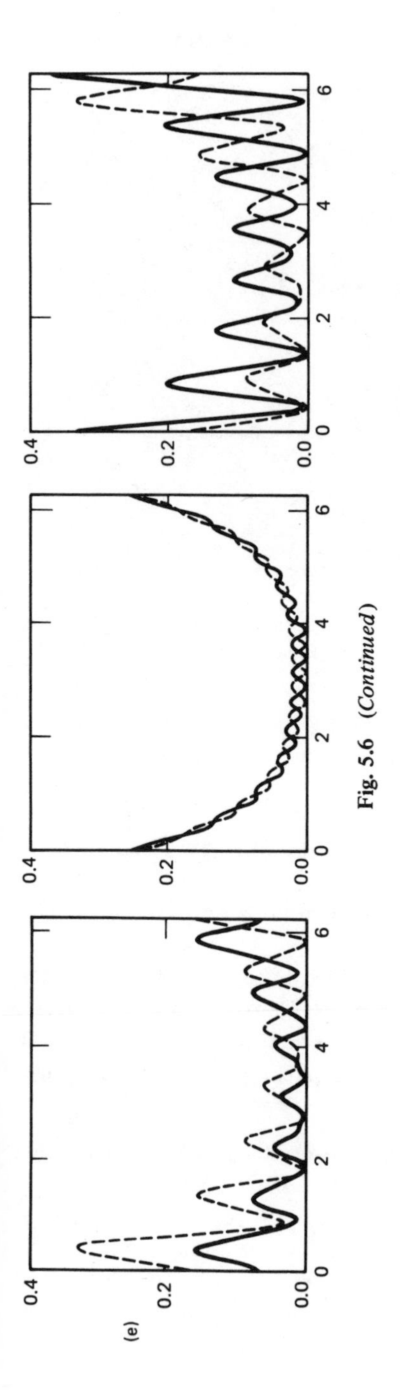

Fig. 5.6 *(Continued)*

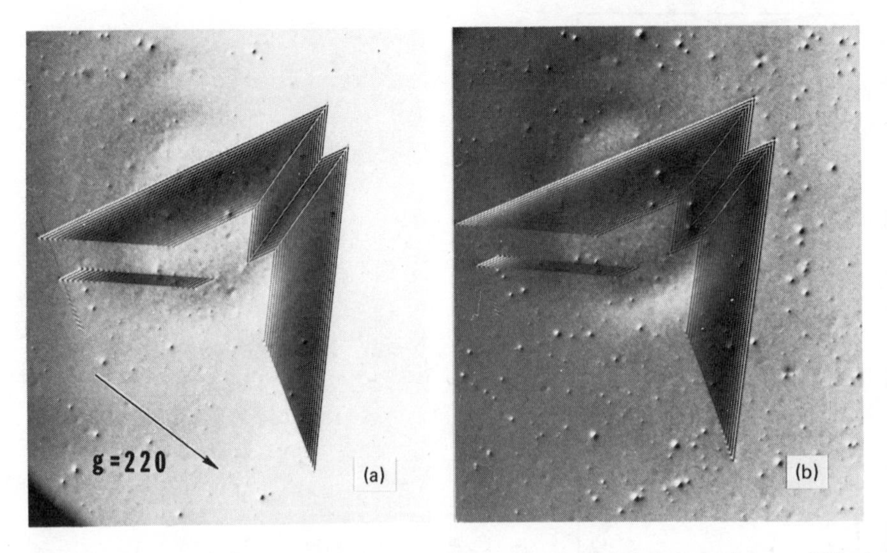

Fig. 5.7 Micrographs of stacking faults in silicon, $\alpha = \pm 2\pi 3$ and 0, imaged in $\mathbf{g} = 220$ at 1 MV: (*a*) bright field, (*b*) dark field. Notice defects from displacement damage.

visibility is greater when the fault is near the top (entrance) surface, and for positive values when the fault is near the bottom (see Figs. 5.6*c* and *d*: $w = s\xi_g = -0.4$ and $+0.4$).

It should be noted that these general rules apply only to crystals that are sufficiently thick and strongly absorbing (Head[2] gives $t = 0.25\xi_g'$ as sufficient) because they are a consequence of the asymmetry in the absorption of the two Bloch waves.

The situation encountered when the absorption is lower (for higher voltages, thinner crystals, lighter elements, elements of higher Debye temperature, etc.) is exemplified by the plots of Figs. 5.8*c* and *d*. Note that the gray region near the center of the fault is no longer present; instead, more complicated double fringes extend right across the profile.

3.1.2 *Other Faults with* $\alpha \neq \pi$. Given that the crystal is thick, though, sign determination is not restricted to the special case of $\alpha(\pm 2\pi/3)$ given above. Contrast of similar nature should be detectable for $0.1\pi < |\alpha| < 0.9\pi$ (e.g., Figs. 5.6*a* and *b*). For α too near zero the contrast becomes vanishingly weak, while for $|\alpha|$ near π the contrast tends to that of the special case $\alpha = \pi$, described below. Again, if the absorption is too small, complicated fringes ensue (Figs. 5.8*a* and *b*).

3.1.3 *Faults with* $\alpha = \pi$. When the phase angle is π a special situation exists, which is shown most clearly at the exact reflecting position, where both bright and dark field profiles are symmetrical but, in the presence of strong absorption, uninterestingly, uniformly gray (Fig. 5.6*e*). Away from the reflecting position, however, fringes are found again, with many properties similar to those for $\alpha \neq \pi$, such as bright and dark fields similar near the entrance surface. However, in this case the sign of the first fringe is bright for negative deviation w, and dark for positive deviation, in contrast to the invariant nature of the sign for $|\alpha| = 2\pi/3$.

If the absorption is less, more complicated fringes occur, an effect that is particularly noticeable at the reflecting position (e.g., Fig. 5.8*e*). As noted in Chapter 3, Section 8.3, the extinction distances for superlattice reflections (which can give rise to $|\alpha| = \pi/2$) are often large and hence specimens may always be "thin." Thus Fig. 5.9*a* may be directly compared with the experimental micrographs of Figs. 5.9*b* and *c*.

3.1.4 *Orientation Faults.* When there is no phase change at the boundary, but an abrupt orientation change, α is zero, but w is different in the two crystal sections. This means that, unlike the stacking fault situation, the intensities of both bright field and dark field images are in general different on opposite sides of the fault. This and the other general characteristics of thick absorbing crystals listed in Chapter 3, Section 8.2, may be seen in the calculated profiles of Fig. 5.10 and the experimental micrographs of Fig. 3.28. Of particular note are the *differences* between the profiles and those of the stacking fault (Fig. 5.6), for example, the symmetry, or lack of symmetry, of the image and the different spacings of the fringes at the upper and lower surfaces. Fringes at boundaries always have spacings corresponding to thickness fringes in the thinner wedge-shaped subcrystal; where the orientations of the two subcrystals are different, the fringe spacings are in general dissimilar.

3.1.5 *Presentation of Results.* The results of the calculations have been presented in Figs. 5.6, 5.8, and 5.10 in the most obvious way, as plots of the intensities as a function of one variable, the depth of the fault below the top surface. Comparison of such line profiles, which are calculated for a relatively small number of columns (say, separated by $0.05\xi_g$ in most of the figures), with the experimental images of Figs. 5.7, 5.9, and 3.15 is straightforward since the images are invariant in one dimension. In this case, therefore, the simplest way is the best. However, if, for example, the faults occur only in a crystal that is bent or tapered, the image will vary in two dimensions.

Since micrographs are "pictures" and the eye is an extremely good instrument for comparing the overall features of pictures, it is reasonable to try to display the results of calculations in a similar form. An example of such a computed

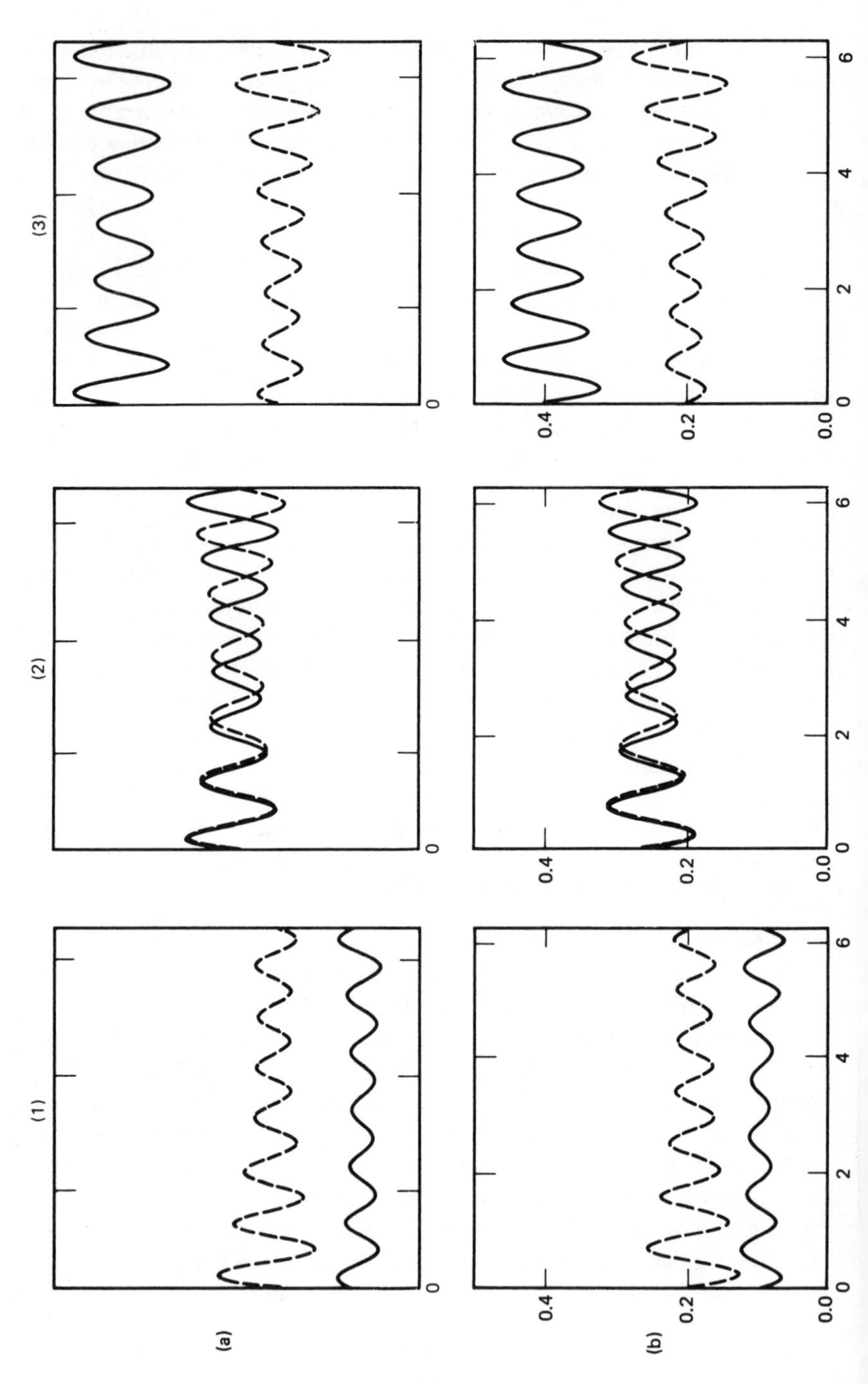

(a)
(b)
(1)
(2)
(3)
0
2
4
6
0.0
0.2
0.4

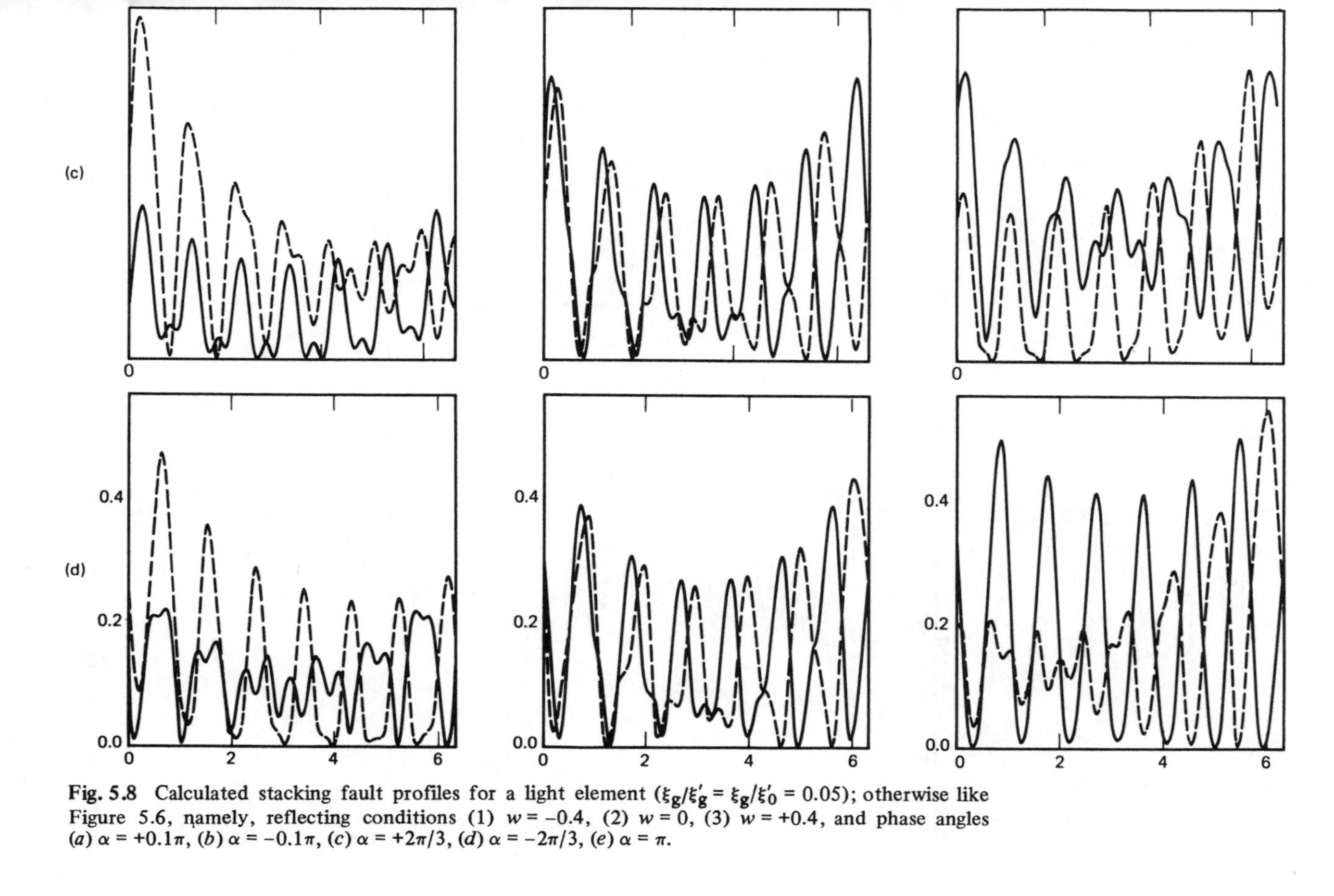

Fig. 5.8 Calculated stacking fault profiles for a light element ($\xi_g/\xi'_g = \xi_g/\xi'_0 = 0.05$); otherwise like Figure 5.6, namely, reflecting conditions (1) $w = -0.4$, (2) $w = 0$, (3) $w = +0.4$, and phase angles (*a*) $\alpha = +0.1\pi$, (*b*) $\alpha = -0.1\pi$, (*c*) $\alpha = +2\pi/3$, (*d*) $\alpha = -2\pi/3$, (*e*) $\alpha = \pi$.

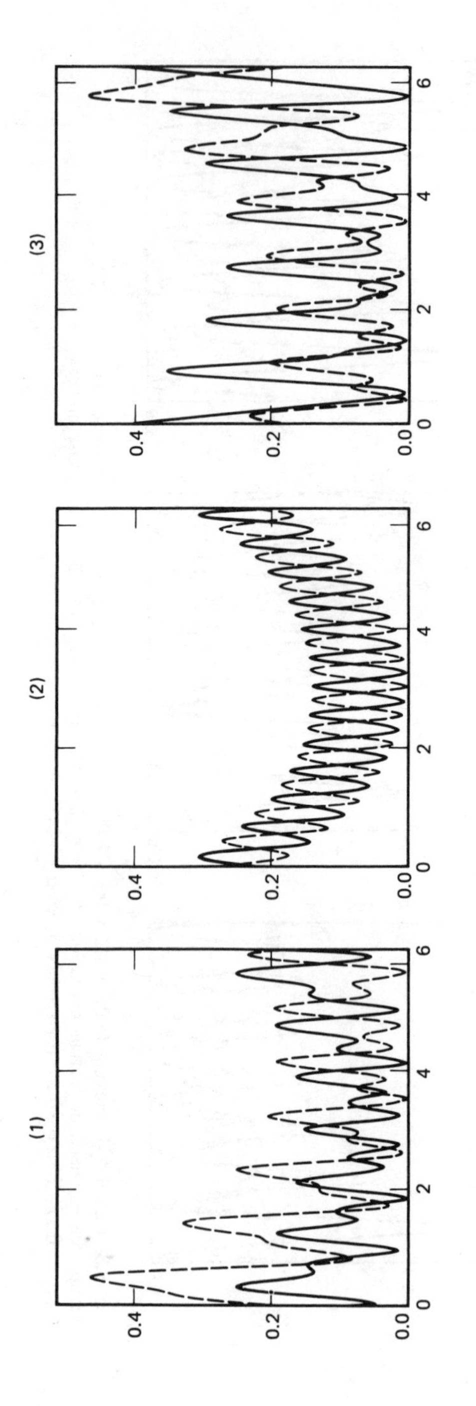

Fig. 5.8 *(Continued)*

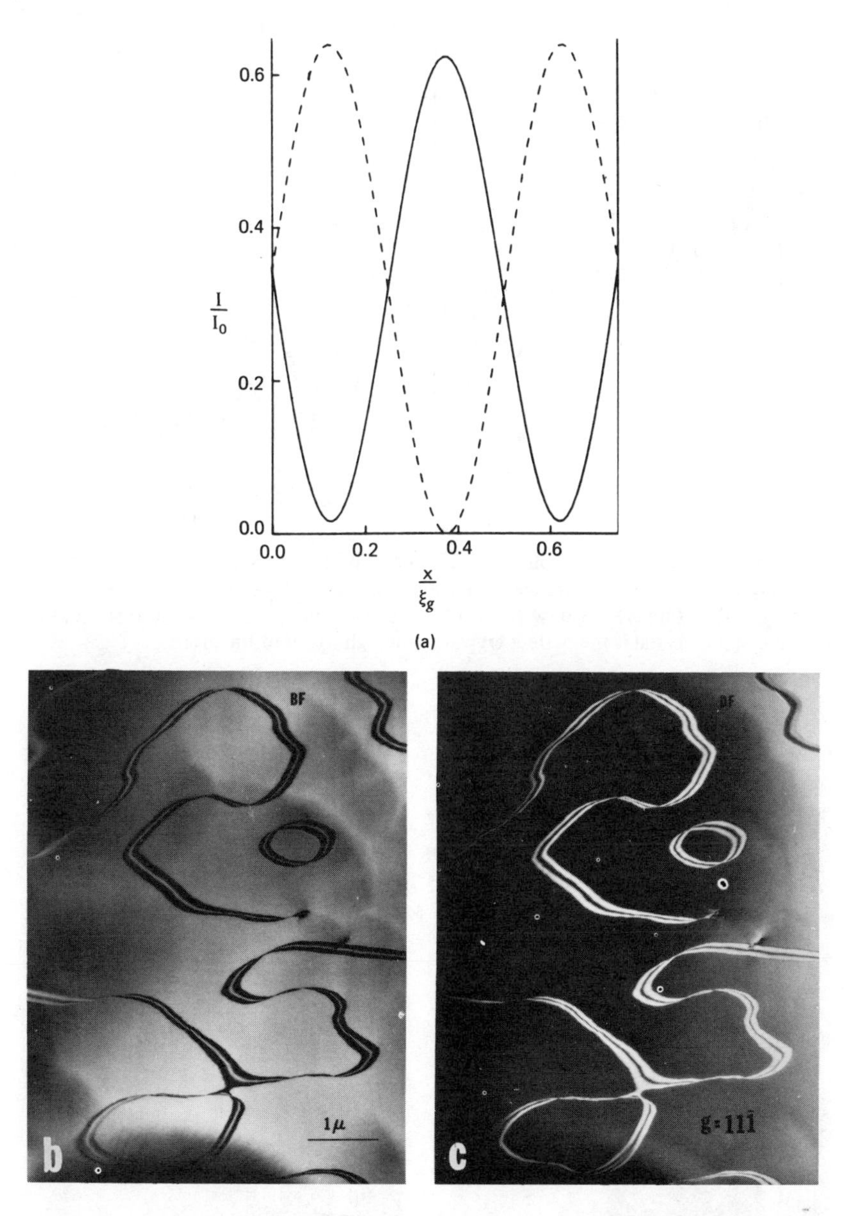

Fig. 5.9 Pi faults in thin crystals. (*a*) Calculated profiles of bright field (solid) and dark field (dashed) intensities ($\xi_g/\xi'_g = \xi_g/\xi'_0 = 0.1$, $t/\xi_g = 0.75$, $\alpha = \pi$). (*b*) Bright field image of antiphase boundaries in anorthite (650 kV electrons). (*c*) Corresponding dark field image. (*b*, *c*) Courtesy Springer-Verlag, *Contributions to Mineralogy and Petrology*. **40,** 70 (1973).

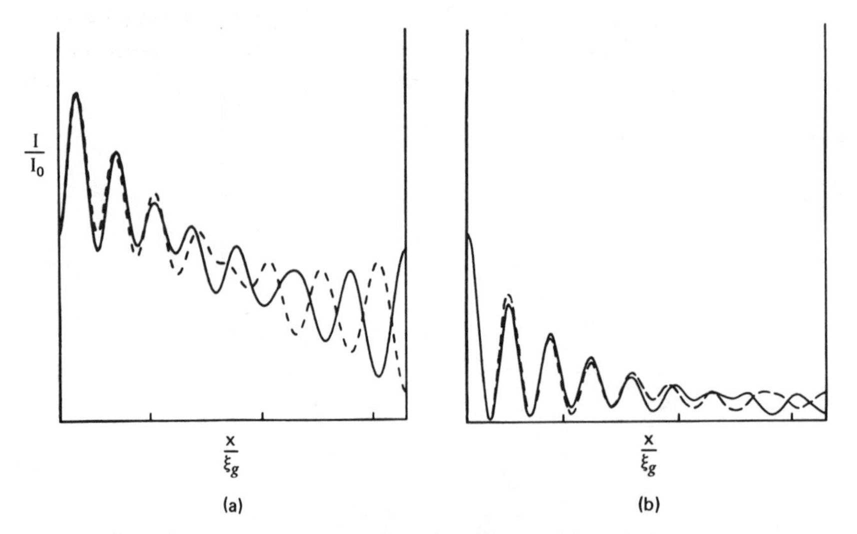

Fig. 5.10 Calculated bright (solid) and dark field (dashed) fringe profiles for single inclined orientation faults in a medium weight element ($\xi_g/\xi'_g = \xi_g/\xi'_g = 0.1$) of reasonable thickness ($6.25\xi_g$). (*a*) w (upper) = 1.0, w (lower) = 0. (*b*) w (upper) = −1.0, w (lower) = 0. In each case the upper crystal is the perfect crystal at the right hand of the profile.

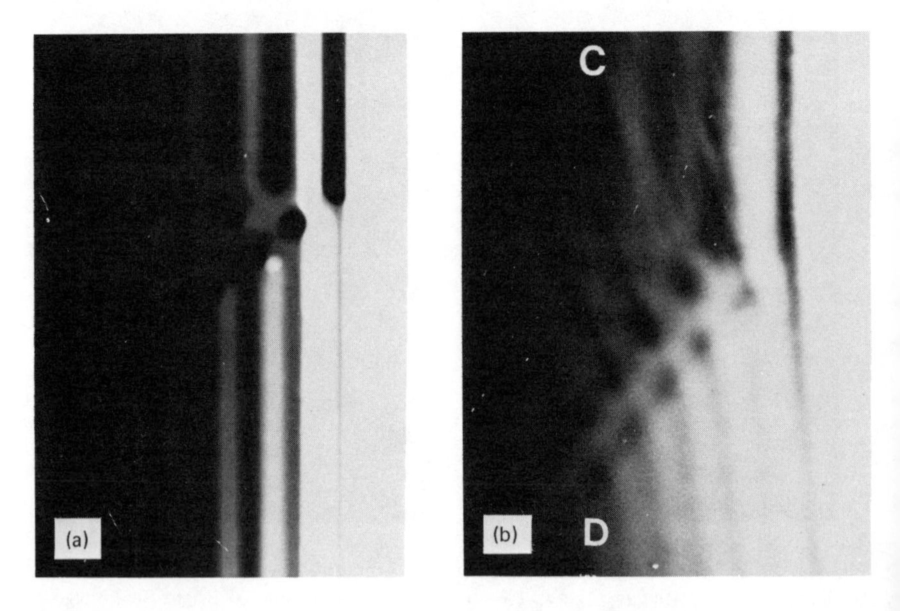

Fig. 5.11 Comparison of calculated four-tone contour map (*a*) and experimental bright-field micrograph (*b*) for an inclined orientation fault in a tapered specimen.

picture (using a limited four-tone gray scale) and the experimental micrograph of an "orientation" planar fault, a coherent twin boundary, are shown in Figs. 5.11*a* and *b*. At this boundary there is no abrupt phase change, but the orientations of the two crystal slabs are different. As noted above, this orientation difference may be deduced from the change in spacing of the thickness fringes on opposite sides of the boundary at *C* and *D*.

3.2 Dislocations

The majority of calculations of the images expected from dislocations have been carried out by integration of the Howie-Whelan equation (eq. 4.21 here) directly, although, as has been noted, alternative formulations may be quicker in many circumstances and therefore preferable, particularly when a large number of calculations are needed. What is required to perform the integration are reasonable values of the extinction and absorption parameters for the material (found, e.g., by comparison of perfect crystal thickness fringes with calculated profiles, although for a large number of materials of medium atomic weight the copper parameters are adequate approximations, See Table 4.1) and an analytical expression for the strain field β'_g (eq. 4.22), which describes the way in which the reflecting planes are bent (i.e., the expression describing the planes shown in the sketches of Fig. 3.6 or 3.8).

3.2.1 *General Features of Contrast.* The principles of the method are illustrated by means of a simple example. It is necessary to set up a convenient coordinate system such as that shown in Fig. 5.12, which is suitable for a dislocation lying parallel to the specimen surfaces. The displacement field for such a dislocation in an isotropic material is then (e.g., ref. 2) given by

$$2\pi \mathbf{R} = \mathbf{b}\Phi + \mathbf{b}\frac{\sin 2\Phi}{4(1-\nu)} + (\mathbf{b} \wedge \mathbf{u})\left[\frac{1-2\nu}{2(1-\nu)}\ln r + \frac{\cos 2\Phi}{4(1-\nu)}\right], \quad (5.19)$$

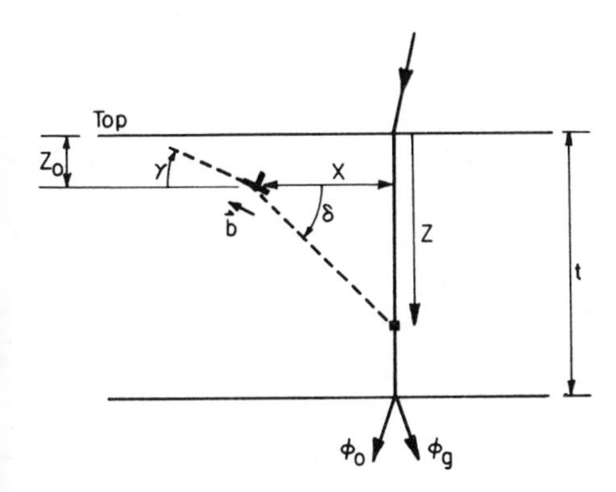

Fig. 5.12 Example of a suitable coordinate system to describe the strain field of a dislocation.

where $\Phi = \delta - \gamma$, **u** is the unit vector describing the dislocation line direction, and ν is the Poisson ratio.

For particular values of dislocation depth (z_0) and column position (x), r and Φ in eq. 5.19 are functions of z only, and hence β_g' for the value of **g** assumed is a known function of z. Numerical integration of eq. 4.21 is then performed for a number of different columns (x changed) to produce the simplest image profile, a trace "across" the dislocation. Examples of such traces for a variety of typical situations[3] are shown in Fig. 5.13. The traces in Figs. 5.13*a–c* and 5.13*e–h* illustrate a number of "rules" of fairly general applicability, provided that the specimen is reasonably thick ($t = 8\xi_g$ here), the dislocation is not too near either surface, and a strong, two-beam reflecting condition is chosen.

1. For $\mathbf{g} \cdot \mathbf{b} = 1$ the bright and dark field images are similar in nature (although the image contrast and background intensity depend on the deviation parameter w; compare Figs. 5.13*a* and *c*), being dark lines on a bright background. The image width (measured as, say, full width at half depth) is approximately $\xi_g/3$ for a screw dislocation. Generally edge dislocation images are wider than those of screw dislocations (compare Figs. 5.13*a* and *b*), and (not shown) image widths decrease if the dislocation line is inclined to the specimen surface. The image position is displaced to one side of the dislocation position for nonzero values of w, the direction of displacement depending on the sign of $(\mathbf{g} \cdot \mathbf{b})w$ (see Fig. 5.13*c*). This effect is more noticeable in the bright field image.
2. For $\mathbf{g} \cdot \mathbf{b} = 0$ the contrast depends on the value of $\mathbf{g} \cdot \mathbf{b} \wedge \mathbf{u}$, being essentially zero for $\mathbf{g} \cdot \mathbf{b} \wedge \mathbf{u} < 0.6$ (curve 1 in Fig. 5.13*f*) but appreciable for values greater than this (curves 2–4). Hence only screw or near-screw dislocations totally disappear for $\mathbf{g} \cdot \mathbf{b} = 0$ (see, e.g. Fig. 3.3*b*).
3. For $\mathbf{g} \cdot \mathbf{b} = 2$ the images are generally double (see Figs. 5.13*g* and *h*) and symmetrical at the reflecting position (Fig. 5.13*g*), but significantly asymmetrical at deviated positions (Fig. 5.13*h*).
4. Partial dislocations, for which $\mathbf{g} \cdot \mathbf{b}$ may be nonintegral, show contrast similar to that of perfect dislocations except that, of necessity (since partials must bound a fault of some sort), the background levels are in general different on opposite sides of the dislocation image (Fig. 5.13*e*). For small values of w, dislocations with $|\mathbf{g} \cdot \mathbf{b}| < \frac{1}{3}$ are effectively invisible, while those for $|\mathbf{g} \cdot \mathbf{b}| > \frac{1}{3}$ may be seen. However, for the high penetration imaging situations frequently used (to view thick specimens with greater screen brightness), say $w \simeq 1$, partials with $\mathbf{g} \cdot \mathbf{b} = -\frac{2}{3}$ may also be invisible,[4] whereas the contrast for $\mathbf{g} \cdot \mathbf{b} = +\frac{2}{3}$ remains strong.

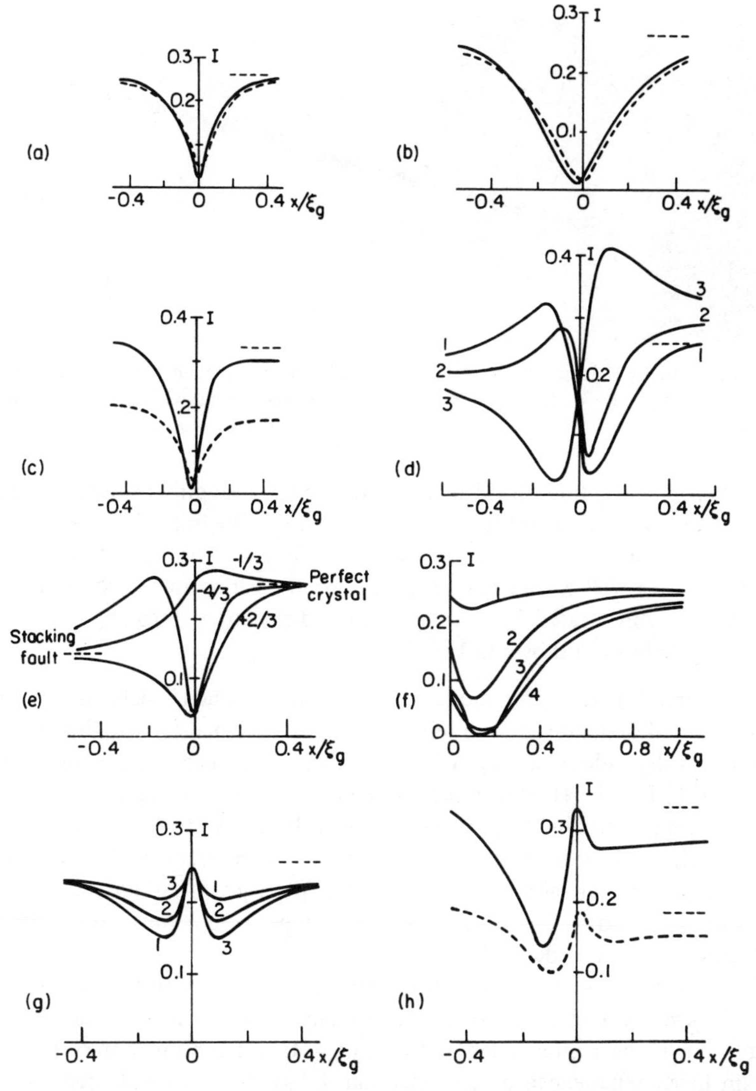

Fig. 5.13 Sketches of typical two-beam images of dislocations based principally on figures from ref. (3), which should be consulted for full details. Bright field, solid lines; dark field, dashed. Unless otherwise stated, the dislocation in question is in the middle of a specimen of thickness $t = 8\xi_g$, $w = 0$, and $\xi_g/\xi'_0 = \xi_g/\xi'_g = 0.1$. (*a*) screw dislocation, $\mathbf{g} \cdot \mathbf{b} = 1$; (*b*) edge, $\mathbf{g} \cdot \mathbf{b} = 1$; (*c*) screw, $\mathbf{g} \cdot \mathbf{b} = 1$, $w = 0.3$; (*d*) screw, $\mathbf{g} \cdot \mathbf{b} = 1$, depths, $z_0/\xi_g = 7.25, 7.5, 7.75$; (*e*) partials, $\mathbf{g} \cdot \mathbf{b} = -\frac{1}{3}, -\frac{4}{3}, +\frac{2}{3}$, $t/\xi_g = 6$, $z_0/\xi_g = 1.7$; (*f*) edge, $\mathbf{g} \cdot \mathbf{b} = 0$, $m = \mathbf{g} \cdot \mathbf{b} \wedge \mathbf{u} = 0.08$, 0.23, 0.40, 0.46; (*g*) screw, $\mathbf{g} \cdot \mathbf{b} = 2$, $z_0/\xi_g = 4.0, 4.25, 4.5$; (*h*) screw, $\mathbf{g} \cdot \mathbf{b} = 2$, $z_0/\xi_g = 4.0$, 4.25, 4.5, $w = 0.3$.

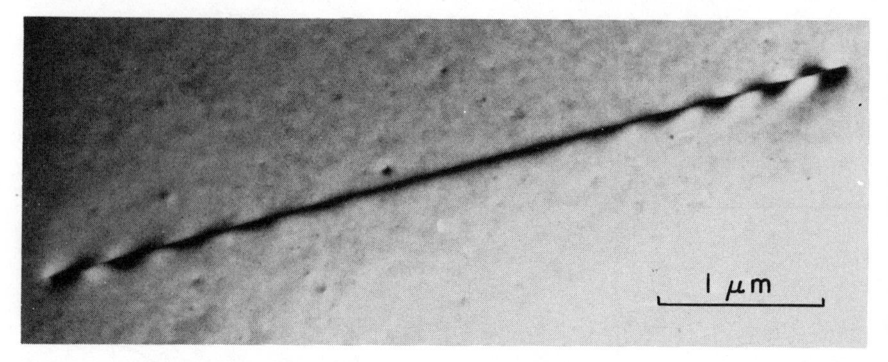

Fig. 5.14 Bright field image of a dislocation threading a specimen from top to bottom. Note the complexity of the image near either surface compared with its simplicity near the center.

If the specimen is thin, or if the dislocation is too near either surface, more complicated images ensue, as illustrated in Fig. 5.13*d*. The image here shows a black-white contrast and zigzags from side to side of the dislocation position. This zigzagging, as well as the constancy of image position in the center of a thick specimen, is illustrated by the image of a dislocation threading the specimen from top to bottom shown in Fig. 5.14.

3.2.2 *Time-Saving Techniques.* As noted above, when specimens are thin, or when the dislocation is near either surface, the image is no longer simple, becoming dependent on depth (see the oscillations near both ends of the dislocation in Fig. 5.14). Under these circumstances or in situations where greater image detail is required (e.g., if the crystal is elastically anisotropic or noncentrosymmetric), a large number of image profiles at different depths are necessary for comparison, or, alternatively, a simulated image. In either case the integration must be carried out for a very large number of columns, and many of the integration steps are identical.

Reference to Fig. 5.15 shows how time is wasted in conventional calculations on dislocations threading the specimen from top to bottom. If calculations are made down the columns denoted as 2 or 3, then (ignoring surface effects) the strain fields experienced over the section A_iB_i (i = 2, 3) are identical and hence the two integrations must have much in common. One way of acknowledging this similarity and making use of it in scattering matrices was mentioned in Section 2.3. The scattering matrix method developed by Thölén[1] implies in its most general form storage of matrices for a very wide range of values of β'_g, many of which are not used in a particular calculation. An alternative variant developed by Head[5] calculates only the matrices pertinent to the problem in hand and thus may be preferable in situations where computer storage is limited.

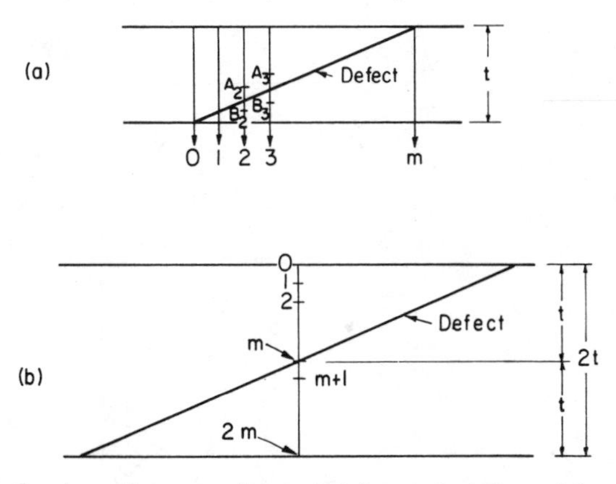

Fig. 5.15 Illustrating how time may be wasted in a calculation of image contrast. (*a*) Conventional arrangement of $(m + 1)$ columns through a crystal of thickness t. (*b*) Head's[5] arrangement of a single column through a crystal of thickness $2t$ used to synthesize the $(m + 1)$ solutions of (*a*).

3.3 Small Defects

The defects so far considered, that is, planar faults and dislocations, are extended in at least one dimension. The term "small defect" may be used to include any defect whose dimensions are all small on the scale of the extinction distance (i.e., less than approximately $\xi_g/3$). For defects in this size range the contrasts from different parts of the defect (e.g., opposite sides of a small dislocation loop) overlap, making detailed interpretation difficult. The general principles of contrast calculation are, however, identical to those used in the case of dislocations, namely, integration of the equations for a suitably constructed strain field, this time over a two-dimensional net since the image varies in this way; see, for example, Fig. 3.4.

The characteristics of the images may be summarized briefly as follows:

1. Defects with zero or small strain fields (e.g., voids, small gas bubbles, inclusions of second phase with zero misfit) are visible only through "scattering factor" contrast—columns through such defects appear to be of different thickness from those through adjacent regions of matrix crystal. They are thus visible only in situations where strong thickness fringes are seen, that is, in thin areas when the crystal is at the reflecting position.
2. Defects with significant strain fields in general produce small "black-dot" contrast (small dark areas on a light background) when imaged at deviated orientations, such contrast being insensitive to differences between the

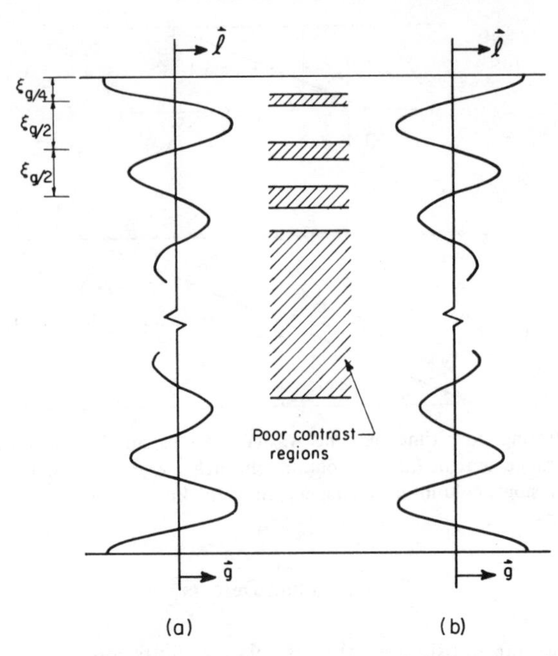

Fig. 5.16 Schematic depth variation in contrast of small defects for orientation near the reflecting position in bright field. The vector **l** is directed from the black to white contrast lobe.

defects. At the reflecting position the contrast becomes much more distinct, characteristically of "black-white" nature, particularly when the defect is near either surface of the specimen. As mentioned in the case of dislocations, such black-white contrast is oscillatory with the depth of the defect (with period equal to that of thickness fringes), the direction of the contrast at any particular depth depending on the sign of **g** · **R**. This direction **l** is usually defined as the vector from black to white. An example of such contrast may be seen in Fig. 5.7, particularly 5.7*b*. The depth oscillation is shown schematically in Fig. 5.16, from which it may be seen that bands of opposite sign are separated by bands of zero contrast, the maximum width of the former being $\xi_g/2$ (except for the surface band of maximum width $\xi_g/4$). Defects of interstitial and vacancy types have opposite displacement fields **R** and hence may be distinguished by their opposite oscillatory contrasts with depth, provided that careful depth measurements are made using stereo techniques. The defects in Fig. 5.7*b* all lie in the same thickness band (having been produced by heavy-ion irradiation with its well-defined depth of defect production). As for any other defect, a column for which **g** · **R** = 0 throughout produces zero contrast. Thus the strong contrast effect near the reflecting position is also characterized by a line of no contrast, where $\mathbf{g} \cdot \mathbf{R} = 0$, which may be seen in Fig. 5.7*b* as the line separating black and white contrast lobes. For spherically symmetrical strain fields (e.g., pure dilatation in isotropic materials) the line of no contrast is perpendicular to **g** for all **g** (and the defect is visible for all **g**), whereas for defects with shear displacements, or in cases where anisotropic elasticity effects are important, the relationships are more complicated.

Contrast from small defects is discussed again in the next section.

4 Image Displays

Image displays have already been mentioned in considering planar faults; the overall picture comparison is much more meaningful in many cases than any number of single-line traces. This is particularly true when detail in oscillatory images is relevant (e.g., thickness fringes near bend contours, dislocations near surfaces), and various techniques have been used. The various types of image displays include the following:

1. Pen and ink contour map (e.g., Hashimoto et al.[6]), which requires cartographic skills both to execute and to decipher.
2. Overlaid contour map (e.g., Fig. 5.11), which is more like a picture, but with a strictly limited gray scale.

Nowadays two computer-generated output formats are available:

3. Overprinted lineprinter output (e.g., Head[5]), where the correct combinations of overprinted characters give a gray scale. When the result is photographed and printed (slightly out of focus for best results), pseudomicrographs of reasonable quality are quickly and cheaply produced, which have made possible close comparisons of the details of oscillatory images of dislocations (see, e.g., the many excellent examples in ref. 7).
4. Similarly (and more expensively), a television output system may be modulated to provide superb "pictures," the TV scan being much more readily modulated than is ink on paper. An example is shown in Fig. 5.17, where dislocations are compared rather directly with their experimental counterparts.[8]

The complexities in the contrast from small defects constitute another area where image simulations may be most helpful. An example is shown in Fig. 5.18, where calculations for four possible Frank loop defects are compared with a

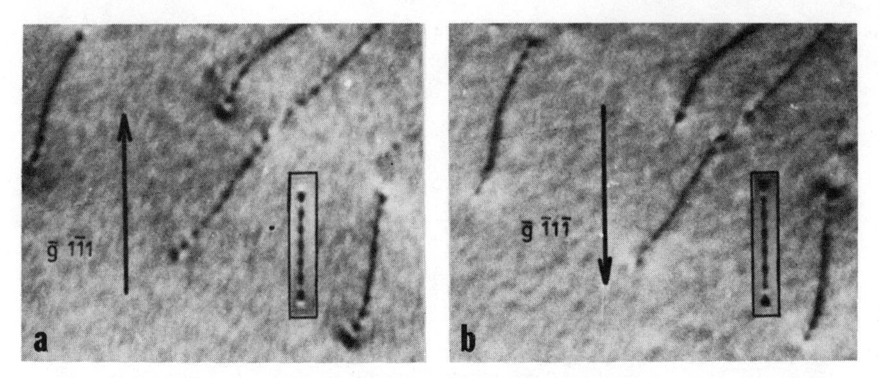

Fig. 5.17 Computer-simulated images compared with the corresponding micrographs of dislocations in Cu-10% Al. Courtesy P. M. Hazzledine, H. P. Karnthaler, and M. S. Spring.

micrograph from ion-damaged copper. Obviously a large amount of information is available from micrographs, so the model of the defect being investigated may be refined to obtain the best comparison. The real difficulty lies in defining the defect correctly and accurately, using anisotropic elasticity, and so on, rather than the simple expression of eq. 5.19; see, for example, ref. 7.

The two-beam theory has now been developed formally to solve the problem of contrast from any postulated defect; it has been shown how, in principle, such things as Burgers vectors of dislocations may be determined, initially by $\mathbf{g} \cdot \mathbf{b} = 0$, but more fully by comparing micrographs under a variety of g's. The accuracy of this comparison is, however, limited by the restriction of the problem to two beams; this restriction is removed in the following section.

5 Solutions for Typical Situations (n-Beam)

In many cases the two-beam situation cannot be obtained experimentally, for example, in the high voltage electron microscope and with crystals of large spacing (i.e., where the reflecting sphere simultaneously cuts many rel-rods). Under these circumstances many-beam calculations are essential.

5.1 Perfect Crystal

The ith Bloch wave of initial amplitude $\epsilon^{(i)}$ propagates through the crystal with wave vector γ $(= \gamma_{re} + i\gamma_{im})$ in the same way as in the two-beam case. Thus the scattering matrix formulation of eqs. 5.1 and 5.4 holds; in particular, the wave amplitudes $\phi_g(z)$ at depth z for a single incident wave of unit amplitude are

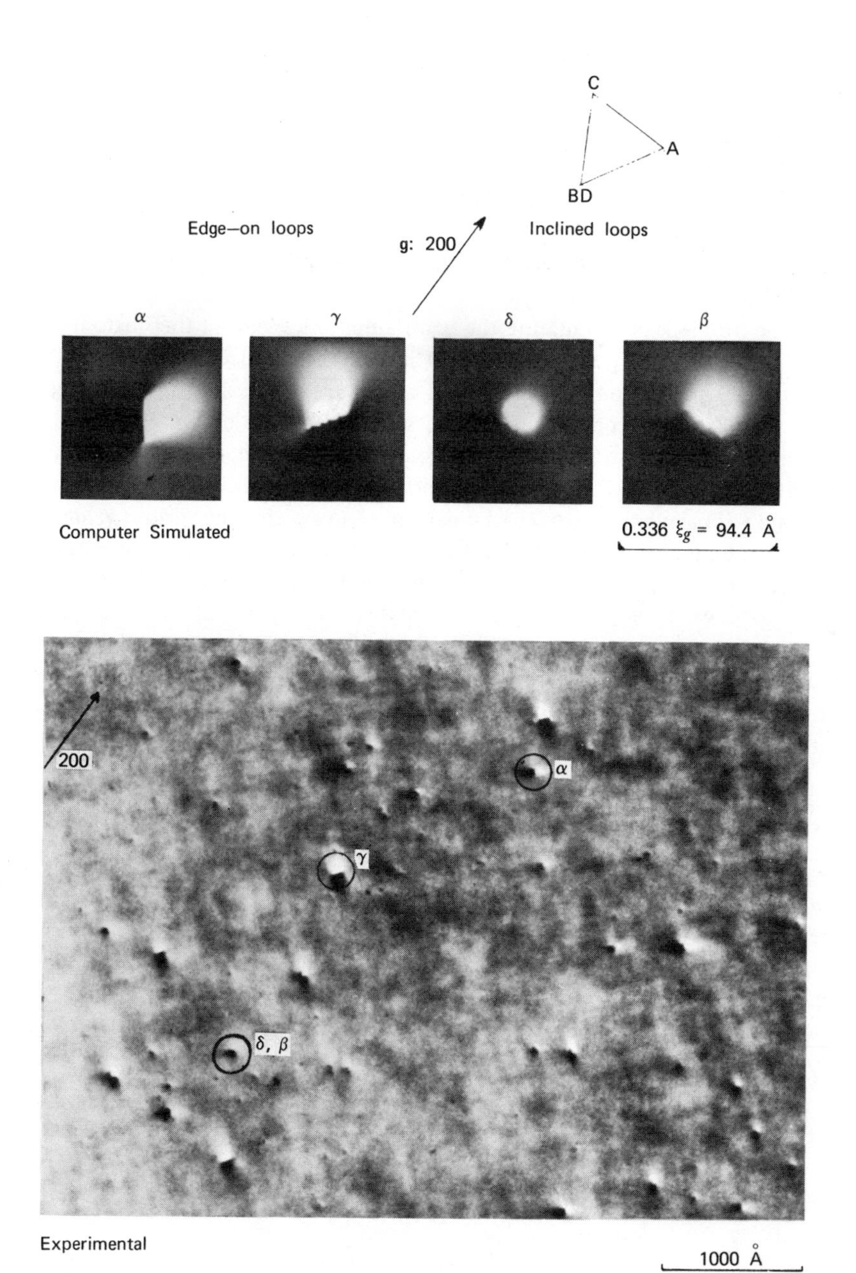

Fig. 5.18 Comparison of images of small Frank loops in ion-bombarded copper. In this particular reflection two out of the four possible configurations may be unambigously identified, other reflections being necessary for the complete analysis. Courtesy D. K. Saldin.

given by

$$\phi_g(z) = \sum_i C_0^{(i)} C_g^{(i)} \exp(2\pi i \gamma^{(i)} z). \tag{5.20}$$

(This result follows immediately from eq. 5.3, using the fact that $\mathsf{C}^{-1} = \tilde{\mathsf{C}}$ and $\epsilon^{(i)} = C_0^{(i)}$.)

Thus the problem of calculating the wave amplitudes leaving the bottom of a slab of perfect crystal consists of (*a*) setting up the matrices A_{re} and A_{im} for the particular orientation and reflections of interest, (*b*) calculating the eigenvectors $\mathbf{C}^{(i)}$ and corresponding eigenvalues $\gamma_{re}^{(i)}$ of A_{re} by one of the standard subroutines available in computer libraries (e.g., Householder's or Jacobi's method), (*c*) calculating the imaginary parts of γ by matrix multiplication (eq. 4.66), and (*d*) calculating the set of emergent beams ϕ_g for the various values of z that are of interest, and hence the beam intensities $|\phi_g|^2$. The same set of calculations may then be carried out for the next orientation of interest except that, if the beams used remain the same, then in step (*a*) only the diagonal elements of A_{re} need be recalculated. Step (*b*) consumes by far the largest amount of computer time for large values of n—hence the advisability of changing orientation the smallest possible number of times during a set of calculations. An outline of the calculation process is shown in Fig. 5.19.

In many situations, however (e.g., in the case of systematic reflections), one is concerned only with the perturbing effect of weakly excited beams on the two principal beams ϕ_0 and ϕ_g, rather than the full solution. Under conditions where only two Bloch waves (i and j—usually 1 and 2) are strongly excited the main features of the contrast are governed by their interference with each other. Thus there is a "principal" extinction distance $\xi\ [=1/(\gamma^{(i)} - \gamma^{(j)})]$ which may be used in place of ξ_g in *two-beam* calculations with a consequent saving in computing time. The value of ξ may be calculated by the n-beam theory of eq. 4.65 and Fig. 5.19 or, under certain circumstances, analytically. An example of such an analytical calculation is given by Howie[9] for the four-beam symmetrical systematic situation with the incident beam at the Bragg angle for the lowest order reflection g, the result being

$$2k\,\Delta k = U_1 + U_3 + \left[\left(g^2 + \frac{U_1 - U_3}{2}\right)^2 + (U_1 + U_2)^2\right]^{1/2} - \left[\left(g^2 - \frac{U_1 - U_3}{2}\right)^2 + (U_1 - U_2)^2\right]^{1/2}, \tag{5.21}$$

where $\xi = 1/\Delta k$ and $\xi_{g-h} = K/U_j$ ($j = |g - h|$), U_j and K being the modified lattice potential and incident electron wave vector, respectively. Equation 5.21 is expected to hold reasonably well when $U_2 > U_3$ and $g^2 > |U_1|$, that is, for

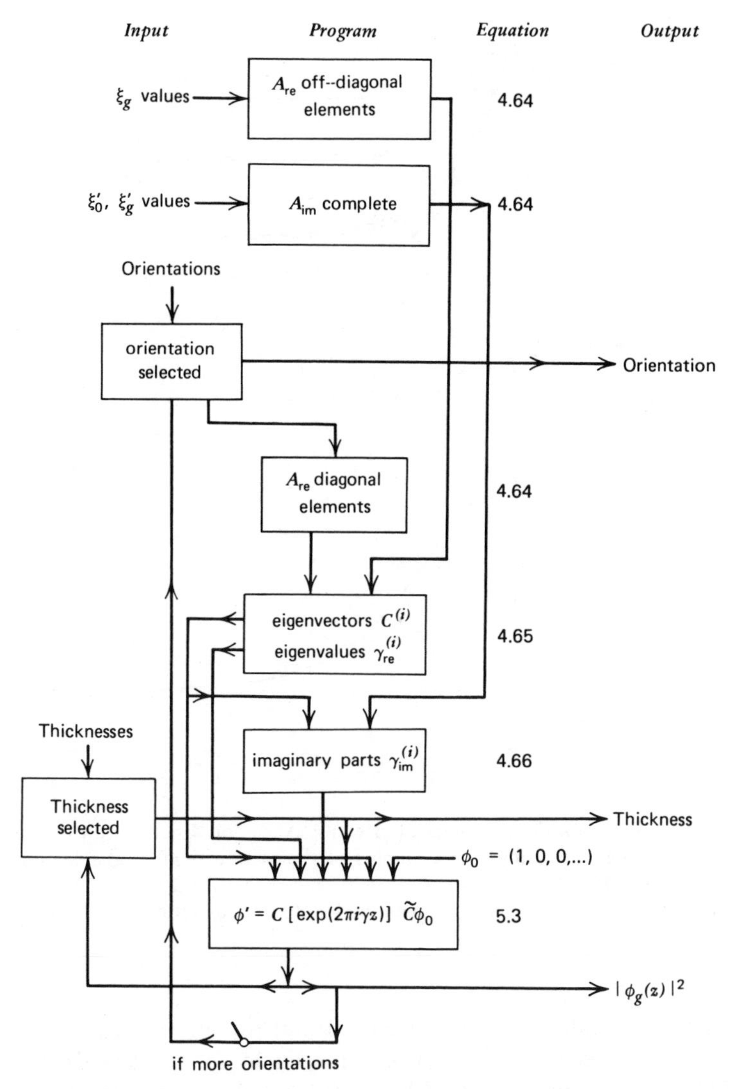

Fig. 5.19 An outline of a computer program to calculate n-beam intensities from perfect crystals. The subscripts re and im stand for real and imaginary respectively.

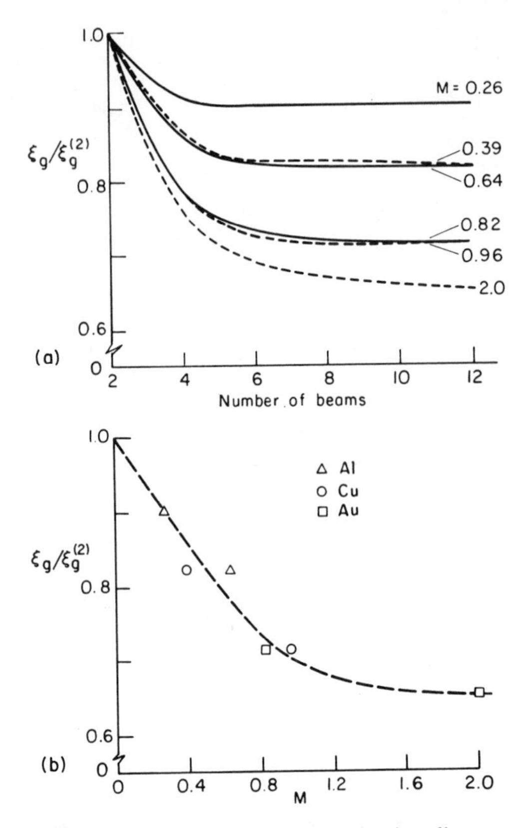

Fig. 5.20 (*a*) Variation in the calculated effective extinction distance with the number of beams used in the calculations for a number of typical situations; (e.g. aluminum, many-beam parameter, M = 0.26; copper, M = 0.39; and gold, M = 2.0, at 100 kV. g = 111). (*b*) Unified plot of terminal many-beam extinction distance as a function of the many-beam parameter M.

structures with uniform structure factors at not too high energies (U_1 increases with energy through m in eq. 4.47). For example, for 100 keV electrons with g = 111 eq. 5.21 gives values of ξ = 51.0 nm and 14.1 nm for aluminum and gold, respectively, while 22-beam theory from eq. 4.65 yields ξ = 51.8 nm and 13.0 nm (see Fig. 5.20 for a typical graph), using ξ_{111} values of 57.3 nm and 15.3 nm (and suitable values for the higher orders required; see Table 4.1). It should be noted that the diffraction conditions envisaged here are the nearest possible to two-beam—the systematic reflections *cannot* be avoided by suitable tilting. The graphs shown in Fig. 5.20 are typical results for the effect of systematic reflections on strong, low order principal reflections. The decrease in ξ with increasing number of beams is not, however, perfectly general. The reverse may occur for higher order reflections or in structures with varying structure factors; for example, for silicon **g** = 111 and two-beam $\xi_g = 60.5$ nm, while the many-beam systematic value is increased to 61.7 nm.

Inspection of eq. 5.21 shows that the inequalities mentioned when taken to stricter limits reduce the solution to the two-beam value. Thus a "many-beam" parameter M may be defined as

$$M = \frac{U_1}{g^2} = \frac{K}{g^2 \xi_g}, \tag{5.22}$$

true two-beam conditions having $M \ll 1$. The pertinent values of M are noted on the plots in Fig. 5.20, where the effect of increasing M by both voltage and atomic weight is shown.

To further illustrate the change from two-beam to many-beam through a transitional "modified two-beam" region, two-beam and n-beam thickness fringes are plotted in Fig. 5.21. For $M \lesssim 0.5$ (Fig. 5.21*b*) the effects of the other

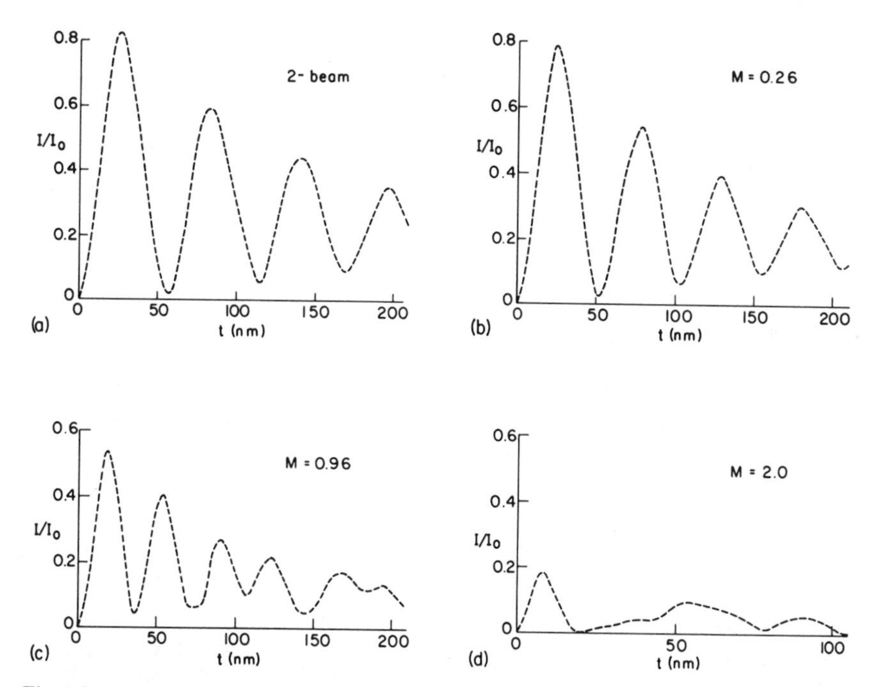

Fig. 5.21 *N*-beam thickness fringes calculated for various typical values of the many-beam parameter *M*. In each case the first-order dark field (**g** = 111) intensity is plotted against thickness measured in terms of the *n*-beam extinction distance, for the systematic case with **g** exactly satisfied. (*a*) Aluminum, two-beam, 100 kV. (*b*) Aluminum, eight-beam, 100 kV, $M = 0.26$. (*c*) Copper, eight-beam, 1 MV, $M = 0.96$. (*d*) Gold eight-beam, 1 MV, $M = 2.0$.

beams excited are negligibly small apart from the reduction in the extinction distance; for $0.5 \lesssim M \lesssim 1$ (Fig. 5.21*c*) the fringes are still well defined, but less regular; for $M > 1$ (Fig. 5.21*d*) the fringes are obviously distorted. The ranges quoted here are, of course, approximate, but they serve to indicate when the more complicated many-beam calculations may have to be undertaken instead of, or to supplement, the simpler two-beam calculations.

A similar trend may be discerned in the behavior of bend-contour profiles. Provided that plots are calculated at the same thickness in terms of the *effective* extinction distance, the two-beam curves are very similar to the many-beam calculations (e.g., Figs. 5.22*a* and *b*: $M = 0.26$) near the reflecting position. Far away from the reflecting position the original two beams assumed are obviously incorrect; for example, in Figs. 5.22*a* and *b* it may be seen that $w = +2$ corresponds to the second-order reflection being satisfied, while $w = -2$ is the symmetry position. For larger values of M the divergence between the two sets of calculations becomes much more pronounced, even near the reflecting position.

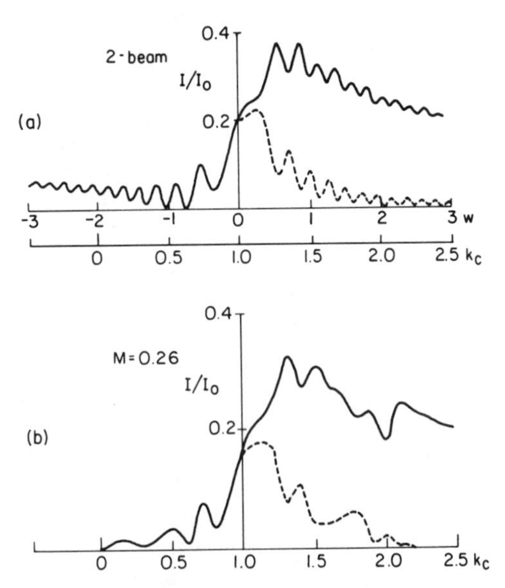

Fig. 5.22 *N*-beam rocking curves calculated for a single typical value of the many-beam parameter M. In each case the bright field (solid) and first-order dark field ($\mathbf{g} = 111$) intensity is plotted against orientation for a specimen of thickness $5.25 \times$ (*n*-beam extinction distance). In all cases the dark field image is symmetrical about the first-order reflecting position, and in (*b*) the bright field about $k_c = 0$. (*a*) Aluminum, two-beam, 100 kV, orientation defined by the deviation w. (*b*) Aluminum, eight-beam, 100 kV, orientation defined by where the Ewald sphere cuts through the systematic row (e.g., $k_c = 1.0$ is the reflecting position for g_1, etc.). Note that $w = -2$ in (*a*) corresponds to the symmetry position, $k_c = 0$, and $w = +2$ to $k_c = 2$, the second-order reflecting position. $M = 0.26$.

5.2 Planar Faults

Here again the equations developed for the two-beam case may be used with increase in their size. Thus, for example, the beams at the lower face of the specimen are described by eq. 5.11 with the fault matrices of eq. 5.8 generalized to

$$\mathsf{F}_{jk} = \{\exp(i\alpha_g)\}_{jk}, \qquad \alpha_g = 2\pi \mathbf{g} \cdot \mathbf{R}_{jk},$$

$\mathbf{R}_{jk}$ being, as before, the displacement of the kth slab of crystal relative to the jth (its immediate predecessor). The method of calculation of an image profile is similar to that shown previously for the two-beam case (Fig. 5.4), with the calculation of $\mathbf{C}_g^{(i)}$ and $\gamma^{(i)}$ replaced by the equivalent sections of Fig. 5.19.

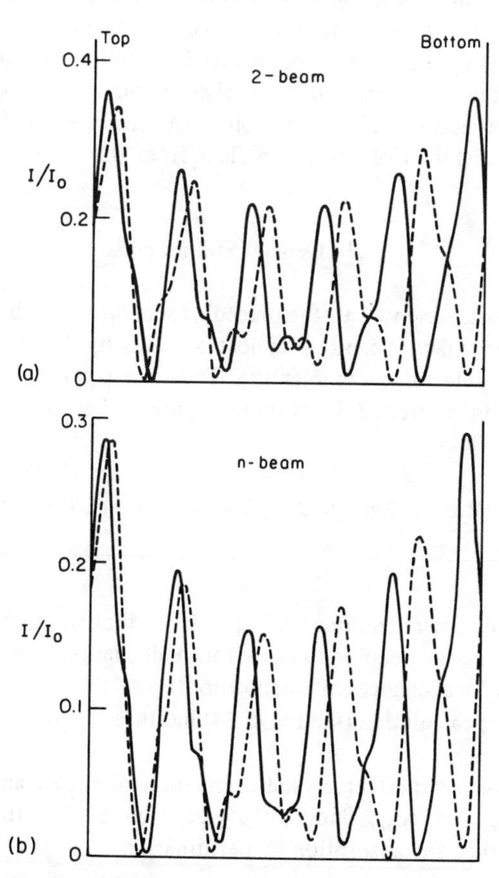

Fig. 5.23 Stacking fault fringe profiles typical of low atomic number elements using data for aluminum at the first-order reflecting position, $\mathbf{g} = 111$, 100 kV, $\alpha = +2\pi/3$: (*a*) two-beam, (*b*) eight-beam, $M = 0.26$. In each case the thickness is $5.25 \times$ (*n*-beam extinction distance); bright field (solid), first order dark field (dashed).

As noted with regard to a perfect crystal, the extent to which many-beam effects modify the fault fringe pattern in the optimum two-beam orientation (systematic reflections with the crystal at the reflecting position for the lowest order reflection) depends on the value of the many-beam parameter M. Figure 5.23 is a plot of the fringe pattern calculated for an inclined stacking fault under both two-beam and many-beam conditions for a value of $M = 0.26$. Here again it is seen that the concept of a modified extinction distance would be sufficient to explain the discrepancies.

It cannot be overemphasized, however, that the situations described above illustrate the *minimum possible* modifications caused by extra reflections being excited. If the crystal orientation is such that a number of nonsystematic reflections (i.e., reflections not in the direction of the principal row) are excited, the complicating effects of these additional reflections may be considerable. An idea of the type of complication may be deduced by noting that for nonsystematic reflections not only the magnitude but also the direction of **g** is different; the nonsystematic reflection will thus respond to a completely different component (**g** · **R** or its derivative) of the strain field from that "seen" by the principal reflection.

5.3 General Strain Fields

From the material presented in the preceding section it may be appreciated that generalization to many beams is almost a formality (apart from the actual increase in computational complexity). Thus the equations developed in the two-beam case in section 2.5 continue to hold, that is, the excitations are governed by

$$\frac{d\boldsymbol{\epsilon}(z)}{dz} = 2\pi i\{\exp(-2\pi i\gamma z)\}\,\mathsf{C}^{-1}\{\Delta\mathsf{A}(z) + \beta_g'(z)\}\,\mathsf{C}\{\exp(2\pi i\gamma z)\}\boldsymbol{\epsilon}(z), \tag{5.23}$$

which is a slight generalization[10] of eq. 5.17 in that the term $\Delta\mathsf{A}(z)$ has been introduced to make it possible to allow for small changes in extinction distances (e.g., because of local changes in composition).

The method of calculation (see Fig. 5.24) has three steps:

1. Using the n-beam perfect crystal calculation discussed above to set up the matrices A_{re} and A_{im}, finding the eigenvectors $\mathbf{C}^{(i)}$, the eigenvalues $\gamma_{re}^{(i)}$ of A_{re}, and then the absorption by perturbation.
2. Integrating the equations for $\boldsymbol{\epsilon}$, that is eq. 5.23, using the proposed strain field for the defect to generate β_g' and any structure factor changes to gen-

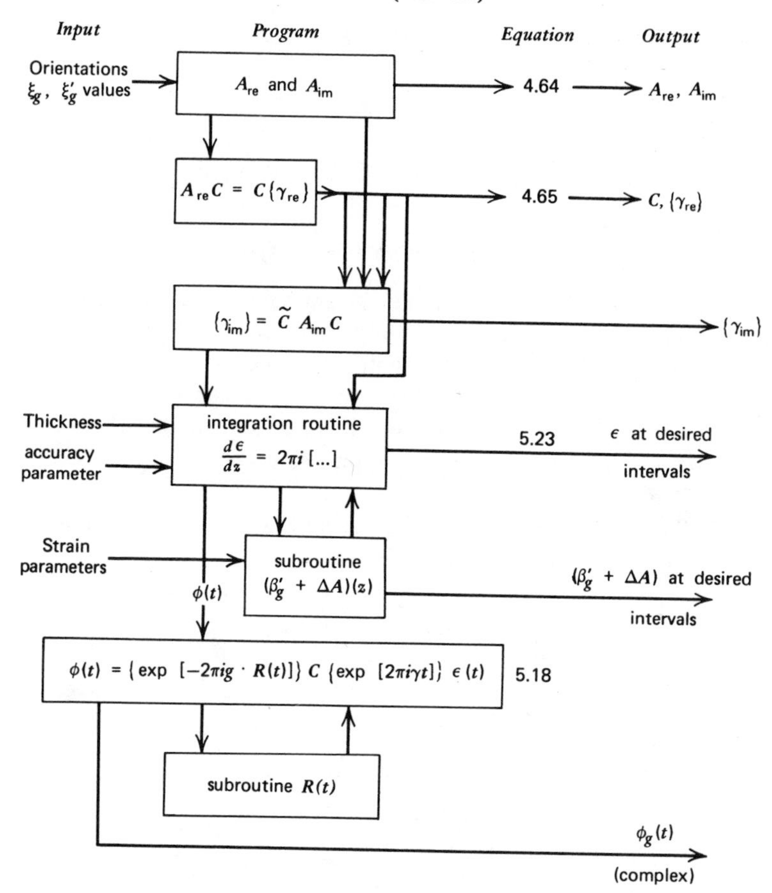

Fig. 5.24 An outline of a computer program to calculate n-beam solutions for continuously varying strain fields.

erate ΔA as a function of z. The initial values of $\boldsymbol{\epsilon}$ are of course the $C_0^{(i)}$ values.

3. After the integration, recomposing the emergent beams from eq. 5.18, the factor $\{Q(z)\}^-$ being relevant only if "lattice images" are to be formed (subroutine $\mathbf{R}(t)$ here); see Section 9.2.

5.3.1 *Integration Routines.* While on the subject of integration it is profitable to consider at greater length what the computer actually does here, and hence appreciate further how computing effort is often wasted. The equations for

$d\boldsymbol{\phi}/dz$ and $d\boldsymbol{\epsilon}/dz$ are of the same general form:

$$\frac{d\boldsymbol{\phi}}{dz} = \mathbf{f}(\boldsymbol{\phi}, z).$$

All numerical integration routines must do something like this crude approximation:

$$\boldsymbol{\phi}(z + \Delta z) = \boldsymbol{\phi}(z) + \Delta\boldsymbol{\phi} = \boldsymbol{\phi}(z) + \mathbf{f}(\boldsymbol{\phi}, z)\,\Delta z, \tag{5.24}$$

that is, calculate the "slopes" **f** at the current values of $\boldsymbol{\phi}$ and add $\Delta\boldsymbol{\phi} = \mathbf{f}\Delta z$. Now for accuracy the $\Delta\boldsymbol{\phi}$'s must be very small; so, if **f** is large, Δz must be correspondingly small, that is, the number of integration steps Δz must be large to go through a thickness z. Thus having **f** small will always provide an advantage; this is the situation in the Bloch wave formulation, where **f** consists of β_g' and ΔA terms and is therefore appreciable *only* near the defect, being small elsewhere.

But whether in $\boldsymbol{\phi}$ or $\boldsymbol{\epsilon}$ notation the machine numerical integration process will be similar and is shown schematically below:

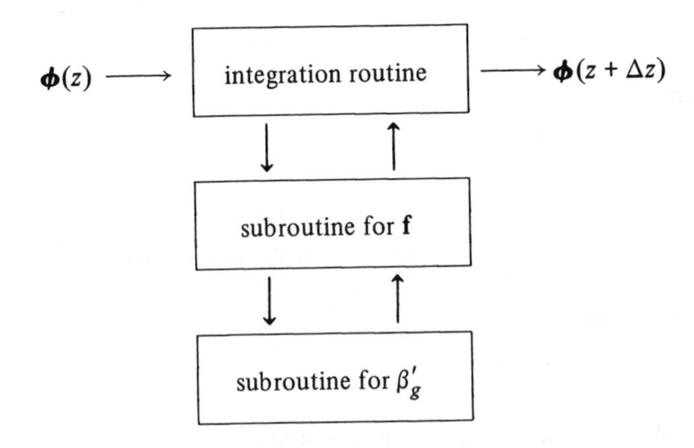

The standard step is to advance $\boldsymbol{\phi}(z)$ [or $\boldsymbol{\epsilon}(z)$] to $\boldsymbol{\phi}(z + \Delta z)$ [$\boldsymbol{\epsilon}(z + \Delta z)$] by a sophisticated version of eq. 5.24, such as the method of Runge-Kutta, Nordsieck, or Krogh.[11] This very general integration routine needs a subroutine for calculating **f** that is general (e.g., a many-beam or two-beam routine) to a number of situations. This subroutine **f** needs β_g', the strain field (and possibly ΔA), and this is most conveniently calculated in another subroutine that is specific to the defect geometry and order of elasticity assumed. Therefore the problem is approached in a modular way—(*a*) a general integration routine, of which several may be available in the computer library; (*b*) a subroutine linking the beams or waves, of sufficient generality to allow simple extension of the number of beams, and so on; and (c) a strain field subroutine applicable to the particular problem of the moment.

The accuracy of the process is obviously of some concern, and although certain library routines have built-in accuracy checks, one should always be prepared to verify the results. Some methods are outlined in Fig. 5.25. In Fig. 5.25a the whole column is integrated with n steps of fixed length h, and then the integration repeated with $2n$ steps of $h/2$. If the results are the same to within

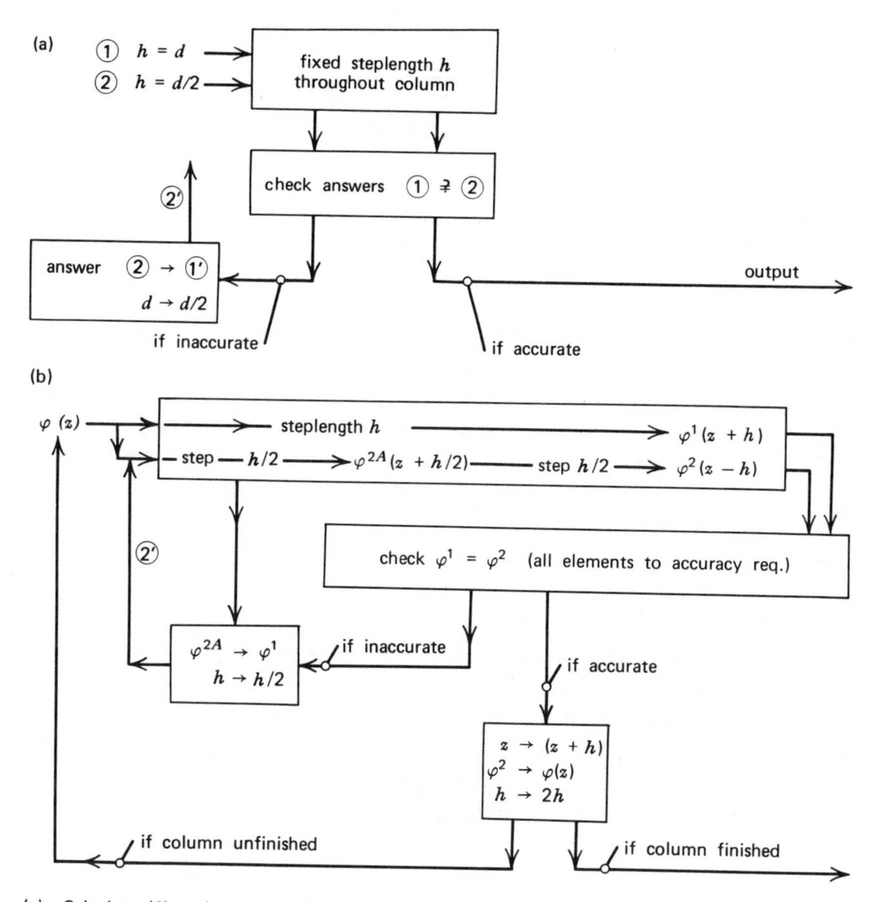

(c) Calculate ($\beta' + s$) in advance of each use of routine for $\varphi(z + h)$ and adjust h according to ≪rules≫ determined by experience (e.g., by method (a)) to be appropriate.

Fig. 5.25 Outline of nonautomatic accuracy checks on numerical integration routines. Courtesy Academic Press; M. J. Goringe.[18]

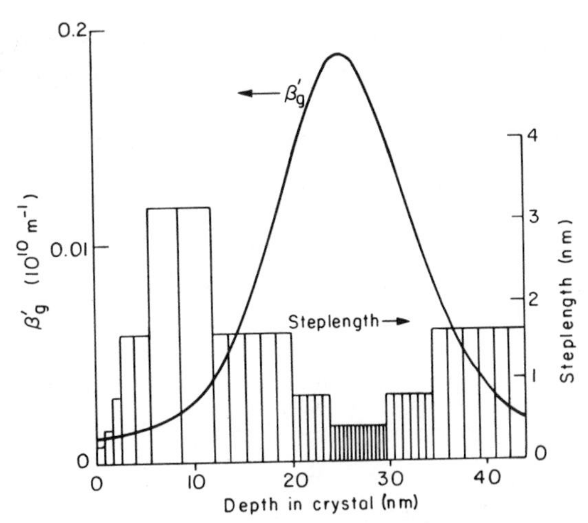

Fig. 5.26 An example of the steplength, L, used by an automatic integration routine (Nordsieck) for a screw dislocation. The histogram indicates the steplength and the smooth curve shows the value of β'. Courtesy D. J. H. Cockayne.

the desired accuracy, the second (slightly more accurate) solution is sufficient. If not, then $4n$ steps of $h/4$, and so on, are used until two successive answers agree. Note the implicit assumption that smaller step lengths produce greater accuracy —a feature common to all numerical solutions of differential equations. In Fig. 5.25*b* a check is made at each small step, and then h is doubled if it was successful, or halved if unsuccessful. This means that the step length decreases near the defect and increases after it. In Fig. 5.25*c* some criterion for h, based on experience, is adopted for the current value of **f**.

In the case of Fig. 5.25*b* or *c* the step length will vary through the integration. An example (from a self-adjusting Nordsieck routine carrying out an n-beam calculation in Bloch wave formulation) of how it varies is shown in Fig. 5.26. From a tentative start at 0.19 nm (a small fraction of the unit of length, which was the principal extinction distance) in strain-free material the step length was doubled to 0.39, 0.78, 1.56, and 3.12 nm. Then the value of β'_g became appreciable (full curve with peak), and the step length was reduced to 1.56, 0.78, and 0.39 nm at maximum β'_g, opening out, as β'_g decreased, to 0.78 and then 1.56 nm.

6 The Critical Voltage, V_c

The terms "critical voltage" and "disappearance voltage" have been used recently in the field of high voltage electron microscopy to define the voltage at which a number of related phenomena occur. Experimentally the voltage is indicated by the disappearance (or minimization of the intensity) of (*a*) a higher order bend contour in the image, (*b*) the corresponding spot in the diffraction pattern, or (*c*) the corresponding Kikuchi line pair in the diffraction pattern, or by a reversal in the asymmetry shown by related Kikuchi lines. Although a full discussion of the behavior of Kikuchi lines is beyond the scope of this book, the last-named symmetry factor gives a clue to the explanation of the phenomenon.

Returning to the *n*-beam Bloch wave solution for perfect crystal of Section 5.1 and to the dispersion surface diagram of Fig. 4.13 for a systematic row of reflections, it may be noted that at all reflecting positions (or Brillouin zone boundaries, i.e., where $-2k_{xy} = ng$, where *n* is an integer—the vertical lines in Fig. 4.13) the Bloch wave solutions must be either symmetric *S* or antisymmetric *A*. This property follows immediately from Bloch's theorem when the value of k_{xy}/g is integral or half integral and is the generalization of the two-beam situation discussed in chapter 4, section 4.3, Figs. 4.11 and 4.13. In the present context it is sufficient to note that at the *n*th-order reflecting position the product $\epsilon^{(j)} \times C_{ng}^{(j)}$ is positive for a symmetric Bloch wave *j* and negative for an antisymmetric wave. Furthermore, at such reflecting positions the waves with the largest excitation, $C_0^{(j)} = \epsilon^{(j)}$, are those with $j = n$ and $j = n + 1$, that is, the two Bloch waves "nearest" the modified free electron sphere. Thus, if all other waves may be neglected, the amplitude $\phi_{ng}(t)$ of the beam *ng* emitted at the bottom of a slab of perfect crystal of thickness *t* will be, from eq. 5.20,

$$\boldsymbol{\phi}_{ng}(t) = \sum_{j=n,n+1} C_0^{(j)} C_{ng}^{(j)} \exp(2\pi i \gamma_{\text{re}}^{(j)} t) \exp(-2\pi \gamma_{\text{im}}^{(j)} t). \qquad (5.25)$$

In general, eq. 5.25 predicts a varying nonzero diffracted amplitude for finite values of *t*, because of the changing phase relationships between the two waves. However, if the two values of γ_{re} are identical, the phase relationship is fixed and if, further, the premultiplying products are opposite in sign, $\phi_{ng}(t)$ will be small for all values of *t* [not necessarily zero, because of contributions from the neglected Bloch waves and the fact that the premultiplying factors $C_0^{(j)} C_{ng}^{(j)}$ and the absorption terms $\exp(2\pi\gamma_{\text{im}}^{(j)} t)$ are not strictly identical]. Hence the condition that would reproduce the experimental observation of a minimum (disappearance) in the intensity of the *n*th diffracted beam is the degeneracy (i.e., same value of γ_{re}) or "touching" of the two branches of the dispersion surface, n and $n + 1$. In the eigenequation (eq. 4.65) from which the values of γ_{re} are derived the matrix A_{re} has off-diagonal elements involving U_{g-h} which depend on the accelerating voltage through the relativistic mass of the electron (eq. 4.46). All terms in A_{re} also depend on voltage through the wave vector **k**, but for constant relative orientation (e.g., at the exact reflecting position) the effect is only that of a scaling factor (since $s \propto 1/K$). Thus it is to be expected that for any material a number of "critical" voltages will exist at which different higher

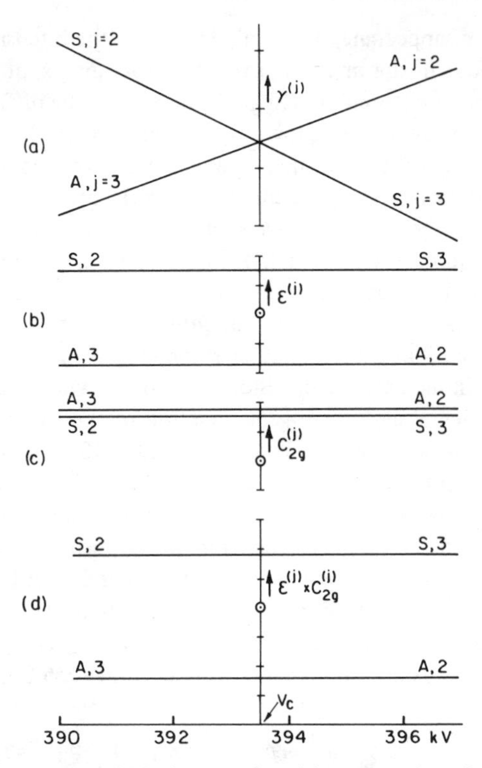

Fig. 5.27 Various parameters of Bloch waves 2 and 3 as a function of voltage, at the exact second-order reflecting position; copper $\mathbf{g} = 111$, five-beam calculation. (*a*) The propagation parameter γ_{re} and symmetry (symmetric S if $C_0^{(j)}$ and $C_{2g}^{(j)}$ have the same sign, antisymmetric A otherwise). (*b*) The excitations $\epsilon^{(j)} = C_0^{(j)}$. (*c*) The second-order coefficient $C_{2\mathbf{g}}^{(j)}$. (*d*) The product $C_0^{(j)} C_{2\mathbf{g}}^{(j)}$.

order Bloch wave degeneracies occur. The second-order effect (i.e., minimum in ϕ_{2g}) was the first to be discovered and is probably the easiest to observe.

To further amplify the points made above Fig. 5.27 illustrates how a number of the relevant parameters of the two principal Bloch waves vary with the energy of the incident electron beam. The propagation factors γ_{re} of the two waves vary almost linearly with voltage, so that the critical voltage V_c can be determined rapidly and accurately by interpolation between voltages on either side. Over the small range of voltage plotted the components of the Bloch waves are essentially constant, except for the abrupt interchange of symmetry which occurs exactly at the critical voltage. A similar variation is seen in the absorption factors.

Whether or not a critical voltage will be found in the range of voltages available on a particular microscope depends on a large number of variables (the value of

Table 5.1 Second-Order Critical Voltages (kV) Calculated (*a*) from Eq. 5.27 and (*b*) by Solution of the *n*-Beam Equations ($n = 9$)[a]

		V_c (2g) in kV	
Material	2*g*	(*a*)	(*b*)
Aluminum	222	653	685 ± 5
	400	1322	1345
Copper	222	695	615
	400	1089	995
Gold	222	103	55
	400	283	245

[a]Note the large discrepancies between this table and Table 2.5; these are produced by quite small changes in the values of the scattering factors used in the calculations.

g and all the Fourier components U_{ng}). The most important terms for the second-order critical voltage (which is the one most commonly determined) are those in U_g and U_{2g}, and an approximate three-beam condition may be found for the degeneracy of branches 2 and 3 (see Exercise 5.3):

$$U_{2g}^2 + U_{2g}g^2 = U_g^2. \tag{5.26}$$

When the relativistic factors are inserted and the equation is rearranged, the formula for the value of V_c is found to be

$$V_c(2g) = \frac{2.26 \times 10^4}{[(\xi_{2g}^{100})^2 - (\xi_g^{100})^2]} (\xi_g^{100})^2 \xi_{2g}^{100} g^2 - 5.12 \times 10^5 \tag{5.27}$$

where, if the extinction distances ξ_g^{100} (two-beam values calculated for 100 keV electrons) are measured in Å and the basic reflecting vector $|\mathbf{g}|$ in reciprocal Å, the value of $V_c(2g)$ is in volts. The accuracy of eq. 5.27 is indicated by the voltages shown in Table 5.1, where values calculated using the three-beam equation are compared with those obtained from a full solution of the *n*-beam equations for the three "standard" materials (see Table 4.1).

The potential usefulness of measurements of the critical voltage as an analytical tool may also be seen by inspection of eq. 5.27, although of course accurate work or measurements at orders higher than the second demand full solution of the *n*-beam equations. Under conditions where the predicted value of $V_c(2g)$ is in the range of present-day high voltage electron microscopes (100 to 1500 kV, say) small changes in the relationship between the first- and second-order scattering factors produce relatively large changes in V_c. Anything that affects this relationship is thus susceptible to detection by measurement of the critical voltage.

Fig. 5.28 Convergent beam electron diffraction pattern of the systematic 111 row from a small area of a specimen of Cu-10 at. % Al at 110°C. The dark central line in the second-order disk indicates that the microscope was at the critical voltage for g = 222 (300 kV) at this temperature. Courtesy D. Imeson.

The accuracy to which the critical voltage may be measured depends on the technique adopted, which in turn may depend on the material under investigation. One simple method is to detect the minimization of the intensity of the diffraction spot, but the accuracy obtained is low. The second method is to observe the minimization of the intensity in the corresponding dark field image, which, of course, allows differences between different areas of the specimen to be seen in a straightforward fashion. Again accuracy is relatively low (~10 kV) because of the slow variation in the intensity near the minimum. The most accurate methods for measurements on areas of specimen of the size of the diffraction selected area are (*a*) observation of the Kikuchi lines, which, in addition to passing through a minimum in intensity, suffer changes in symmetry,[12] and (*b*) minimization of the second-order diffraction spot in a convergent beam diffraction pattern. An example of a determination using the latter technique is shown in Fig. 5.28, where the characteristic dark central line of the rocking curve may be clearly seen in the second-order diffraction disk. The voltage determined by any of these methods is measured most accurately by a Kikuchi line method (see Chapter 2, Section 5.5.5).

The effects discussed in this section occur for the simplest orientations of the specimen, that is, a single, systematic row of reflections; and, as previously noted, the many-beam effects leading to critical voltages are inevitable. Similar effects occur when other nonsystematic reflections operate; the changes in Kikuchi lines that occur under such conditions have been discussed in some detail by Gjønnes and Høier.[13]

Current areas of research using critical voltages include the following: (*a*) more accurate determination of electron scattering factors and hence new knowledge of electron distribution and bonding (by refining the relative magnitudes of the extinction distances);[14,15] (*b*) determination of Debye-Waller factors—$V_c(2g)$ is

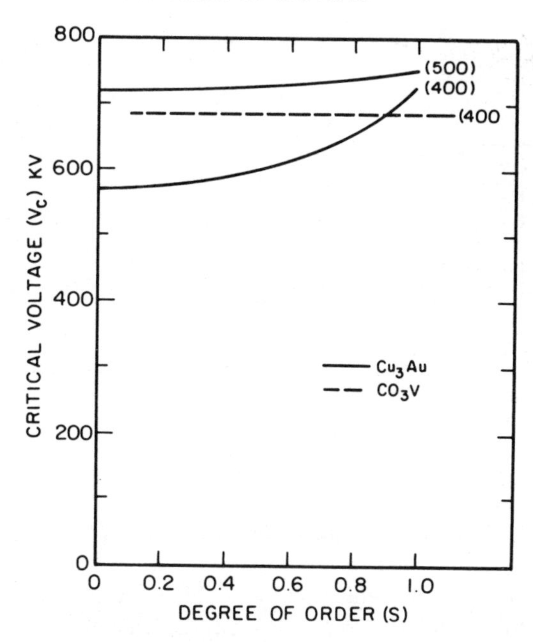

Fig. 5.29 Calculated variation in critical voltage with degree of order S for Cu_3Au and Co_3V. The critical voltage is sensitive to order in the case where scattering factors for the two species are dissimilar. Courtesy Taylor and Francis.[17]

in general reduced by atomic vibrations that increase ξ_{2g} more than ξ_g (see footnote to eq. 4.13), that is, V_c is reduced by an increase in temperature, as in the example shown in Fig. 5.28; (*c*) detection of local inhomogeneities or segregation of atoms with different scattering factors;[16] and (*d*) determination of the degree of order[17] (possible when the systematic row includes both matrix and superlattice reflections, the relationship between them being sensitive to the degree of order; see the example in Fig. 5.29). This field of research activity is in a state of flux, and it remains to be seen in how many systems useful results will be achieved in areas such as those just mentioned.

7 Factors Affecting Resolution of Defects

In this section consideration is given to some general principles of the way in which strain fields produce contrast and hence as to how the operating conditions of the microscope may be set to produce optimum resolution. In general, the images produced under diffraction contrast conditions (e.g., the dislocation images of Fig. 5.13) are very much larger than the ultimate resolving power of

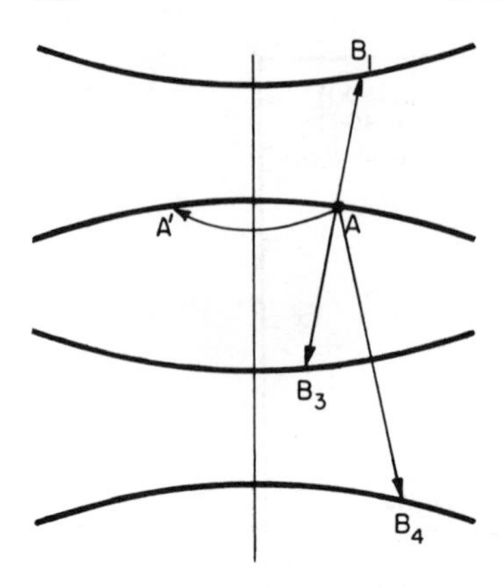

Fig. 5.30 Schematic diagram of Bloch wave scattering.

the microscope, and hence the latter factor will be neglected. For crystalline materials the resolving power of the modern instrument is fully utilized only in the lattice and structure imaging modes discussed in Sections 9.3 and 9.4, and in Chapter 3, Sections 12 and 13.

The Bloch wave scattering equation (eq. 5.23) for a strained material may be written in a slightly different form to pick out the jth component:

$$\frac{d\epsilon^{(j)}}{dz} = 2\pi i \left\{ \sum_l \epsilon^{(l)}(z) \exp\,[2\pi i(\gamma^{(l)} - \gamma^{(j)})z] \sum_g C_g^{(j)*} C_g^{(l)} \beta_g' \right\}. \tag{5.28}$$

This equation describes how the electrons scatter from branch to branch under the influence of the strain field and is shown schematically in Fig. 5.30. Now, by a suitable phase change, intraband scattering (that, is $j \to j$, such as AA') may be removed completely. Then, if the weak scattering approximation is made, that is, if one sets

$$\epsilon^{(l)}(z) = \epsilon^{(l)}(0) = C_0^{(l)} = \text{const},$$

eq. 5.28 integrates to yield

$$\Delta\epsilon^{(j)} = \epsilon^{(j)}(z) - \epsilon^{(j)}(0) =$$

$$2\pi i \sum_{l \neq g} \sum_g \epsilon^{(l)} C_g^{(j)*} C_g^{(l)} \int_0^t \beta_g' \exp\,[2\pi i(\gamma^{(l)} - \gamma^{(j)})z]\; dz. \tag{5.29}$$

If, furthermore, the limits of integration may be extended to infinity (i.e., if $\beta_g' = 0$ outside a small volume), then $\Delta\epsilon^{(j)}$ (which, representing a difference from what pertains in perfect crystals, has something to do with "contrast") is given principally by the Fourier transform of β_g' with respect to the $\Delta\gamma$ between the wave in question and the wave with the largest excitation, that is, with the largest term on the right-hand side of eq. 5.29. Thus image contrast will be strong where β_g' is appreciable over a distance of the order of the corresponding extinction distance ξ^{eff} $(=1/\Delta\gamma)$. Given that dislocation strain fields are approximately cylindrical, this implies that dislocation image widths are on the order of ξ^{eff}. In practice the width is usually taken to be $\xi^{\text{eff}}/3$. It also follows that image widths should be smaller if the important $\Delta\gamma$ is larger (that is, with a smaller effective extinction distance).

In the following sections possible ways of reducing the effective extinction distance between the important Bloch waves are considered.

7.1 Weak-Beam Technique

One of the easiest ways of decreasing the effective extinction distance (thinking for the present in two-beam terms) is to increase the deviation parameter s_g. Equation 4.43 is rewritten for ease of reference:

$$\xi_g^{\text{eff}} = \xi_g/\sqrt{1 + s_g^2\xi_g^2}\,. \tag{5.30}$$

If, in addition, the initial value of ξ_g is small (e.g., because a low index reflection is used), the desired effect is obtained. To obtain an estimate of the resulting image width (and hence the "resolution" of closely spaced, similar defects) the criterion $\xi^{\text{eff}}/3$ may be used. For a typical (copper $\mathbf{g} = 111$) value of $\xi_g = 30$ nm eq. 5.30 predicts that to obtain a width of 2 nm or less the value of the deviation parameter should be $s_g \geqslant 2 \times 10^{-8}\ \text{m}^{-1}$. Now, if s_g is large the kinematical theory is a reasonable approximation, and this approximation allows a very simple explanation of the image peak position to be given, as outlined below.[19] Again for convenience the kinematical integral of eq. 4.19 is rewritten as

$$\phi_g = \frac{\pi i \phi_0}{\xi_g} \int_0^t \exp\left[-2\pi i(sz + \mathbf{g}\cdot\mathbf{R})\right] dz, \tag{5.31}$$

The parts of the integration over which the argument of the exponential (the phase) is varying produce oscillatory factors whose contributions are negligibly small if there is a nonoscillatory part. The condition for the presence of a nonoscillatory section is that the "phase" be constant over a range of z, that is,

$$\text{phase} = -2\pi(sz + \mathbf{g}\cdot\mathbf{R}) = \text{const}$$

or

$$s + \frac{d}{dz}(\mathbf{g}\cdot\mathbf{R}) = 0. \tag{5.32}$$

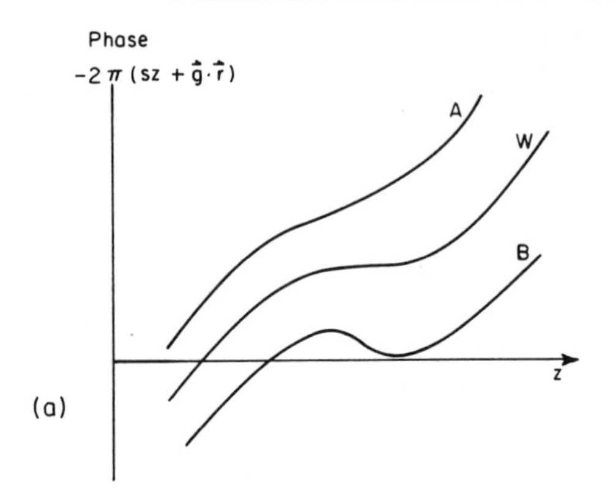

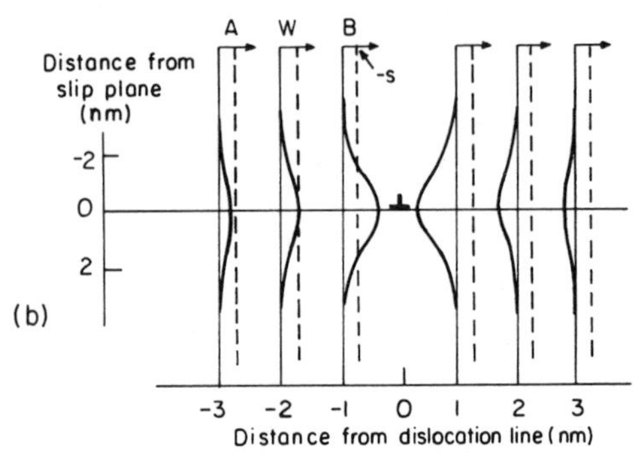

Fig. 5.31 (*a*) Schematic variation of the phase in the kinematical integral (eq. 5.32) for columns at various positions A, W, B relative to an edge dislocation; see (*b*). (*b*) Variation of β' for columns near an edge dislocation with a typical value of s denoted by the dashed line. The rate of change in the phase—see (*a*) – is proportional to $(s + \beta')$ and so is zero when $\beta' = -s$. Courtesy D. J. H. Cockayne.

This is a standard "stationary phase" condition—the phase should be at a turning point. Furthermore, the contribution to ϕ_g will be largest from the column of crystal that has stationary phase over the greatest distance (see Fig. 5.31). This is achieved if

$$\frac{d^2}{dz^2}(\mathbf{g} \cdot \mathbf{R}) = 0 \tag{5.33}$$

as well—a condition of "most stationary phase." The image will be on one side of the dislocation where β_g' $[=(d/dz)\,\mathbf{g} \cdot \mathbf{R})]$ and s are opposed. This is illustrated in Fig. 5.31*b*; the weak-beam image peak would be predicted to be near W (2 nm from the actual dislocation position).

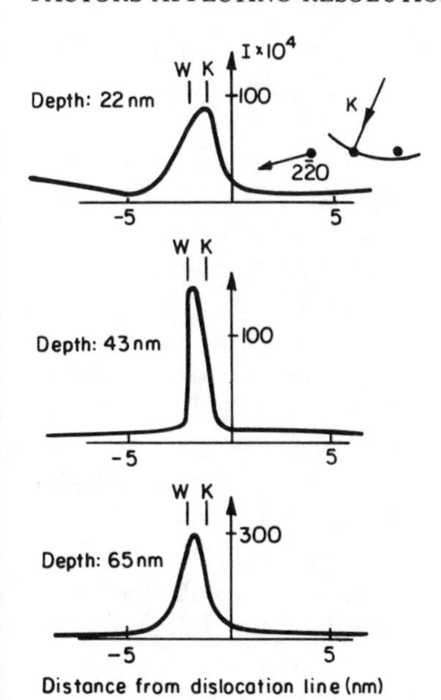

Fig. 5.32 Many-beam calculated intensity in the weak dark field beam $\mathbf{g} = 2\bar{2}0$ for the orientation indicated in the inset for a dislocation at different depths in a specimen, with the weak beam, W, and kinematical, K, image positions indicated. Courtesy D. J. H. Cockayne.

These very simple criteria (eq. 5.32 and 5.33) predict the image position W by simple differentiation of the strain field. Hence no lengthy computation is required.

The approximate kinematical theory could also be used to perform the whole integration of eq. 5.31 and thus work out the image shape (see, e.g., Fig. 7.17 of ref. 20). The image peak position so obtained is denoted as K. The third (and most accurate) way to predict the image contrast is to integrate the many-beam or many-wave equations (probably the latter since large s values are intrinsically slow in beam notation integration).

The sort of comparisons found are typified by Fig. 5.32. Here the calculated dark field intensity is presented for $\mathbf{g} = 2\bar{2}0$ (with $\mathbf{g} = \overline{2}20$ not quite satisfied) for a dislocation at different depths in a foil. It should be noted that (*a*) the "true" image is very narrow (~2 nm), (*b*) the peak oscillates in position by much less than this width, (*c*) the peak is displaced to somewhere roughly between the kinematical K and simple weak-beam W peak positions, (*d*) the contrast is bright on a dark background, (*e*) the contrast is large because the background intensity is very low, and (*f*) the intensity is low ("weak"). Note also the convenient notation ($\mathbf{g}$, $n\mathbf{g}$) for a systematic row, the first parameter, $\mathbf{g}$ being the reflection used to form the image, and the second parameter, $n\mathbf{g}$ being the position at which the Ewald sphere cuts through the systematic row (other than the origin). For the example in Fig. 5.32 the conditions are ($-\mathbf{g}$, 1.1 $\mathbf{g}$) with $\mathbf{g} = \bar{2}20$.

In principle it may be seen from the above discussion that the larger the deviation parameter used the narrower is the image (eq. 5.30) and the nearer is the image peak to the true defect position (Fig. 5.31), that is, the better is the resolution in the senses of both separation from other images and definition of defect geometry. However, it also follows that the image peak intensity is lower (the distance over which the phase is zero is shorter–eq. 5.31), falling roughly as s_g^{-2} (cf. eq. 4.15). Since there inevitably is additional background intensity in the experimental micrograph which is insensitive to s_g, the contrast is in practice reduced as s_g becomes very large. Consequently the minimum value of s_g consistent with the required resolution ($\sim\xi^{\text{eff}}/3$, eq. 5.30) should be used in practice. The appropriate experimental conditions are illustrated in Fig. 1.21.

A typical example of the use of the weak beam technique is shown in Fig. 5.33. In Fig. 5.33*a* the standard bright field image of a specimen of Cu-10 at. % Al is shown, while Fig. 5.33*b* shows the same area under weak beam dark field conditions. In the latter picture the dissociated nature of the dislocations may be clearly seen, with an image width of less than 2 nm. Note in particular the clarity with which the dissociated and constricted segments of the dislocations are shown.

If the problem is reexamined by means of *n*-beam theory, it follows from eq. 5.20 that the change in the amplitude of the beam *g* at the bottom of the crystal slab of thickness *t* caused by any defect is given by

$$\Delta\phi_g(t) = \sum_j \Delta\epsilon^{(j)} C_g^{(j)} \exp\,(2\pi i \gamma^{(j)} t), \tag{5.34}$$

where the summation is over the significant Bloch waves j, $\Delta\epsilon^{(j)}$ being given by eq. 5.29. The wave or waves which are significant in their contribution to $\Delta\phi_g(t)$ are those with *both* $C_g^{(j)}$ and $\Delta\epsilon^{(j)}$ large. From eq. 5.29 the dominant contribution to $\Delta\epsilon^{(j)}$ is most likely to come from the waves that have the largest excitation $\epsilon^{(l)}$; that is, if differences in the absorption of the Bloch waves are neglected, contrast is produced by scattering *from* the waves of largest excitation *into* the Bloch waves of largest beam component in the direction of interest. In general, the waves with the largest excitation $\epsilon^{(l)}$ are those nearest the modified free electron sphere (e.g., $l = 4, 5$ at the fourth-order reflecting position C in Fig. 4.13; $l = 5$ only between the fourth and fifth reflecting positions at D), while the waves of largest component $C_g^{(j)}$ are (as a consequence of the periodicity of Bloch waves, discussed in Chapter 4, Section 6.3) those nearest the sphere when the order of the reflection satisfied is decreased by 2g (e.g., $j = 2, 3$ at C for $g = 1$; $j = 3$ at D). Thus near a reflecting position there are four possible transitions of importance $l \rightarrow j$, each of which in a sense gives rise to an "image." However, away from an exact reflecting position, only one transition occurs. As the order of reflecting involved increases, the $\Delta\gamma$ values for all these transitions

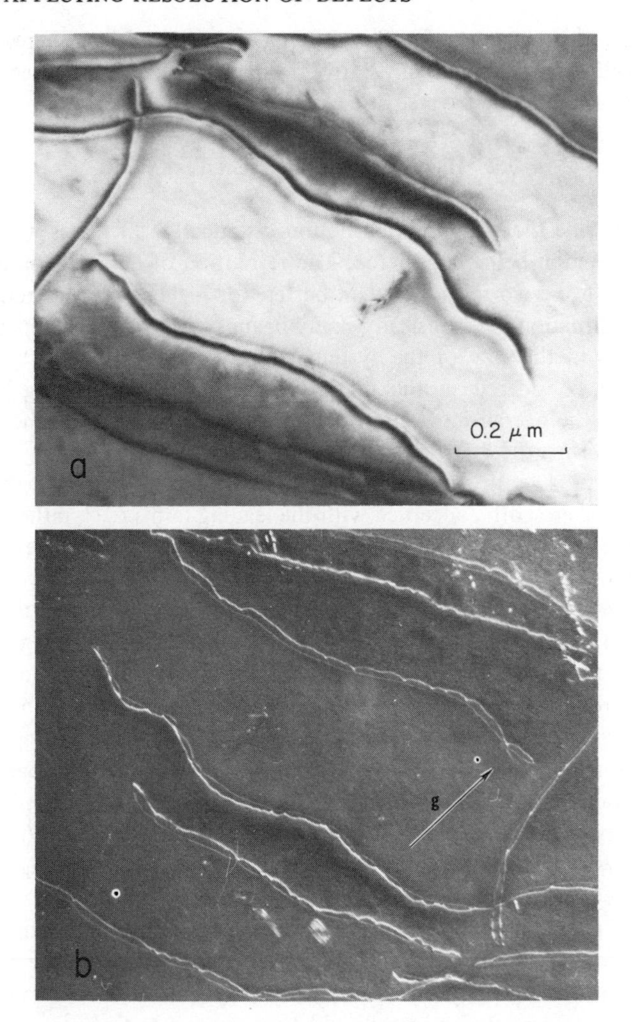

Fig. 5.33 Images of disociated dislocations in Cu-10% Al, $\mathbf{b} = \frac{1}{2}$ [110] → $\frac{1}{6}$ [21$\bar{1}$] + $\frac{1}{6}$ [121] imaged in $\mathbf{g} = 220$, that is, $\mathbf{g} \cdot \mathbf{b} = 2$ with $\mathbf{g} \cdot \mathbf{b}^p = 1$ for each partial dislocation. (*a*) Bright field near the reflecting position. (*b*) Weak beam dark field, showing image width of less than 2 nm. Courtesy C. B. Carter.

increase as the free electron sphere "tilts over," that is, the effective extinction distance decreases and the Fourier transform in eq. 5.29 implies improved resolution for larger values of $\Delta\gamma$. Thus the best resolution, in the sense of narrowest and least complicated images, is achieved with large values of s_g with no reflection exactly satisfied (i.e., between high order reflecting positions). The stricture on no other reflection being satisfied is, of course, more severe for nonsystematic reflections than for the systematic case considered here, since even more complication is potentially introduced by the different direction of the extra reflection in the nonsystematic case.

7.2 High Order Bright Field Technique

As a general rule the two-beam extinction distance ξ_{ng} increases with the order n of the reflection being considered. The resolution of defects would therefore be expected to be worse, rather than better, for higher order reflections. However, if the situation is intrinsically many-beam (large many-beam parameter M, eq. 5.22), as in the case of the systematic row of reflections with small ξ_g at high voltage considered in the preceding section, the scattering behavior is governed by the overall interaction of the waves rather than by the weak two-beam effect. Thus the contrast may again be considered as being given in some way by eqs. 5.34, with $g = 0$ for the bright field image. However, the terms in $C_g^{(j)}$ and $\Delta\epsilon^{(j)}$ are both concerned with the leading terms $C_0^{(j)}$, rather than being two separate quantities. This means that the significant scattering events cannot be picked out by consideration of the terms $C_g^{(j)}$ alone; the overall Fourier transforms of eq. 5.29 and the different absorption coefficients of the waves are relevant. With only excitation and absorption taken into account, the principal mechanism producing contrast (dark compared with background) would be scattering from waves of large excitation and small absorption into waves of significantly larger absorption. Wave 1 is always in the latter category, and at any reflecting position the differences between the absorption parameters of adjacent branches are accentuated in comparison with the differences found when the crystal orientation is away from the reflecting position (see, e.g., the parameter γ_{im} shown schematically for the two-beam case in Fig. 4.11*c*). As already noted, two waves at least are strongly excited at any reflecting position, while away from such orientations one wave only is dominant. Thus, as in the weak beam case, the image is, in principle, more complicated if the crystal is set at an exact reflecting position rather than in any general orientation. Also as in the weak-beam case, the images should be narrower as the intersection of the Ewald spheres moves out along the systematic row, since the values of $\Delta\gamma$ between the highly excited waves and *all* other waves (with the exception of the adjacent wave) increase (see Fig. 4.13).

7.3 Images at Critical Voltages

Another way of improving resolution would appear to be to remove the interference between a pair of Bloch waves by making them coincident, that is, $\Delta\gamma = 0$. The resultant infinite effective extinction distance should have no effect on the image, and hence the contrast would be dominated by scattering to more distant branches of the dispersion surface, with consequently improved resolution. The coincidence condition is attained at a critical voltage (see Section 6) at the exact reflecting position for the higher order reflection $n\mathbf{g}$ involved. As shown previously, the background intensity in this reflection is, in reasonably thick crystals, extremely low, and thus any changes introduced by defects should produce images of high contrast. However, the images are of necessity complex in principle for the reasons discussed above, since it is implicit in the method that the crystal be set at an exact reflecting position. The scattering mechanisms discussed above are thus consistent with the experimental observation of rather wide, high contrast images in the critical voltage dark field.

7.4 Comparison of Methods

The preceding discussions of Bloch wave scattering and image resolution and contrast make it possible to compare the methods proposed for achieving high resolution. In principle the best resolution is attained in either weak beam or high order bright field (in the latter provided that the situation is intrinsically n-beam) at very large deviations, when no other reflection is exactly satisfied. The problem with either method is that the contrast attained in practice, whether positive as in the weak beam case or negative as in bright field, is low. In the dark field case inelastic scattering events not considered at all in the theory may increase the background intensity and decrease the signal-to-noise ratio unacceptably, while in the bright field case the contrast is intrinsically small. In both cases the observed image contrast increases as the deviation is decreased until an acceptable compromise orientation is reached at which the image brightness and contrast are such that focusing is possible (and exposure not unacceptably long) and reasonable resolution is obtainable; see, for example, the comparison made in ref. 21. From published material the two methods appear similarly successful in practice in obtaining clear-cut images at high voltages, but at conventional (100 kV) voltages the weak beam method is obviously superior. At critical voltages, images at the exact reflecting position are of high contrast but poor resolution.

Discussion of another method recently proposed to improve resolution at high voltages, multibeam imaging,[22] is deferred until the general properties of images formed by using more than one beam are discussed in Section 9.

8 General Symmetry Principles

In common with other forms of wave propagation electrons obey a fundamental relationship generally known as reciprocity, which was first applied to diffraction phenomena in the electron case by von Laue. In a strict sense the principle may be applied only to elastically scattered electrons, but recently Howie[23] has shown that, if the energy lost is not too great, inelastically scattered electrons may also be considered. This means that, as well as being applicable to images in the electron microscope, certain properties of Kikuchi diffraction patterns may also be deduced. A formal statement of the principle is as follows: "If a certain signal is detected at a point $\mathbf{r}_2$ when a source is at point $\mathbf{r}_1$, the same signal in amplitude and phase would be detected at $\mathbf{r}_1$ if the source were at $\mathbf{r}_2$." For the Fraunhofer diffraction situation usually present in the electron microscope the positions $\mathbf{r}_j$ of source and detector outside the specimen may be interpreted as directions, and hence the properties of either bright field or any dark field images discussed.

An application of the principle to deduce the relationship between images of deformed crystals is shown schematically in Fig. 5.34. At the left of each sequence the standard orientation of source S, detector D (the objective aperture selecting the beam considered–heavy line), and crystal, containing defects near its upper surface, is shown. The deformation along the reference column (•) of the crystal traversed by the beam is defined as usual by the displacement $\mathbf{R}(z)$. Application of the reciprocity relation R yields the second figure (2) by interchange of source and detector. The next step depends on the symmetry of the crystal structure of the specimen; in Fig. 5.34 the crystal is assumed either to be centrosymmetric (and therefore invariant under an inversion I through its center) or to possess a horizontal mirror plane M_H, or possibly both. Such symmetries have been applied in moving from (2) to (3). The final step (if required) is to transform the position of the defects, which are now near the bottom of the specimen, to the same side of the reference column as were the original defects, having regard to the symmetry of their strain fields. This operation, denoted as D, may or may not change the "sign" of the defect, depending on the strain field; compare (3) and (4). Alternatively a vertical mirror plane M_v has been applied. Comparing Figs. 5.34a(1) and a(4), and noting that a constant displacement $\mathbf{R}_0$ of the whole crystal may be made without altering intensities, it may be formally stated that bright field images from columns with displacement fields defined by $\mathbf{R}_1(z)$ and $\mathbf{R}_2(z)$ are identical if

$$\mathbf{R}_2(z) = \mathbf{R}_0 - \mathbf{R}_1(t - z), \tag{5.35}$$

the relationship deduced by Howie and Whelan[24] after consideration of scattering matrices. Note that this relationship holds under n-beam conditions and all orientations.

A somewhat more restricted relationship is apparent in dark field [compare Figs. 5.34c(1) and c(3) and take the exact orientation of the reflecting planes into account], namely, that under two-beam conditions and at the exact reflecting position ($\phi = \theta_g$) the dark field images from columns with displacement

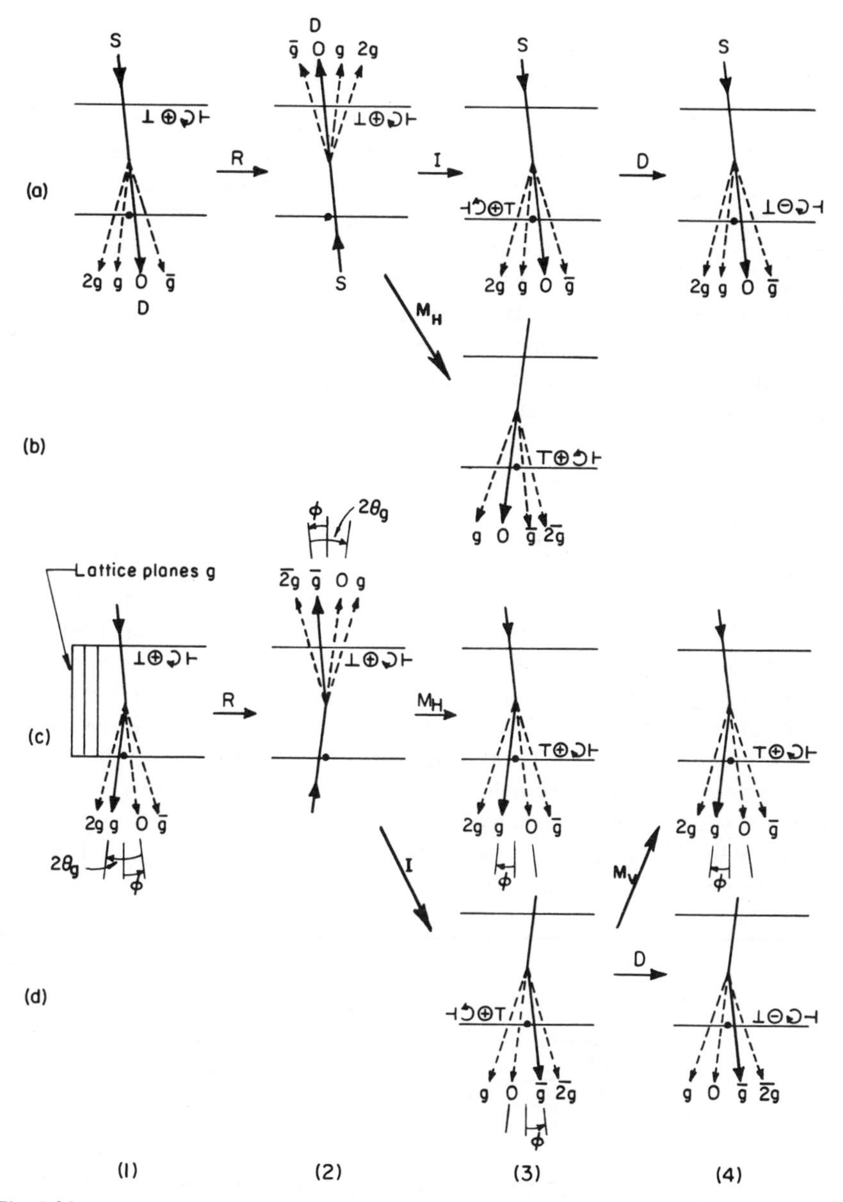

Fig. 5.34 Application of reciprocity (columns 1 and 2) followed by operation of specimen symmetry (columns 2 and 3) and defect strain-field symmetry (columns 3 and 4), for a number of defects and reflecting conditions (*a–d*).

fields $\mathbf{R}_1(z)$ and $\mathbf{R}_2(z)$ are identical if

$$\mathbf{R}_2(z) = \mathbf{R}_0 + \mathbf{R}_1(t - z), \tag{5.36}$$

as deduced by Ball.[25] Figs. 5.34*a* and *b* show, via the alternative paths M_H and I, the identity of images formed in bright field when both incident beam and defect strain field are reversed [i.e., images depend on $(\mathbf{g} \circ \mathbf{R})$ and are the same for $-\mathbf{g}, -\mathbf{R}$]; compare Figs. 5.34*a*(4) and *b*(3). Similarly, Figs. 5.34*c* and *d* show the invariance in dark field at the exact reflecting position; compare Figs. 5.34*c*(3) and *d*(4).

Some of these principles may be applied by considering Fig. 5.14—the bright field image of the dislocation is symmetrical with respect to the center of the specimen as shown by Figs. 5.34*a*(1) and *a*(4). Further application of the reciprocity relation will be made in Section 10 on scanning transmission electron microscopy.

9 Images Formed by More Than One Beam

As mentioned in Chapter 1, Section 2.1, the resolution of the microscope is affected by the compromise chosen between large objective aperture (to achieve better resolution in the Abbé sense) and small aperture (to reduce aberrations or to select only one beam with which to form the image, or both). The purpose of the present section is to show how diffraction contrast image calculations that have been carried out predominantly with one emitted beam in mind (bright field, or one of several dark field, images), may be easily extended to cover the situation where more than one beam contributes to the image.

The general situation is shown schematically in Fig. 5.35. The incoming electron beam is modified by the specimen to produce an angular spread of

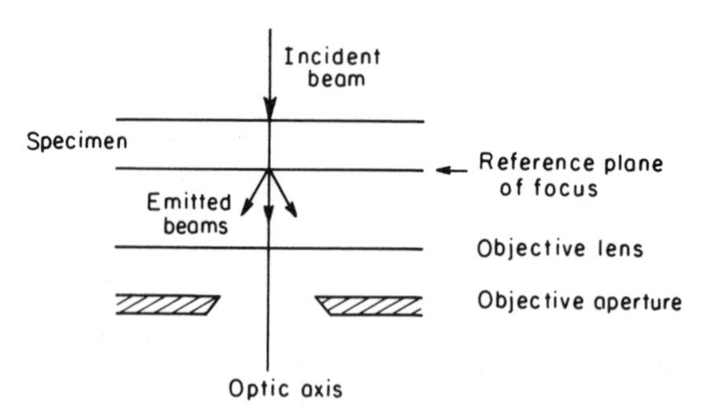

Fig. 5.35 Schematic diagram pertinent to the effect of objective lens aberration and defocus.

emitted electrons at the exit surface; in the case of crystalline specimens this angular spread is confined to discrete Bragg reflection directions **g**. The lenses of the microscope form the image using the beams that pass through the objective aperture. In the absence of any lens aberrations the in-focus image bears an exact correspondence to the beams emitted from the specimen. In this case the amplitude $\phi(\mathbf{r})$ at the point **r** [neglecting the phase factor $\exp(2\pi i\mathbf{k}\cdot\mathbf{r})$ common to all beams] is given by the sum of the individual beam amplitudes ϕ_g:

$$\phi(\mathbf{r}) = \sum_g \phi_g \exp\{2\pi i[\theta_g(\mathbf{r}) + \mathbf{g}\cdot\mathbf{r}]\}, \tag{5.37}$$

where $\theta_g(\mathbf{r})$ are the phase factors corresponding to the beams **g** *produced by the specimen* (see, e.g., the factors F_n^- in eq. 5.11 and Q^- in eq. 5.18), and the factor $\mathbf{g}\cdot\mathbf{r}$ takes account of the different beam directions outside the specimen. Thus in the absence of lens aberrations (or in circumstances where such aberrations may be neglected) eq. 5.37 shows that a simple sum should be performed to calculate the image amplitude.

9.1 Effects of Lens Aberrations: Contrast Transfer Functions

As noted in Chapter 1, Section 4.8, for work at high resolution the effects of aberrations and defocus of the objective lens must be taken into account. The reasons for this and, in particular, the factors relevant to lattice imaging are considered briefly in this section. For a fuller discussion see ref. 26.

For axial illumination the phase χ, relative to that of an undeviated axial beam, of a beam emitted from the specimen at an angle α to the optic axis is (taking only spherical aberration C_s and objective defocus D into account)

$$\chi(\alpha) = \frac{\pi}{\lambda}\left(\frac{C_s\alpha^4}{2} + D\alpha^2\right). \tag{5.38}$$

For reflection from a particular set of lattice planes **g** this may be rewritten, using Bragg's law $\lambda g = 2\theta$ and noting that the angle α is, for an incident beam along the optic axis, equal to 2θ, as

$$\chi(g) = \pi\lambda g^2\left(\frac{C_s\lambda^2 g^2}{2}\right) + D. \tag{5.39}$$

A beam **g** is imaged as $\exp[i\chi(\mathbf{g})]$ by the lens, that is, as a complex quantity, which may be plotted as a function of $|\mathbf{g}|$ for the applicable values of C_s and λ. Such a plot is known as the contrast transfer function (CTF) of the lens, and typical curves are shown in Fig. 5.36. These curves have been plotted for three voltages, assuming $C_s\lambda$ = constant (see Fig. 1.5), to illustrate the improvement in resolution reasonably anticipated with increasing voltage. If the contrast in

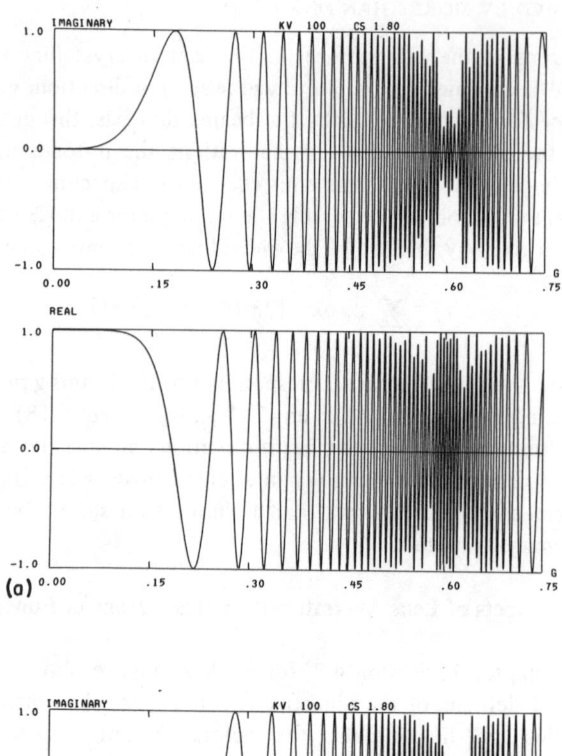

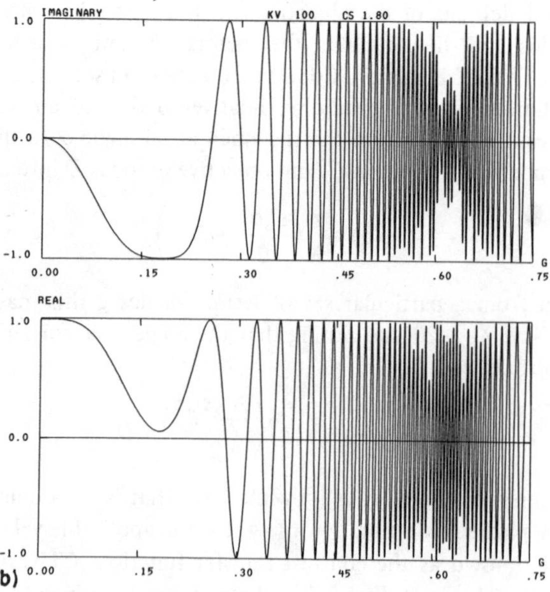

Fig. 5.36 See page 319 for legend.

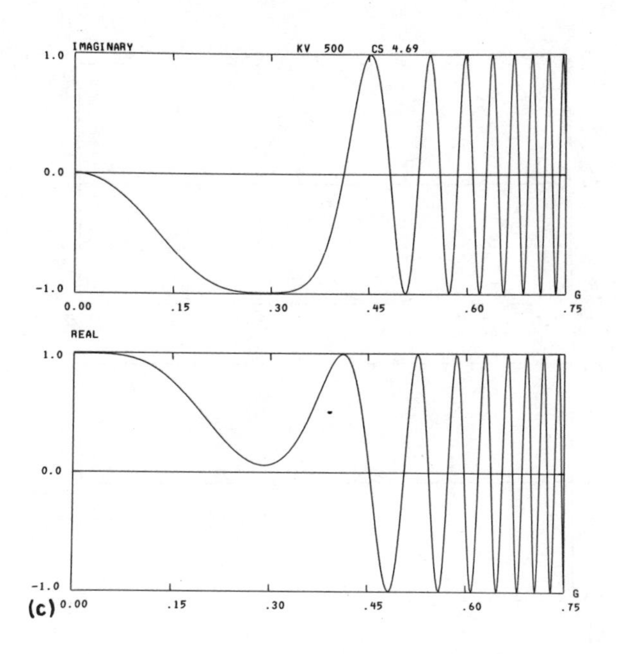
IMAGINARY
KV 500
CS 4.69
1.0
0.0
-1.0
0.00
.15
.30
.45
.60
.75
G
REAL
1.0
0.0
-1.0
(c) 0.00
.15
.30
.45
.60
.75
G

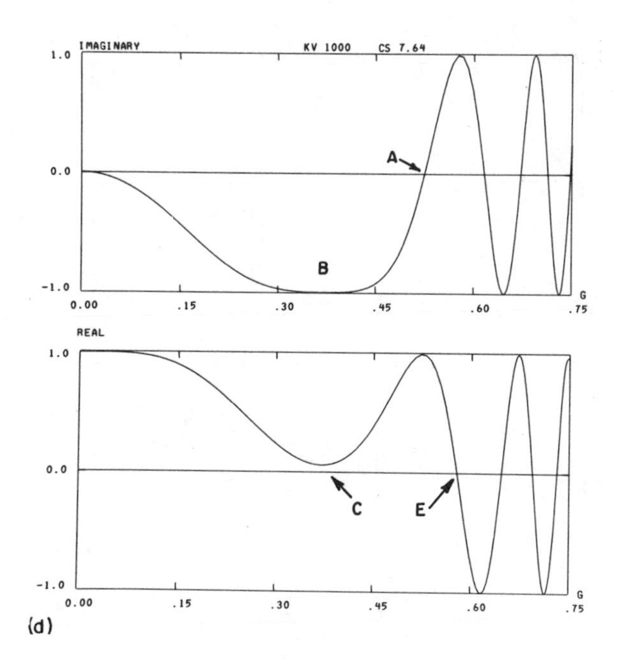
IMAGINARY
KV 1000
CS 7.64
1.0
0.0
-1.0
A
B
0.00
.15
.30
.45
.60
.75
G
REAL
1.0
0.0
-1.0
C
E
0.00
.15
.30
.45
.60
.75
G
(d)

Fig. 5.36 *cont.*

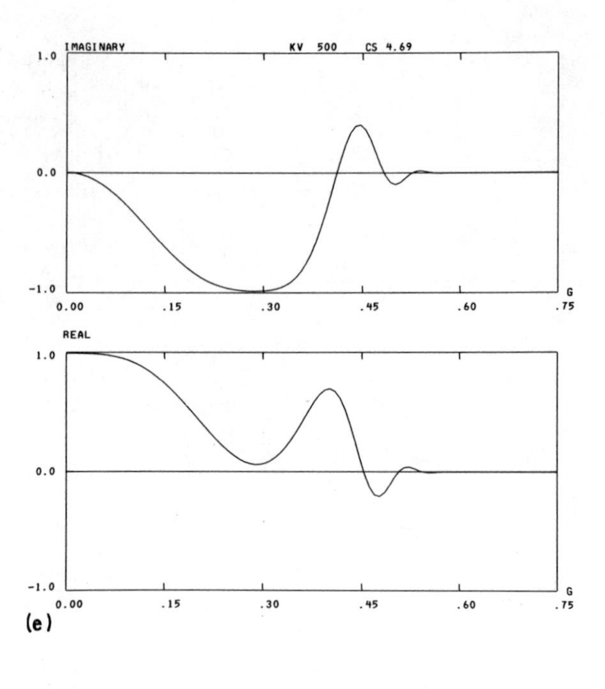
IMAGINARY
KV 500
CS 4.69
1.0
0.0
-1.0
0.00
.15
.30
.45
.60
.75
G
REAL
1.0
0.0
-1.0
0.00
.15
.30
.45
.60
.75
G
(e)

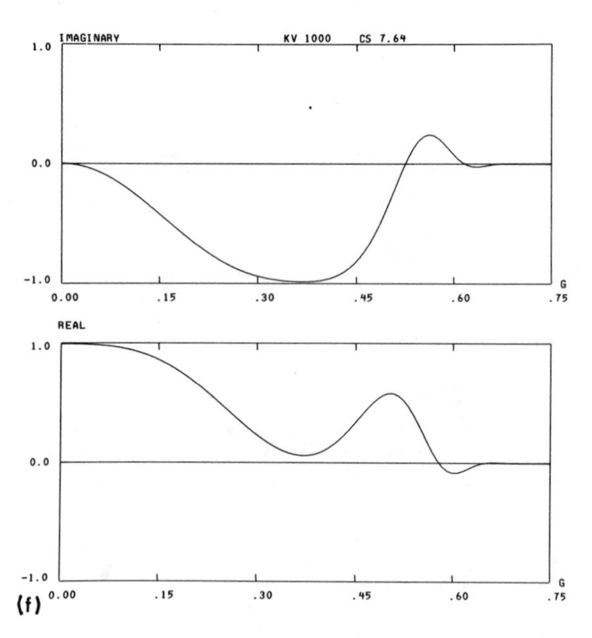
IMAGINARY
KV 1000
CS 7.64
1.0
0.0
-1.0
0.00
.15
.30
.45
.60
.75
G
REAL
1.0
0.0
-1.0
0.00
.15
.30
.45
.60
.75
G
(f)

Fig. 5.36 *cont.*

the features under consideration is amplitude contrast, the real part of the CTF (amplitude CTF) is relevant. However, contrast at high resolution in the electron microscope is essentially *phase contrast* (see, e.g., Chapter 4, Section 2.2, and Exercise 4.1), and thus it is the *imaginary part* of the CTF (phase CTF) which must be considered. Under these circumstances information on all spatial frequencies (reciprocal of g) down to $d = 1/g$ is transferred by the lens when the CTF is purely imaginary, or a reasonable approximation when the imaginary part of the CTF is negative, for example, out to point A in Fig. 5.36*d*. Inspection of the curves indicates resolution of all periodicities down to 0.5, 0.3, and 0.25 nm in Figs. 5.36*b–d*, respectively, for example. Maximum contrast over the widest range of periodicities is achieved if the CTF is "flat" and purely imaginary, as around B in Fig. 5.36*d*. The widest ranges are found for negative values of defocus D, particularly at the largest negative value for which the real part first becomes zero for $\chi(g) = +\pi/2$, rather than for $\chi(g) = -\pi/2$ (i.e., at point E rather than C or, strictly, where the curve just touches the axis at C). Setting $\chi(g) = -\pi/2$ in eq. 5.39 and differentiating with respect to g yields a turning value (maximum) at

$$D_{\text{Sch}} = -(C_s\lambda)^{1/2}. \tag{5.40}$$

This value of optimum defocus [or its almost numerical equivalent $-(2.5/\sqrt{2\pi})(C_s\lambda)^{1/2}$] is known as the Scherzer[27] defocus. The graphs in Figs. 5.36*b–d* are plotted for defocus almost equal to the Scherzer value, and thus the resolution limits quoted above are the best available for axial illumination using the lenses assumed in the calculations.

The analysis so far has calculated only the conditions for the best "first" flat in the phase CTF. Further inspection of eq. 5.39 shows that similar turning

---

Fig. 5.36 Plots of the contrast transfer function exp $(i\chi(g))$ for a number of related situations. (*a*) 100 kV, $C_s = 1.8$ mm, $D = 0$; (*b*) like (*a*) except $D = -80$ nm; (*c*) 500 kV, $C_s = 4.7$ mm, $D = -80$ nm; (*d*) 1 MV, $C_s = 7.6$ mm, $d = -80$ nm; (*e*) like (*c*) except smoothed for energy losses, etc.; (*f*) like (*d*) except smoothed. For these curves $C_s\lambda$ is constant, and the values of defocus in (*b–d*) are approximately the Scherzer[27] defocus values, that in (*a*) the Gaussian focus. In each case the ordinate of the upper curve is the imaginary part and the lower the real part, the abscissa being g in units of Å^{-1}. The truncation effects [particularly (*a*) and (*b*) for $g > 0.40\,\text{Å}^{-1}$] should be ignored. In (*e*) and (*f*) the smoothing (see, e.g., ref. 28) was effected by setting the energy loss $\Delta E = 2$ eV, $C_c = 5.3$ and 8.9 mm, respectively, and divergence $\Delta\alpha = 0.001$. These diagrams should be compared with the optical diffractograms of Fig. 1.41. Courtesy P. Rez.

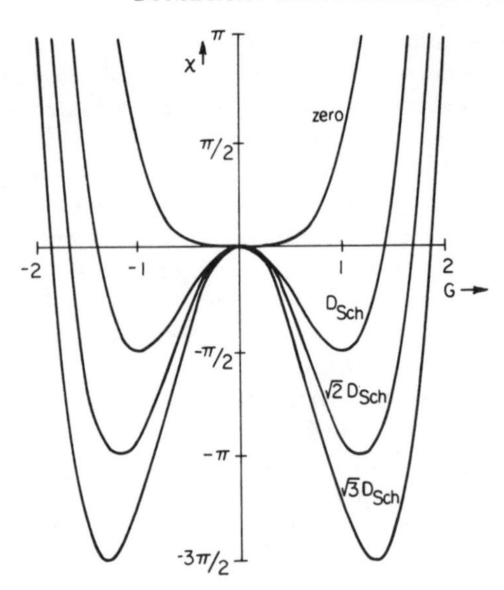

Fig. 5.37 Plot of phase χ of beam at angle α to the optic axis, plotted in terms of the dimensionless parameter $G = C_s^{1/4}\lambda^{-1/4} (= gC_s^{1/4}\lambda^{3/4}$ for axial illumination) for specific values of defocus D; see eqs 5.38 to 5.40.

values occur at defocus values

$$D = D_{\text{Sch}}\sqrt{n}\,, \tag{5.41}$$

where n is odd (and corresponding best transfers in the amplitude CTF with n even) and that they occur at larger values of g (i.e., better spatial resolution). This is best seen in a plot of χ against g for different values of defocus D, as shown schematically in Fig. 5.37. The contrast transfer flats occur when $\delta\chi/\delta g = 0$ and $\chi = -n\pi/2$ together. However, other factors usually prevent utilization of these higher frequency windows, such as variations in high tension, fluctuations in lens supplies, intrinsic energy spread in the illumination (1 to 2 eV for heated filaments), energy losses in the specimen, and beam convergence. These problems may be taken into account by means of envelope functions,[28] whose principal effect is to reduce contrast transfer at high frequencies. Beam divergence smears out all rapid oscillations in the transfer function. Allowance for these factors has been made in parts of Fig. 5.36 (compare Fig. 5.36*c* with *e* and Fig. 5.36*d* with *f*). Figure 5.37 illustrates the general principal that $\delta\chi/\delta g$ should be zero for the directions of the emitted beams considered; in the case of axial illumination $g = 0$ is one of these directions. By this means the deleterious effects of beam divergence, electrical instabilities, and the like are minimized—the resultant changes in the value of χ are smallest near a turning value.

If instead of using axial illumination the incident beam is directed at an angle α to the optic axis, the analysis becomes somewhat more complicated, but the principles noted above remain applicable. In particular, beams passing through the aperture suffer least effect from instabilities if the value of χ for each is near a turning value. The situation where both the incident beam and a diffracted beam are at minima in the $\chi(\alpha)$ curve of Fig. 5.37 (one at the left, the other at the right) is especially favorable since second-order effects are of the same sign for both ($d^2\chi/dg^2$ is positive for both, in contrast to the axial condition, in which it has opposite signs). In the case of symmetrical tilted illumination contrast transfer is of course amplitude in nature (since the phase difference between the two beams is zero), but it is of particular importance in lattice fringe imaging, where the beams involved are essentially in phase (see eqs. 4.26 to 4.35 and Section 9.3).

9.2 Moiré Fringes

Moiré fringes may be seen in images of moderate resolution (as well as in high resolution situations) when the aberration factor χ can be neglected. The simplest situation then envisaged is that of two overlapping crystals, each perfect and of uniform thickness and orientation, producing diffracted amplitudes in two closely adjacent diffraction spots $\mathbf{g}_1$ and $\mathbf{g}_2$, one from each crystal. If only these two diffracted beams pass through the objective aperture, eq. 5.37 becomes

$$\phi(\mathbf{r}) = \phi_1 \exp[2\pi i(\theta_1 + \mathbf{g}_1 \cdot \mathbf{r})] + \phi_2 \exp[2\pi i(\theta_2 + \mathbf{g}_2 \cdot \mathbf{r})]$$

and the image intensity $I(\mathbf{r})$ is

$$I(\mathbf{r}) = \phi(\mathbf{r})\phi^*(\mathbf{r}) = \phi_1^2 + \phi_2^2 + 2\phi_1\phi_2 \cos 2\pi(\Delta\theta + \Delta\mathbf{g} \cdot \mathbf{r}), \qquad (5.42)$$

where $\Delta\theta = \theta_1 - \theta_2$ and $\Delta\mathbf{g} = \mathbf{g}_1 - \mathbf{g}_2$. Equation 5.42 represents a set of fringes with a period in the direction $\Delta\mathbf{g}$ of $1/\Delta g$ and contrast C, given by

$$C = \frac{I_{max} - I_{min}}{I_{max} + I_{min}} = \frac{2\phi_1\phi_2}{\phi_1^2 + \phi_2^2}, \qquad (5.43)$$

which is maximized for $\phi_1 = \phi_2$, that is, if the beams are of equal amplitudes. Such fringes are known as moiré fringes and, provided that the phase relationships between the two interfering beams are as assumed (θ_1 and θ_2 constants) and their amplitudes are constant, are a set of straight, parallel fringes of constant visibility. Changes in thickness, for example, may alter both phase (θ_i) and amplitude (ϕ_i) of either or both contributing beams, resulting in a local bending of the fringes (by changing $\Delta\theta$ and hence shifting the origin of the pattern), as well as altering their visibility.

The fringes produced in this way by overlapping crystals are called parallel moiré fringes if $\mathbf{g}_1$ and $\mathbf{g}_2$ are parallel (and hence $\Delta\mathbf{g}$ is parallel to both), and rotational moiré fringes if $g_1 = g_2$. Parallel moirés may be produced, for example, during epitaxial growth of materials with slightly different lattice spacings, and rotational moirés by rotational misorientations of layers of the same materials. A general moiré pattern may contain elements of both types. An example of such fringe patterns may be seen in Fig. 2.13.

9.3 Lattice Plane Images

If instead of two diffracted beams from different, superimposed crystals, the objective aperture allows through two beams from the exit surface of a single crystal of uniform thickness, for simplicity ϕ_0 and ϕ_g, then, under the same conditions as assumed for moiré fringes, the intensity distribution is given by the analogue of eq. 5.42:

$$I(\mathbf{r}) = \phi_0^2 + \phi_g^2 + 2\phi_0\phi_g \cos 2\pi(\Delta\theta + \mathbf{g} \cdot \mathbf{r}). \tag{5.44}$$

Equation 5.44 represents fringes having *the same spacing* $(1/g)$ as the physical crystal planes $\mathbf{g}$, but because of the factor $\Delta\theta$, not necessarily coincident with them. Such fringes, known as lattice fringes or lattice plane images, are most visible when $\phi_0 = \phi_g$, that is, at thicknesses $t = (n \pm \frac{1}{4})\,\xi_g$—see Fig. 4.6, with the crystal set exactly at the reflecting position. Because of inelastic scattering processes not taken into account, the best visibility is usually in the thinnest crystal at $t = \xi_g/4$. Additionally, at the reflecting position for perfect crystal the factor $\Delta\theta$ is zero, and under these conditions the fringes appear to be coincident with the crystal lattice planes from which they derive.

However, lattice images are an essentially high resolution, thin crystal phenomenon, particularly if the lattice under consideration is that of a metal, with spacing near the resolution limit of the instrument. Under these circumstances the phase factor χ of eq. 5.38 may not be omitted. As already noted, however, its deleterious effects on the image may be minimized by positioning the optic axis to bisect the angle between the two beams (the "symmetric orientation" or tilted beam illumination, as opposed to axial). Examples of such lattice fringe images are shown in Figs. 1.2*c* and 3.29 to 3.31.

9.4 Crystal Structure Projections

If, instead of restricting the beams passing through the aperture to two, a large number are allowed to form the image, the full summation of eq. 5.37 must be used once more. If the aberration factor is ignored for the present, the expression has the form of a straightforward Fourier series. Thus the effect of including higher order reflections in the same row as the original pair is to "sharpen up" the fringe pattern, that is, to give it more detail. A similar row of reflections at an angle to the first produces an inclined set of fringes, and so on. Thus the effect of including a large number of reflections, symmetrically placed about the optic axis and symmetrically excited, is to produce a two-dimensional lattice structure image with detail down to spacings corresponding to the highest order, appreciably excited reflections included. (See Chapter 3, Sections 12 and 13.) Furthermore, if the crystal is sufficiently thin, the amplitudes ϕ_g and the phases θ_g of the beams are given by the phase grating approximation (see Chapter 4, Section 2.2). Under these conditions the integrated potential experienced by the electrons passing through the specimen is directly mirrored by the emitted beams. Hence, for example, individual heavy atom impurities or vacant lattice sites can be directly "resolved" in the micrograph. Once again, though, resolution at this high level means that the lens aberration factor must be fully taken into account; in particular, spurious deductions as to the nature of a defect from a micrograph at one particular defocus must be avoided.

A number of examples of analyses made using lattice and crystal structure images were shown in Chapter 3 (in particular, in Figs. 3.29 and 3.35).

9.5 Multibeam Images at High Voltages

One of the problems in imaging thick specimens is low intensity. If more beams are allowed through the aperture, the intensity is obviously increased overall, and, provided that the contrast of a particular defect is of the same nature (e.g., dark on a light background), image visibility is unaltered. The reason why this technique is not normally used is that the bright field and different dark field images are differently displaced by the phase distortion factor of eq. 5.38, and for a nonperiodic object such as a dislocation these displacements wash out the image completely. The image displacements are proportional to θ_B^3 and θ_B for spherical and chromatic aberrations, respectively, where θ_B is the Bragg angle. At high voltages Bragg angles are much reduced [e.g., $\theta_B(1\ \mathrm{MV}) \simeq \theta_B(100\ \mathrm{kV})/4$] and the displacements therefore decreased the spherical component by a very large factor, even if the aberration constants of the lens are somewhat increased ($C_s\lambda$ = constant assumed in Section 9.1). Thus at high voltages the different images will be more nearly superimposed and hence the gain in image intensity (or in specimen thickness for comparable intensity) will be worthwhile.[22]

At the resolution level of interest in thick specimens–very much poorer than "lattice" images–the interference terms (e.g., the final term of eq. 5.44) average to zero; the image intensity may be calculated as the sum of the individual beam intensities (shifted if necessary by the small aberration and defocus displace-

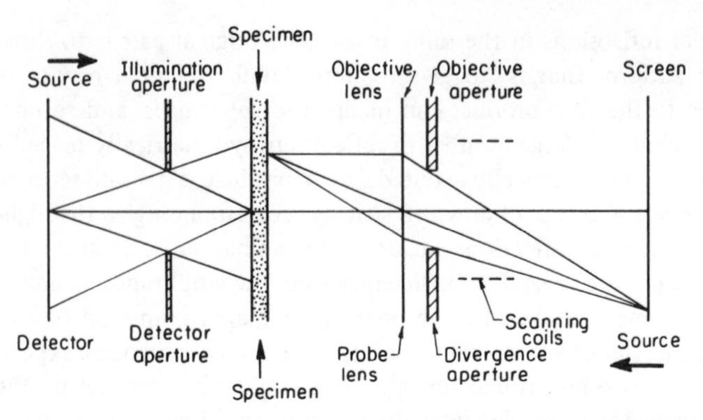

Fig. 5.38 Schematic ray diagram illustrating the reciprocity between TEM (upper row, read left to right) and STEM (lower row, read right to left).

ments). The contrast at a particular defect is thus more complicated than that of a normal single-beam image, but little extra effort is required for its calculation since the individual images will have been computed in any case during the many-beam integration necessary at high voltages.

10 Scanning Transmission Electron Microscopy

As mentioned in Section 8, the general principle of reciprocity may be applied to the propagation of electrons through materials. Inspection of Fig. 5.38 indicates how the symmetry principles noted in Fig. 5.34, parts (1) and (2), are used in practice. Any defect will produce the same image contrast whether viewed in a microscope that operates from left to right (conventional TEM) or one that operates from right to left (STEM). An example of this equivalence of image contrast was shown in Fig. 1.32.

The principal advantage originally claimed for the scanning transmission mode of operation was that, since no postspecimen lenses need be used, resolution should be superior in thick specimens (or, equivalently, that thicker specimens could be viewed at the same resolution), because chromatic aberration (which limits resolution in thick specimens in conventional TEM) is absent. The limitation on resolution in the STEM mode (provided that sufficient signal is available) is the dispersion of the initially small probe produced by both intrinsic beam convergence and the inevitable elastic and inelastic scattering events in the crystal. Present indications are that such factors limit any improvement in resolution or penetration to a rather small factor. Thus, as noted in Chapter 1, the main advantage of the scanning mode is the improved microdiffraction and microanalytical capability implicit in operations with small, scanned probes.

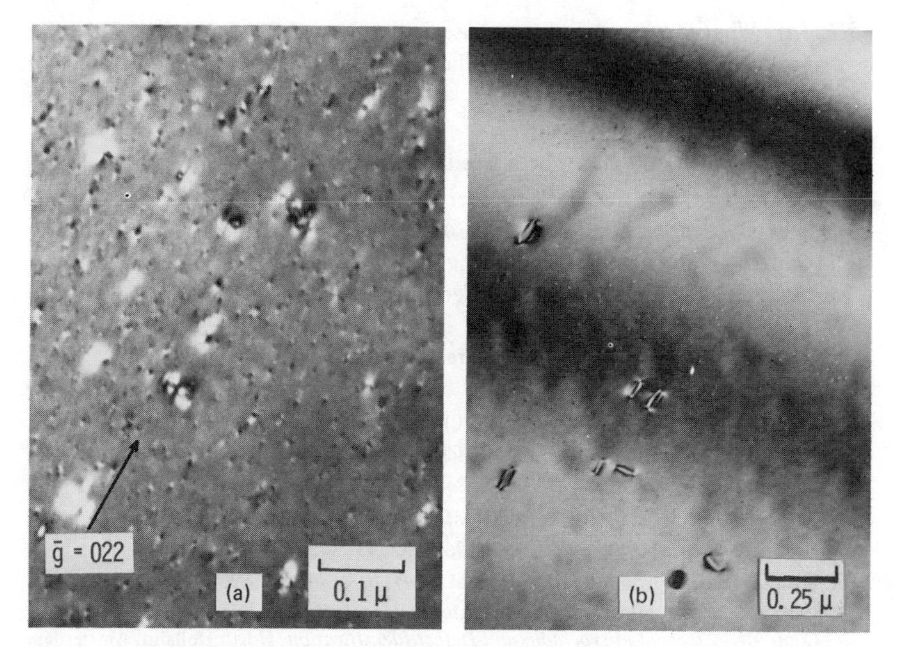

Fig. 5.39 Defect contrast features pertinent to Exercise 5.1. Courtesy K. H. Westmacott.

Exercises

5.1 Comment on the small defect contrast features in Fig. 5.39.

5.2 Show how to derive an effective two-beam extinction distance from a systematic four-beam situation, that is, derive eq. 5.21.

5.3 Using a suitable diffraction geometry for the Ewald sphere and a systematic row of reflections, deduce the "three beam" condition for the critical voltage of the second-order reflection (eq. 5.27).

5.4 Using the result from Exercise 5.3 and data from Table 4.1, estimate the accuracy to which the stoichiometry of an alloy near Cu_3Au may be determined by measurement of a convenient critical voltage (which may be measured to ±5 kV). Assume that the lattice parameter of Cu_3Au = 0.360 nm.

5.5 Confirm, in the two-beam restricted case, that intraband scattering may be removed from eq. 5.28 by a suitable substitution, that is, that scattering of type AA' in Fig. 5.30 may be eliminated.

5.6 With the aid of Fig. 5.34 show that the reciprocity relation implies that weak beam dark field images of defects taken under conditions (**g**, 3**g**) and (**g**, −**g**) are identical.

5.7 (*a*) What is the value of the phase factor χ (eq. 5.39) for the $\mathbf{g} = 200$ beam from a nickel foil using axial illumination, 500 kV electrons, $C_s = 5$ mm, and Scherzer defocus? (The lattice parameter of nickel = 0.352 nm, $\lambda = 1.42$ pm.)

(*b*) Will the 200 planes be imaged in the axial mode?

5.8 How large are the defocus increments between Fourier images of (*a*) $Nb_{12}O_{29}$: $a = 2.89$ nm, $b = 0.334$ nm, $c = 2.07$ nm in [010] projection, and (*b*) gold: $a = 0.408$ nm, using the three beams $\bar{2}00$, 000, and 200? In both cases assume that 500 kV electrons are used ($\lambda = 1.42$ pm).

References

1. Thölén, A. R., *Phil. Mag.*, **22,** 175 (1970).
2. Read, W. R., *Dislocations in Crystals*, McGraw-Hill, New York, 1953.
3. Howie, A. and Whelan, M. J., *Proc. R. Soc.*, **A267**, 206 (1962).
4. Silcock, J. M. and Tunstall, W. J., *Phil. Mag.*, **10,** 361 (1964).
5. Head A. K., *Austr. J. Phys.*, **20,** 557 (1967); **22,** 43, 345 (1969).
6. Hashimoto H., Howie A., and Whelan, M. J., *Proc. R. Soc.*, **A269,** 80 (1962).
7. Head, A. K., Humble, P., Clarebrough, L. M., Morton, A. J., and Forwood, C. T., *Computed Electron Micrographs and Defect Identification*, North Holland, Amsterdam, 1973.
8. Karnthaler, H. P., Hazzledine, P. M., and Spring, M. S., *Acta Met.*, **20,** 459 (1972).
9. Howie, A., *Phil. Mag.*, **14,** 223 (1966).
10. Cockayne, D. J. H., Doctoral thesis, University of Oxford, 1970.
11. Mayers, D. F., in *Numerical Solution of Ordinary and Partial Differential Equations* (Ed L. Fox), Pergamon, London, 1962, p. 16. See also Krogh, F. T., Jet Propulsion Laboratory, Section 314, *Tech. Mem.* 275, 1971.
12. Thomas, L. E., *Phil. Mag.*, **26,** 1447 (1972).
13. Gjønnes, J. and Høier, R., *Acta Crystallogr.*, **A27,** 313 (1971).
14. Thomas, L. E., Shirley, C. G., Lally, J. S., and Fisher, R. M., in *High Voltage Electron Microscopy* (Eds P. R. Swann, M. J. Goringe, and C. J. Humphreys), Academic, London, 1974; p. 38. See also Hewat, E. A. and Humphreys, C. J., *ibid.*, p. 52; Jones, I. P. and Tapetado, E. G., *ibid.*, p. 48.
15. Rocher, A., Sinclair, R., and Thomas, G., *Microscopie Electronique a Haute Tension 1975*, Société Francaise de Microscopie Electronique, Paris, 1976, p. 125.
16. Butler, E. P., *Micron*, **5,** 293 (1975).
17. Sinclair, R., Goringe, M. J., and Thomas, G., *Phil. Mag.*, **32,** 501 (1975).
18. Howie, A., in *Electron Microscopy in Material Science* (Ed. U. Valdrè), Academic, New York, 1971, p. 274.
19. Cockayne, D. J. H., Ray, I. L. F., and Whelan, M. J., *Phil. Mag.*, **20,** 1265 (1969).
20. Hirsch, P. B., Howie, A., Nicholson, R. B., Pashley, D. W., and Whelan, M. J., *Electron Microscopy of Thin Crystals*, Butterworths, London, 1965 (revised edition published by Knieger, New York, 1977).

21. Goringe, M. J., Hewat, E. A., Humphreys, C. J., and Thomas, G., *Proceedings of the 5th European Congress on Electron Microscopy*, Manchester, England, 1972, p. 538.
22. Hashimoto, H., U.S.-Japan Seminar on High Voltage Electron Microscopy, Honolulu, 1971 (unpublished).
23. Howie A., *Proceedings of the 5th European Congress on Electron Microscopy*, Manchester, England, 1972, p. 408.
24. Howie, A. and Whelan, M. J., *Proc. R. Soc.*, **A263**, 217 (1961).
25. Ball, C. J., *Phil. Mag.*, **9**, 541 (1964).
26. Hanszen, K. J., *Adv. Opt. Electron Microsc.*, **4**, 1 (1971).
27. Scherzer, O. J., *Appl. Phys.*, **20**, 20 (1949).
28. Frank, J., *Optik*, **38**, 519 (1973).

APPENDIX A

PREPARATION OF THIN FOILS FOR TEM FROM BULK SAMPLES

For most applications of TEM to materials characterization it is necessary to prepare thin foils from bulk samples on which property measurements have already been made (e.g., the undeformed parts of a K_{1c} fracture toughness specimen). In such cases the standard techniques involve chemical thinning, electrolytic polishing (for conductors), and ion thinning (especially for non-conductors or for conductors with a contamination layer). The polishing techniques and recipes for chemical and electrolytic thinning have now been well established,* and it seems unnecessary to repeat them here. Instead, references for specialized techniques of polishing are given in Section A.2, and for new data in general at the end of the appendix (refs 1-38).

A.1 Specific Materials

In Table A.1 the newer techniques have been listed under the specific material. The terms employed for electrolytic thinning are "window" (electropolishing a sample protected by a lacquer except for a window, which is the part to be polished).[39] and "jet" (the sample is automatically polished by a jet of solution).[39] Commercial jet polishing apparatuses are available from several sources,†

*See, for example, Appendix 1 of ref. 39a, ref. 1, and ref. 40.

†Jet thinners: South Bay Technology, 4900 Santa Anita Ave., El Monte, CA 91731; Precision Scientific Co., Dept. M, 3737 Cortland, Chicago, IL 60647; Materials Science North/West, 55 Cocker St., Blackpool, Lancashire, England; Hacker Instruments, Inc., Box 646, West Caldwell, NJ 07006; Ernest F. Fullam, Inc., P.O. Box 444, Schnectady, NY 12301; C. W. French Div., Ebtec Corp., 120 Shoemaker La., Agawam, MA 01001; E. A. Fischione Instrument Manufacturing, 7925 Thon Dr., Verona, PA 15147.

Table A.1 Polishing Techniques

Material	Technique	Solution	Conditions[a]	Reference[b]
Ag–Zn	Jet	9% KCN, 91% H_2O		1
Al	Window	1 part nitric acid, 3 parts methanol	-70°C, 40 V	2
Al–Cu	Chemical	94 parts phosphoric acid, 6 parts nitric acid		3
Al–Au alloys	Window	20% perchloric acid, 80% methanol	16–18 V, 100 cm^{-2}, -55 to -65°C	4
Au	Window (3 mm disks)	100 ml CH_3COOH, 20 ml HCl, 3 ml H_2O	Stainless steel cathode, 32 V, 15°C	5
	Electrolytic			
		(*a*) 15% glycerol, 35% 50% HCl	30 V, -30°C, 1.5 A cm^{-2}–fast polish	
		(*b*) 133 ml glacial CH_3COOH, 25 g CrO_3, 7 ml H_2O	Finish after (*a*) at 22 V, 0.8 A cm^{-2}, 0°C	
Be, Be–Cu	Window	20% perchloric acid, 80% ethanol	-30°C, 40 V	6
Co–Fe	(*a*) Chemical	50% phosphoric acid, 50% hydrogen peroxide		7
	(*b*) Jet	20% perchloric acid, 80% methanol	-20°C	
Cu–Ni–Fe alloys	Chemical	20 ml acetic acid, 10 ml nitric acid, 4 ml HCl		8

Cu–Ti	Jet	750 ml acetic acid, 300 ml phosphoric acid, 150 g chromic acid, 30 ml distilled H_2O	30–40 V, 45–55 A	9
Fe–Al–C	Chemical	HF, H_2O, H_2O_2 (1 : 3 : 16)		10
Fe_3Si	Window	1% perchloric acid, 2.5% hydrofluoric acid in methanol	$-77^\circ C$	11
Gd–Ce alloys	Window	1% perchloric acid, 99% methyl alcohol	$-77^\circ C$	12
Hf	(*a*) Jet	45 parts nitric acid, 45 parts H_2O, 8 parts HF		13
	(*b*) Jet	2 parts perchloric acid, 98 parts ethanol	40–45 V, 16–20 mA, $-77^\circ C$	
In	Window: from 0.1 mm thick	33% HNO_3, 67% methanol	$-40^\circ C$	14
Mg		(*a*) 20% $HClO_4$, 80% ethanol	Fast polish, $0^\circ C$	
		(*b*) 375 ml H_3PO_4, 625 ml ethanol	Finish at 15 V, $0^\circ C$, lowest current	
	Chemical	6% nitric acid, 94% H_2O	$3^\circ C$	15
Mo	Jet	1 part sulfuric acid, 7 parts methanol	10 V	16
Nb	Chemical	60% HNO_3, 40% HF	$0^\circ C$	17
Nb alloys	Window	100 ml lactic acid, 100 ml sulfuric acid, 20 ml HF	$40^\circ C$, $\sim 0.4\ A\ cm^{-2}$	18

Table A.1 (Continued)

Material	Technique	Solution	Conditions[a]	Reference[b]
Ni steels Stainless steels	Chemical	50 ml 60% H_2O_2, 50 ml H_2O, 7 ml HF	Place specimen in H_2O_2 solution, then add HF until reaction starts; produces 1000 Å foils which can then be finished by electron polishing by standard chromic-acetic electropolishing (courtesy Republic Steel Corp.)	
Ni alloys Cu–Ni–Fe	Jet	1 part HNO_3, 3 parts CH_3OH	6–10 V, 18–30 mA, −30°C	
Ni	Jet	20% perchloric acid in ethanol		19
Ni–Al alloys	Jet	20% sulfuric acid, 80% methanol	Dished at 150 V, final polish in same solution at 10–12 V	20
Ni and Ni–Fe	Jet	8 parts 50% sulfuric acid, 3 part glycerine	Less than 10°C, 1.3 A cm^{-2}	21
Rene 95	Jet	250 ml methanol, 12 ml perchloric acid, 150 ml butylcellusolve	−35°C, 65 V	22
Th	Window	5% perchloric acid, 95% methanol	−77°C, 3–10 V	23
Pt	Electrolytic	H_2SO_4 (96%), HNO_3 (65%), H_3PO_4 (80%)	0.2–0.5 A cm^{-2}, 20–30°C	24

Re	Window	Ethyl alcohol, perchloric acid, butoxyethanol (6 : 3 : 1)	$-40^\circ C$, 35 V	25
Ta	Chemical	3 parts HNO_3, 1 part HF	$-10^\circ C$, immerse in petri dish after lacquering	26
Ta	Jet	1 part sulfuric acid, 5 parts methanol	$-5^\circ C$, current density of 0.5 A cm^{-2}	27
Ti	Jet	1 part HF, 9 parts sulfuric acid		28
Ti–Nb	Chemical	4 parts nitric acid, 1 part HF		29
TiC	Jet	6 parts nitric acid, 2 parts HF, 3 parts acetic acid	100 V	30
Ti and Ti alloys	Jet	30 ml $HClO_4$, 175 ml *n*-butanol, 300 ml CH_3OH	15 V, 0.1 A cm^{-2}, may need ion thinning at end to remove contamination layer	
U	Jet	1 part perchloric acid, 10 parts methanol, 6 parts butylcellusolve	$-20^\circ C$, 35–45 V	31
UO_2 (sintered)	Chemical	10 ml HOAc, 20 ml sat. CrO_3, 5 ml 40% HF, 7 ml HNO_3 ($d = 1.42$)		32
	or			
	Chemical	20 ml H_3PO_4 ($d = 1.75$), 10 ml HOAc, 2 ml HNO_3 ($d = 1.42$)		32
V	Chemical	2 parts HF, 1 part nitric acid	Less than $10^\circ C$	33

Table A.1 (Continued)

Material	Technique	Solution	Conditions[a]	Reference[b]
V	Jet	20% sulfuric acid in methanol		
Zn–Al	Chemical	50–70% nitric acid in H_2O		35
Zn–Al	Window	90% methanol, 10% perchloric acid		36
Zr–25% Ti	Chemical	50 ml nitric acid, 40 ml H_2O, 10 ml HF		37
Zr alloys	Jet	5% perchloric acid, 90% ethanol	70 V, below −50°C	38

[a] Where no conditions are listed it means they were not given in the reference.
[b] Listed at the end of Appendix A.

as are temperature control baths and voltage-current density control units.* The conditions achieving a plateau in the current versus voltage plot are those required for polishing. In some cases polishing occurs chemically without an applied current.

A.2 "Universal" Polishing Techniques: Special Techniques

Both chemical and electrolytic techniques are covered here; the titles of the references are self-explanatory.

1. Controlled Jet Polishing, C. K. H. DuBose and J. O. Stiegler, *Rev. Sci. Instrum.*, **38** (5), 694 (1967).
2. Improved Jet Polishing, A. F. Rowcliffe, *J. Inst. Metals,* **94** (7), 263 (1966).
3. Automatic Jet Thinning, R. D. Schoone and E. A. Fischione, *Rev. Sci. Instrum.*, **37** (10), 1351 (1966).
4. Jet Polishing, A. R. Davis, *J. Inst. Met.*, **96,** 61 (1968).
5. Jet Thinning of GaAs, M. J. Hill, D. B. Holt, and B. A. Unvala, *J. Sci. Instrum.*, **2** (1), 301 (1968).
6. Jet Cutting and Thinning, P. G. Merli and U. Valdrè, *J. Sci. Instrum.*, **5** (9), 933 (1972).
7. Modified Double-Jet Thinning, A. Mastenbroek and G. Hamburg, *J. Sci. Instrum.*, **5** (1), 10 (1972).
8. Universal Electropolishing, E. N. Hopkins, D. T. Peterson, and H. H. Baker, *Ames Lab.* (AEC) *Contrib.* 1746, 1965.
9. Universal Electropolisher, K. L. Maurer and H. Schaffer, *Prakt. Metallogr.*, **4** (18), 396 (1967).
10. Polishing of Disc Specimens, G. G. Shaw and C. Q. Bowles, *Rev. Sci. Instrum.*, **37** (4), 516 (1966).
11. Preparation of Disc Foils at Subzero Temperatures, A. S. Pearce and G. Wood, *J. Sci. Instrum.*, **5** (10), 984 (1972).
12. Low Temperature Electropolishing, W. N. Roberts, *J. Sci. Instrum.*, **5** (5), 416 (1972).
13. Wide Area Electrothinning of Foils, C. Herrmann and R. Courtel, *J. Sci. Instrum.*, **7** (5), 350 (1974).
14. Electropolishing of *in situ* E. M. Tensile Samples, D. S. Lashmore, R. N. Gardner, and W. A. Jesser, *J. Sci. Instrum.*, **9** (3), 172 (1976).
15. Foils from Small Specimens, M. Nagumo and M. Eudo, *J. Sci. Instrum.*, **44** (7), 550 (1967).
16. Thinning of Wires, E. S. Meieran and D. A. Thomas, *Trans. AIME*, **227,** 284 (1963).
17. Micro-electropolishing, L. I. van Torne and G. Thomas, *Rev. Sci. Instrum.*, **36** (7), 1042 (1965).

*Temperature control baths: Cole-Parmer Instrument Co., 7425 N. Oak Park Ave., Chicago, IL 60648; The Lab Apparatus Co., 18901 Cranwood Parkway, Cleveland, OH 44128. Voltage-current density control units: Power Designs Pacific, 3381 Miranda Ave., Palo Alto, CA 94304; Brinkmann Instruments, Cantiague Rd., Westbury, NY 11590; Kiethly Instruments, 28775 Aurora Rd., Cleveland, OH 44139.

18. Electrolytic Thinning of Aluminum, Single Crystals, R. E. Medrano, *J. Sci. Instrum.*, 6 (11), 1088 (1973).
19. Fe and Alloys, A. T. Davenport and B. Belyakov, *J. Iron Steel Insti. (London)*, p. 610, June 1968.
20. Titanium Alloys, M. J. Blackburn and J. C. Williams, *Trans AIME*, **239** (2), 289 (1967).
21. Yttrium Aluminum Garnet, D. J. Keast, *J. Sci. Instrum.*, **44,** 862 (1967).
22. Ni–Cr Eutectic, R. Kossowdky, W. C. Johnston, and B. J. Shaw, *Trans. AIME*, **245** (6), 1219 (1969).
23. Reduced Spurious Deformation in EM Thin Foils, B. Junkin and T. Malis, *J. Sci. Instrum.*, 9 (2), 83 (1976).
24. Specimen Preparation Technique of Surface Layers, S. Hoemark, H. Swahn, and O. Vingsbo, *Ultramicroscopy*, **1** (2), 113 (1975).
25. Ultramicrotomy for Metals (deforms the samples), V. A. Phillips, *Prakt. Metallogr.*, **4** (12), 637 (1967).
26. Powders and Finely Divided Samples, L. N. Kalitina and L. D. Skrylev, *Zavod. Lab.*, **33** (1), 50 (1967). (English translation: A foam is produced with added gelatin and then dried. This prevents agglomaration and does not require support films.)

In summary, good foils can be prepared from bulk samples of most materials (conductors) by either chemical or electrolytic methods. If these fail, the now generally accepted method is ion thinning, which is also useful when selective etching can be a nuisance (e.g., different layers in compounds such as GaAs etch or polish at different rates, and this is undesirable).

Since the ion thinning technique[34] has become important for studies of ceramics and minerals in recent years, an outline of the method is given in the following section.

A.3 Ion Thinning (Especially for Nonconductors: Minerals, Ceramics)

It is interesting that ion thinning techniques were first tried over 20 years ago when transmission electron microscopy of crystals was just emerging; but because of the success of chemical and electrochemical thinning techniques, especially for metals, ion thinning was not properly utilized until fairly recently, when nonconducting solids such as ceramics and minerals began to be examined in detail. It has also proved most successful in preparing multiphase and sandwich specimens, where thinning rates by other techniques are nonuniform.

The principles of the latest developments in ion thinning have been described by Barber.[34] A schematic diagram of the apparatus is shown in Fig. A.1, which is a modification of a commercial sputtering device.*

*Several commercial instruments are available from, for example, the following sources: Commonwealth Scientific, 500 Pendleton, Alexandria, VA 22314; Edwards High Vacuum, 3279 Grand Island Blvd., Grand Island, NY 14072; Materials Research Corp., Orangeburg, NY 10962; Polysciences, Paul Valley Industrial Park, Warrington, PA 18976; Technics, 80 N. Gordon St., Alexandria, VA 22304; Veeco Instrument, Terminal Dr., Plainview, NY 11803.

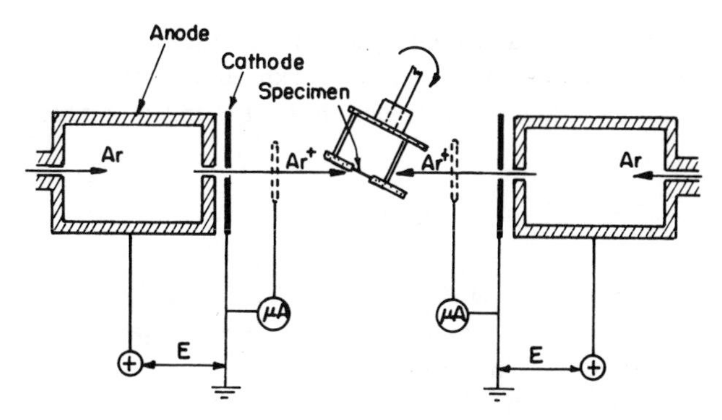

Fig. A.1 Schematic diagram of a double-beam ion thinning apparatus.

When preparing thin foils, it is important to keep their temperature sufficiently low to avoid structural changes. Consequently, ion current densities in excess of $\simeq 200\ \mu A\ cm^{-2}$ are unnecessary. This is a factor of 10^{-1} to 10^{-2} lower than the ion density in normal sputtering systems. Moreover, if the ion energy is increased beyond $\simeq 10$ keV, most of the increase is dissipated in deeper penetration (which is clearly undesirable) and not in greater efficiency of sputtering. With care, damage to the surfaces by ion bombardment can be kept to a minimum, for example, by thinning by 80% at 10 kV and finishing at a much lower voltage. The ion guns are of the hollow anode type, and each produces one beam. Each anode is connected through a large ballast resistor to a well-stabilized 10 kV, 10 mA direct current supply. The bombardment chamber is mounted over a 6 in. diffusion pump and liquid nitrogen trap, backed by a two-stage rotary pump. The vacuum system maintains a working vacuum of between 10^{-3} and 10^{-4} torr while high purity argon gas is entering the anode-cathode interspace through fine needle valves. The main disadvantage of the cold cathode ion source is that it necessitates a relatively large flow of gas into the system. With a typical gas flow rate of $\simeq 5 \times 10^{-2}$ torr sec^{-1} there is an ion current of about 70 μA per gun at 6 kV. The ion current density at the sample is about $200\ \mu A\ cm^{-2}$. The ion currents can be measured and balanced by means of retractable probes.

The sample, with its periphery sandwiched in a thin stainless steel holder, is rotated in the ion beams. The inclination of its exposed surfaces with respect to the beams can be preset at angles between 0 and 30°. It is important that the ion beams "see" both surfaces and erode them at equal rates if no provision is made to stop the backdiffusion of sputtered atoms from the sample and its holder. To minimize the redeposition of sputtered material on the sample, one can place around it a partial shield which is cooled by liquid nitrogen. The shield also prevents the occurrence of arc discharges in the ion guns when materials that give off volatile constituents are thinned. Eventually, the ion sources become electrically unstable to glow discharge because of enlargement of the holes in the cathodes. In normal use, however, they have a life of about 250 hours.

It should be noted that ion thinning can also be used to clean foils from contamination, for example, after continued exposure in the micrscope, or to remove surface films formed during electrolytic or chemical thinning.

References

1. Clark, H., Merriman, E. A., and Wayman, C. M., *Acta Met.*, **17**, 719–734 (June 1969).
2. Garmong, G., Rhodes, C. G., Spurling, R. A., *Met. Trans.*, **4**, 707–714 (1973).
3. Phillips, V. A. and Hugo, J. A., *Micron*, **3**, 212–223 (1972).
4. Toda, T. and Maddin, R., *Met. Trans.*, **245**, 1045–1054 (1969).
5. Meakin, J. E., *Rev. Sci. Instrum.*, **35** (6), 763 (1964).
6. Hazif, R., Edelin, G., and Dupuoy, J. M., *Met. Trans.*, **4**, 1275–1281 (1973).
7. Mahajan, S., *Met. Trans. A*, **6A**, 1880 (1975).
8. Butler, E. P. and Thomas, G., *Acta Met.*, **18**, 347–365 (1970).
9. Cornie, J. A., Datta, A., and Soffa, W. A., *Met. Trans.*, **4**, 727–733 (1973).
10. Watanabe, M. and Wayman, C. M., *Met. Trans.*, **2**, 2221–2227 (1971).
11. Lakso, G. E. and Marcinkowski, M. J., *Met. Trans.*, **245**, 1111–1120 (1969).
12. Koch, C. C., Mardon, P. G., and McHargue, C. J., *Met. Trans.*, **2**, 1095–1100 (1971).
13. Das, G. and Mitchell, T. E., *Met Trans.*, **4**, 1405–1413 (1973).
14. Remaut, G., Lagasse, A., and Amelinckx, S., *Phys. Status Solidi*, **6**, 723–731 (1964).
15. Stevenson, S. and Vander Sande, J. B., *Acta Met.*, **22**, 1079–1086 (1974).
16. Ericksen, R. H. and Jones, G. J., *Met. Trans.*, **3**, 1735–1741 (1972).
17. Van Torne, L. and Thomas, G., *Acta Met.*, **12**, 601 (1964).
18. Klein, M. J. and Gulden, Mary Ellen, *Met. Trans.*, **1**, 105–110 (1970).
19. Olsen, R. J., Judd, G., and Ansell, G. S., *Met. Trans.*, **2**, 1353–1357 (1971).
20. Chakravorty, S. and Wayman, C. M., *Met. Trans. A*, **7A**, 569–582 (1976).
21. Luton, M. T. and Sellars, C. M., *Acta Met.*, **17**, 1033–1045 (1969).
22. Menon, M. N. and Reimann, W. H., *Met. Trans. A.*, **6A**, 1075–1085 (1975).
23. Despres, T. A., *Symposium on Electron Metallography*, 24–25, American Society for Testing Materials, Philadelphia, 1963.
24. Tousek, J., *Collect. Czech Chem. Commun.*, **36** (6), 2348 (1967) (in German). See also C. T. J. Ahlens and R. W. Baluffi, *J. Appl. Phys.*, **38** (2), 910 (1967).
25. Vandervoort, R. R. and Barmore, W. L., *Met. Trans.*, **245**, 825–829 (1969).
26. Rao, P. and Thomas, G., *Acta Met.*, 23, 309 (1975).
27. Liu, C. T., Inuoye, H., and Carpenter, R. W., *Met. Trans.*, **4**, 1839–1850 (1973).
28. Van Landuyt, J. and Wayman, C. M., *Acta Met.*, **16**, 803–814 (1968).
29. Balcerzak, A. T. and Sass, S. L., *Met. Trans.*, **3**, 1601–1605 (1972).
30. Venables, J. D., *Met. Trans.*, **1**, 2471–2476 (1970).
31. Giraud-Heraud, F. and Guillaumin, J., *Acta Met.*, **21**, 1243–1252 (1973).
32. Manley, A. J., *J. Nucl. Mater.*, **15** (2), 143 (1965).
33. Wanagel, J., Sass, S. L., and Batterman, B. W., *Met. Trans.*, **5**, 105–109 (1974).

34. Barber, D. J., *J. Mater. Sci.*, **5**, 3 (1970).
35. Naziri, H. and Pearce, R., *Acta Met.*, **22**, 1321–1330 (1974).
36. Livingston, J. D., Cline, H. E., Kock, E. F., and Russell, R. R., *Acta Met.*, **18**, 399–404 (1970).
37. Sass, S. L., *Acta Met.*, **17**, 813–820 (1969).
38. Carpenter, G. J. C. and Watters, J. F., *Acta Met.*, **21**, 1207–1214 (1973).
39. (a) Hirsch, P. B., Nicholson, R. B., Howie, A., Pashley, D. W., and Whelan, M. J., *Electron Microscopy of Thin Crystals*, Butterworths, London, 1965, Chapter 2 and Appendix 1. (b) Thomas, G., *Transmission Electron Microscopy of Metals*, John Wiley, New York, 1962, Chapter 4.
40. *Material on Electron Metallographic Techniques*, ASTM Technical Publication 547, American Society for Testing Materials, Philadelphia, 1973.

APPENDIX B

THE FOUR-AXIS HEXAGONAL RECIPROCAL LATTICE*

B.1 Basic Definitions

Okamoto and Thomas[1] pointed out that, since most of the important relationships can be written in the form of vector products of direct and reciprocal lattice vectors, the advantages of the Miller-Bravais notation can be fully exploited only if four-axis reference bases are used for both lattices.

In the case of the direct lattice shown in Fig. B.1*a*, this is done by introducing a third symmetric axis, defined by $\mathbf{a}_3 = -(\mathbf{a}_1 + \mathbf{a}_2)$. With respect to this four-vector basis, an arbitrary crystal vector $\mathbf{r}$ may be written in the form

$$\mathbf{r} = u\,\mathbf{a}_1 + v\,\mathbf{a}_2 + t\,\mathbf{a}_3 + w\,\mathbf{c},$$

with $u + v + t = 0$ as usual.

On the other hand, the conventional reciprocal lattice basis vectors $\mathbf{a}_1^*$, $\mathbf{a}_2^*$, and $\mathbf{c}^*$ shown in Fig. B.1*b* do not define a conventional hexagonal unit cell, so that the introduction of a third axis, $\mathbf{a}_3^* = -(\mathbf{a}_1^* + \mathbf{a}_2^*)$, will not have the same geometrical significance as its analogue in the direct lattice. With this end in mind, a far more convenient choice is the basis $\mathbf{A}_1^*$, $\mathbf{A}_2^*$, $\mathbf{A}_3^*$, and $\mathbf{C}^*$, chosen in such a way that the usual Miller to Miller-Bravais transformation of reciprocal lattice points is retained, $[hkl]^* \rightarrow [hkil]^*$ or

$$h\,\mathbf{a}_1^* + k\,\mathbf{a}_2^* + l\,\mathbf{c}^* \longrightarrow h\,\mathbf{A}_1^* + k\,\mathbf{A}_2^* + i\,\mathbf{A}_3^* + l\,\mathbf{C}^*, \qquad \text{(B.1)}$$

where $h + k + i = 0$ as usual.

*The content of this appendix appeared in a slightly different form as an article by P. R. Okamoto and G. Thomas, *Phys. Status Solidi*, **25**, 81 (1968).

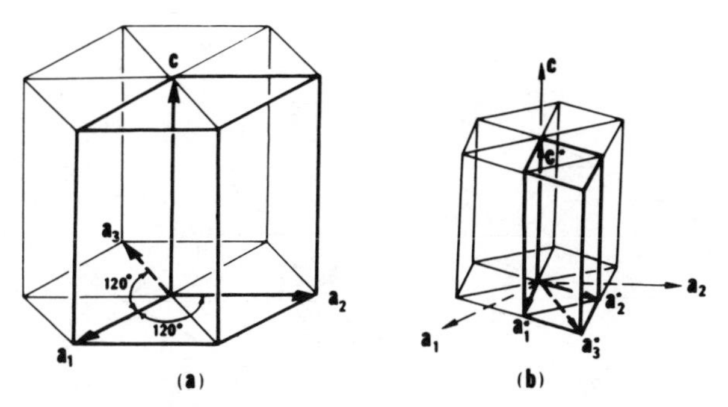

Fig. B.1 (*a*) The conventional unit cell of the direct hexagonal lattice. The third axis $\mathbf{a}_3$ of the four-axis hexagonal notation is also shown. (*b*) The conventional reciprocal lattice is similarly hexagonal but is rotated by 30° relative to the direct lattice about their common sixfold symmetry axis. Courtesy *Physica Status Solidi.*[1]

Otte and Crocker[2] have pointed out that the basis which satisfies the above constraint will not generate a lattice that is the reciprocal of the direct lattice in the mathematical sense (see Chapter 2, Section 1). In other words, the basis vectors $\mathbf{A}_1^*$, $\mathbf{A}_2^*$, $\mathbf{A}_3^*$, $\mathbf{C}^*$ and those of the direct lattice $\mathbf{a}_1$, $\mathbf{a}_2$, $\mathbf{a}_3$, and $\mathbf{c}$ do not satisfy the usual definition $\mathbf{A}_i^* \cdot \mathbf{a}_j = \delta_{ij}$. However, it can be shown that the lattice generated by $\mathbf{A}_1^*$, $\mathbf{A}_2^*$, $\mathbf{A}_3^*$, and $\mathbf{C}^*$ contains as a subset all the points of the conventional reciprocal lattice. Points not belonging to this subset have no physical significance in the sense that their indices are not integral and therefore do not correspond to real crystal planes. Since these points cannot register as spots in an actual diffraction pattern, the two lattices are physically indistinguishable. Therefore in a practical sense the four-index system just developed may be used as a "reciprocal" lattice.

The points discussed above can be more clearly seen by referring to Fig. B.2*a*. The conventional reciprocal lattice points have been indexed using Miller indices based on the primitive reciprocal lattice vectors $\mathbf{a}_1^*$, $\mathbf{a}_2^*$, and $\mathbf{c}^*$ (Fig. B.1*b*). According to the usual definition they can be written as

$$\left.\begin{aligned}
\mathbf{a}_1^* &= \frac{\mathbf{a}_2 \times \mathbf{c}}{V} = \frac{2(2\mathbf{a}_1 + \mathbf{a}_2)}{3a^2},\\
\mathbf{a}_2^* &= \frac{\mathbf{c} \times \mathbf{a}_1}{V} = \frac{2(\mathbf{a}_1 + 2\mathbf{a}_2)}{3a^2},\\
\mathbf{c}^* &= \frac{\mathbf{a}_1 \times \mathbf{a}_2}{V} = \frac{\mathbf{c}}{c^2},\\
V &= \mathbf{a}_1 \cdot \mathbf{a}_2 \times \mathbf{c} = \frac{3a^2c}{2},
\end{aligned}\right\} \qquad \text{(B.2)}$$

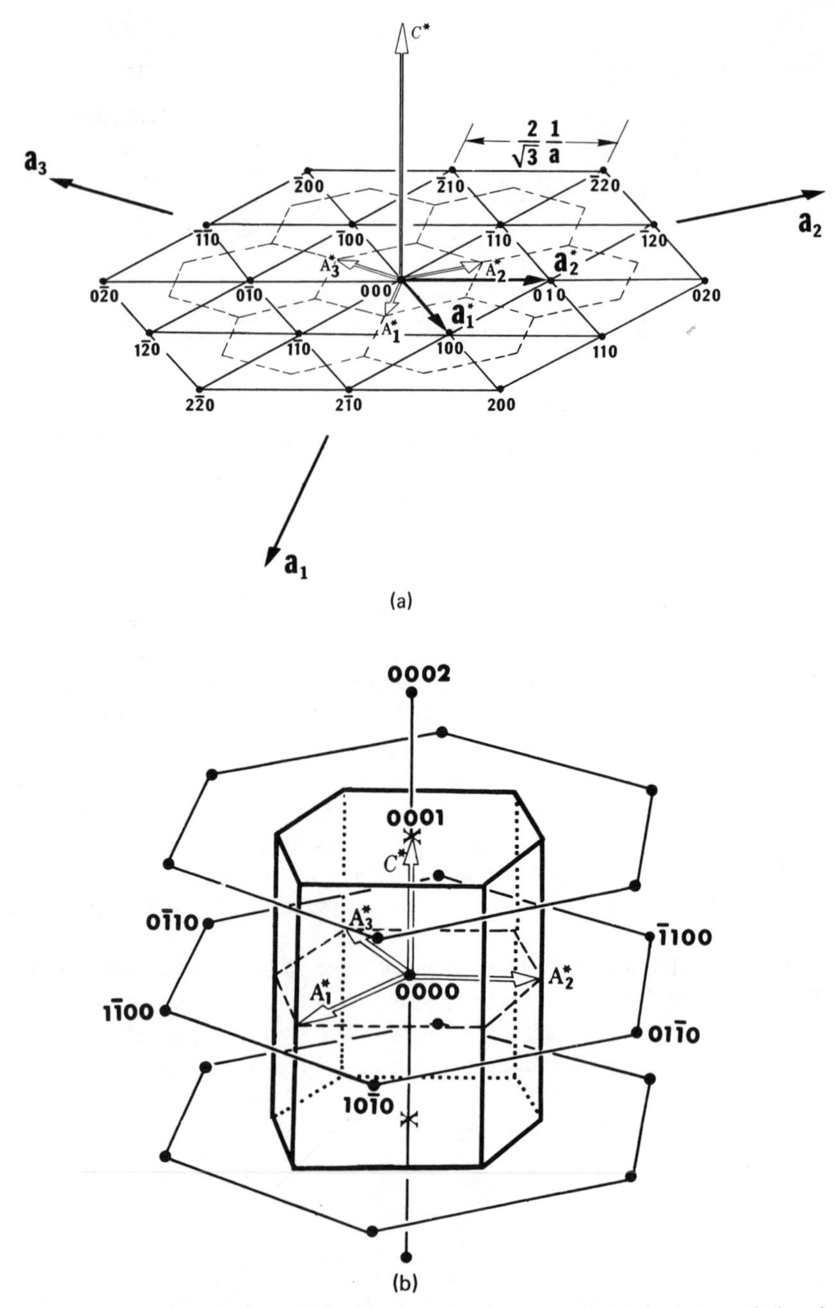

Fig. B.2 (*a*) The [0001] reciprocal lattice plane. The reciprocal lattice points are indexed using Miller indices based on the conventional reciprocal basis vectors. The broken lines indicate the sixfold symmetry of the lattice generated by the four-axis reference basis. The basis vectors of the direct lattice are also shown for reference purposes. (*b*) The first Brillouin zone is shown in bold lines. Only points coincident with the conventional reciprocal lattice are shown, and these are indexed in terms of the four-axis reference basis. ● indicates an allowed reflection and X a forbidden reflection, which may arise from double diffraction effects. Courtesy *Physica Status Solidi.*[1]

where $\mathbf{a}_1$, $\mathbf{a}_2$, and $\mathbf{c}$ are the primitive vectors of the direct lattice, and V is the volume of the unit cell. Substituting eqs. B.2 into the left-hand side of relation B.1 and using the definitions $i = -(h + k)$ and $\mathbf{a}_3 = -(\mathbf{a}_1 + \mathbf{a}_2)$ yields

$$\left.\begin{aligned} \mathbf{A}_1^* &= \frac{2}{3a^2}\,\mathbf{a}_1, & \mathbf{A}_2^* &= \frac{2}{3a^2}\,\mathbf{a}_2, \\ \mathbf{A}_3^* &= \frac{2}{3a^2}\,\mathbf{a}_3, & \mathbf{C}^* &= \frac{1}{c^2}\,\mathbf{c}. \end{aligned}\right\} \tag{B.3}$$

These basis vectors define a hexagonal lattice whose sixfold symmetry, shown in broken lines in Fig. B.2*a*, is identical in orientation with that of the direct lattice. Since these lines can be obtained by bisecting all vectors joining two conventional reciprocal lattice points, they also represent the basal projections of the first Brillouin zones.

In fact, Fig. B.2*b* shows that the hexagonal prism forming the first Brillouin zone serves as a natural choice for the unit cell of the four-axis reciprocal lattice. If one includes as part of the conventional reciprocal lattice the forbidden reflections that can arise from double-diffraction effects, then only one out of every three of the new lattice points coincides with a conventional reciprocal lattice point. Moreover, the basis vectors given by eqs. B.3 are so defined that only these points have integral indices; furthermore, their indices automatically satisfy relation B.1, so that the vector $h\,\mathbf{A}_1^* + k\,\mathbf{A}_2^* + i\,\mathbf{A}_3^* + l\,\mathbf{C}^*$ is normal to the crystal plane $(hkil)$. All other points have nonintegral indices and hence cannot represent real crystal planes. For example, the three basis vectors $\mathbf{A}_1^*$, $\mathbf{A}_2^*$, and $\mathbf{A}_3^*$ are themselves of the form $\frac{1}{3}\langle 2\bar{1}\bar{1}0\rangle^*$.

It is therefore emphasized once again that, although the four-axis lattice defined by eqs. B.3 and the conventional reciprocal lattice are mathematically distinct, they are physically indistinguishable as far as diffraction patterns are concerned. Consequently, indexing procedures may be carried out consistently, using the Miller-Bravais notation at all times. Some of the more useful crystallographic relationships are derived below to illustrate this point.

B.2 Crystallographic Relationships for Hexagonal Systems

In the following derivations the four-axis hexagonal indexing system is used throughout. The basis vectors for the direct and reciprocal lattices are those shown in Figs. B.1 and B.2, respectively.

B.2.1 The Relation Between the Crystal Plane $(hkil)$ and Its Normal $[uvtw]$

The direction $[uvtw]$ normal to the plane $(hkil)$ is simply the reciprocal lattice vector $[hkil]^*$ expressed in terms of the direct lattice coordinate system. This can be done by writing

$$u\,\mathbf{a}_1 + v\,\mathbf{a}_2 + t\,\mathbf{a}_3 + w\,\mathbf{c} = h\,\mathbf{A}_1^* + k\,\mathbf{A}_2^* + i\,\mathbf{A}_3^* + l\,\mathbf{C}^* \tag{B.4}$$

and by substituting relation B.3 into the right-hand side to obtain

$$u\,\mathbf{a}_1 + v\,\mathbf{a}_2 + t\,\mathbf{a}_3 + w\,\mathbf{c} = \frac{2}{3a^2}\left(h\,\mathbf{a}_1 + k\,\mathbf{a}_2 + i\,\mathbf{a}_3 + \frac{1}{\lambda^2}\,\mathbf{c}\right),$$

where $\lambda^2 = \frac{2}{3}(c/a)^2$. Therefore, in the usual notation,

$$[uvtw] = [hkil\lambda^{-2}]. \tag{B.5}$$

Conversely, the plane $(hkil)$ normal to the direction $[uvtw]$ is given by

$$(hkil) = (uvt\lambda^2 w). \tag{B.6}$$

For the special case where λ^2 is unity or $c/a = \sqrt{\frac{3}{2}}$ one has a "cubic" hexagonal lattice in the sense that a plane and its normal have identical indices. As Frank[3] pointed out, a lattice with any other axial ratio can be considered as arising from a deformation of the "cubic" hexagonal lattice corresponding to an elongation by the factor λ along the c-axis.

B2.2 The Dot Product of a Reciprocal Lattice Vector g and a Direct Lattice Vector r

Let $\mathbf{g} = [hkil]^*$ and $\mathbf{r} = [uvtw]$. Then the product

$$\mathbf{g}\cdot\mathbf{r} = (h\,\mathbf{A}_1^* + k\,\mathbf{A}_2^* + i\,\mathbf{A}_3^* + l\,\mathbf{C}^*)\cdot(u\,\mathbf{a}_1 + v\,\mathbf{a}_2 + t\,\mathbf{a}_3 + w\,\mathbf{c}) \tag{B.7}$$

may be expanded and written in the form

$$\mathbf{g}\cdot\mathbf{r} = [hkil]\begin{Vmatrix} \mathbf{A}_1^*\cdot\mathbf{a}_1 & \mathbf{A}_1^*\cdot\mathbf{a}_2 & \mathbf{A}_1^*\cdot\mathbf{a}_3 & \mathbf{A}_1^*\cdot\mathbf{c} \\ \mathbf{A}_2^*\cdot\mathbf{a}_1 & \mathbf{A}_2^*\cdot\mathbf{a}_2 & \mathbf{A}_2^*\cdot\mathbf{a}_3 & \mathbf{A}_2^*\cdot\mathbf{c} \\ \mathbf{A}_3^*\cdot\mathbf{a}_1 & \mathbf{A}_3^*\cdot\mathbf{a}_2 & \mathbf{A}_3^*\cdot\mathbf{a}_3 & \mathbf{A}_3^*\cdot\mathbf{c} \\ \mathbf{C}^*\cdot\mathbf{a}_1 & \mathbf{C}^*\cdot\mathbf{a}_2 & \mathbf{C}^*\cdot\mathbf{a}_3 & \mathbf{C}^*\cdot\mathbf{c} \end{Vmatrix}\begin{bmatrix} u \\ v \\ t \\ w \end{bmatrix}. \tag{B.8}$$

When the matrix elements are evaluated by using relations B.3, the matrix can be written as

$$(\mathbf{A}_1^*\cdot\mathbf{a}_j) = \frac{1}{3}\begin{Vmatrix} 2 & \bar{1} & \bar{1} & 0 \\ \bar{1} & 2 & \bar{1} & 0 \\ \bar{1} & \bar{1} & 2 & 0 \\ 0 & 0 & 0 & 3 \end{Vmatrix},$$

and eq. B.7 becomes

$$\mathbf{g}\cdot\mathbf{r} = h\,u + k\,v + i\,t + l\,w. \tag{B.9}$$

Since $\mathbf{g} \cdot \mathbf{r}$ is proportional to the projection of $\mathbf{r}$ onto $\mathbf{g}$, the two vectors are perpendicular when eq. B.9 vanishes.

B.2.3 The Dot Product of Two Direct Lattice Vectors and of Two Reciprocal Lattice Vectors

These relationships can be expressed in the form of eq. B.8. For the case of two direct lattice vectors, $\mathbf{r}_1 = [u_1 v_1 t_1 w_1]$ and $\mathbf{r}_2 = [u_2 v_2 t_2 w_2]$,

$$\mathbf{r}_1 \cdot \mathbf{r}_2 = [u_1 v_1 t_1 w_1]\,(\mathbf{a}_i \cdot \mathbf{a}_j)\begin{bmatrix} u_2 \\ v_2 \\ t_2 \\ w_2 \end{bmatrix},$$

where

$$(\mathbf{a}_i \cdot \mathbf{a}_j) = \frac{a^2}{2}\begin{Vmatrix} 2 & \bar{1} & \bar{1} & 0 \\ \bar{1} & 2 & \bar{1} & 0 \\ \bar{1} & \bar{1} & 2 & 0 \\ 0 & 0 & 0 & 3\lambda^2 \end{Vmatrix}.$$

Therefore

$$\mathbf{r}_1 \cdot \mathbf{r}_2 = \frac{3a^2}{2}(u_1 u_2 + v_1 v_2 + t_1 t_2 + \lambda^2 w_1 w_2). \tag{B.10}$$

For the case of two reciprocal lattice vectors, $\mathbf{g}_1 = [h_1 k_1 i_1 l_1]^*$ and $\mathbf{g}_2 = [h_2 k_2 i_2 l_2]^*$,

$$\mathbf{g}_1 \cdot \mathbf{g}_2 = [h_1 k_1 i_1 l_1]\,(\mathbf{A}_i^* \cdot \mathbf{A}_j^*)\begin{bmatrix} h_2 \\ k_2 \\ i_2 \\ l_2 \end{bmatrix},$$

where

$$(\mathbf{A}_i^* \cdot \mathbf{A}_j^*) = \frac{2}{9a^2}\begin{Vmatrix} 2 & \bar{1} & \bar{1} & 0 \\ \bar{1} & 2 & \bar{1} & 0 \\ \bar{1} & \bar{1} & 2 & 0 \\ 0 & 0 & 0 & 3\lambda^{-2} \end{Vmatrix}.$$

Therefore

$$\mathbf{g}_1 \cdot \mathbf{g}_2 = \frac{2}{3a^2}(h_1 h_2 + k_1 k_2 + i_1 i_2 + \lambda^{-2} l_1 l_2). \tag{B.11}$$

B.2.4 The Magnitude of Vectors

The magnitude of vectors follows directly from relations B.10 and B.11. For the case of a direct and a reciprocal lattice vector, **r** and **g**, respectively,

$$|\mathbf{r}|^2 = \frac{3a^2}{2}(u^2 + v^2 + t^2 + \lambda^2 w^2), \tag{B.12}$$

$$|\mathbf{g}|^2 = \frac{2}{3a^2}(h^2 + k^2 + i^2 + \lambda^{-2} l^2). \tag{B.13}$$

The latter is also inversely proportional to d^2, where d is the interplanar spacing for the $(hkil)$ planes. It is also useful to remember that, since the direct lattice is itself the reciprocal of the reciprocal lattice, the crystal vector $\mathbf{r} = [uvtw]$ is normal to the reciprocal lattice plane $(uvtw)^*$, whose interplanar spacing d^* is inversely proportional to $|\mathbf{r}|$. This information is often useful when indexing diffraction patterns containing spots from more than one Laue zone or when determining the orientation of unknown patterns by the use of Kikuchi maps (see Chapter 2, Section 5.4).

B.2.5 Interplanar and Interdirectional Angles

These follow directly from eqs. B.10 to B.13. For example, the angle between two vectors, $\mathbf{r}_1 = [u_1 v_1 t_1 w_1]$ and $\mathbf{r}_2 = [u_2 v_2 t_2 w_2]$, is given by

$$\cos(\mathbf{r}_1, \mathbf{r}_2) = \frac{\mathbf{r}_1}{|\mathbf{r}_1|} \cdot \frac{\mathbf{r}_2}{|\mathbf{r}_2|} = \frac{u_1 u_2 + v_1 v_2 + t_1 t_2 + \lambda^2 w_1 w_2}{(u_1^2 + v_1^2 + t_1^2 + \lambda^2 w_1^2)^{1/2} (u_2^2 + v_2^2 + t_2^2 + \lambda^2 w_2^2)^{1/2}}. \tag{B.14}$$

Relation B.14 also gives the angle between the reciprocal lattice planes $(u_1 v_1 t_1 w_1)^*$ and $(u_2 v_2 t_2 w_2)^*$. Similarly, the angle between two planes corresponding to the reciprocal lattice vectors $\mathbf{g}_1 = [h_1 k_1 i_1 l_1]^*$ and $\mathbf{g}_2 = [h_2 k_2 i_2 l_2]^*$ is given by

$$\cos(\mathbf{g}_1, \mathbf{g}_2) = \frac{\mathbf{g}_1}{|\mathbf{g}_1|} \cdot \frac{\mathbf{g}_2}{|\mathbf{g}_2|} = \frac{h_1 h_2 + k_1 k_2 + i_1 i_2 + \lambda^{-2} l_1 l_2}{(h_1^2 + k_1^2 + i_1^2 + \lambda^{-2} l_1^2)^{1/2} (h_2^2 + k_2^2 + i_2^2 + \lambda^{-2} l_2^2)^{1/2}}. \tag{B.15}$$

These examples illustrate the use of Miller-Bravais notation in the derivation of crystallographic relationships needed to index diffraction patterns. A table listing some of the above formulas and other useful relationships can be found in the papers by Otte and Crocker[2] and by Nicholas.[4]

References

1. Okamoto, P. R. and Thomas, G., *Phys. Status Solidi*, **25**, 81 (1968).
2. Otte, H. and Crocker, A. G., *Phys. Status Solidi*, **9**, 441 (1965).
3. Frank, F. C., *Acta Crystallogr.*, **18**, 862 (1965).
4. Nicholas, J. F., *Acta Crystallogr.*, **21**, 880 (1966).

APPENDIX C

RELATIVE RECIPROCAL LATTICE SPACINGS

Relative Reciprocal Lattice Spacings for Face-Centered and Diamond Cubic Lattices[a]

	111	200	220	311	331	420	422	511	531
111	1*								
200	1.155	1							
220	1.63*	1.41	1*						
311	1.92*	1.66	1.17*	1*					
222	2.00	1.73	1.225	1.045					
400	2.31*	2.00	1.415*	1.21*					
331	2.52*	2.18	1.54*	1.31*	1*				
420	2.58	2.235	1.58	1.35	1.027	1			
422	2.85*	2.45	1.73*	1.48*	1.124*	1.096	1*		
333,511	3.00*	2.60	1.84*	1.57*	1.19*	1.16	1.06*	1*	
440	3.27*	2.83	2.00*	1.71*	1.30*	1.217	1.156*	1.09*	
531	3.42*	2.96	2.09*	1.785*	1.36*	1.32	1.21*	1.14*	1*
442	3.46	3.00	2.12	1.81	1.38	1.34	1.225	1.157	1.014
620	3.66*	3.16	2.24*	1.91*	1.45*	1.42	1.29*	1.22*	1.07*
533	3.79*	3.28	2.32*	1.98*	1.503*	1.47	1.34*	1.26*	1.11*
622	3.82	3.32	2.34	2.00	1.52	1.48	1.355	1.28	1.12
444	4.00*	3.47	2.45*	2.09*	1.59*	1.55	1.415*	1.33*	1.17*
711,551	4.12*	3.57	2.52*	2.15*	1.64*	1.595	1.458*	1.374*	1.207*

[a]Asterisks indicate spacings for diamond cubic lattice.

Relative Reciprocal Lattice Spacings for Body-Centered Cubic Lattices

	110	200	211	310	222	321	411	420	332	510 431	521	530 433
110	1											
200	1.415	1										
211	1.73	1.225	1									
220	2.00	1.415	1.155									
310	2.235	1.58	1.29	1								
222	2.45	1.73	1.415	1.095	1							
321	2.645	1.87	1.53	1.185	1.08	1						
400	2.83	2.00	1.63	1.265	1.155	1.07						
411,330	3.00	2.12	1.73	1.34	1.225	1.135	1					
420	3.16	2.235	1.825	1.415	1.29	1.195	1.055	1				
332	3.315	2.345	1.915	1.485	1.355	1.255	1.105	1.05	1			
422	3.465	2.45	2.00	1.55	1.415	1.31	1.155	1.095	1.045			
510,431	3.605	2.55	2.08	1.61	1.47	1.365	1.20	1.14	1.09	1		
521	3.875	2.74	2.235	1.73	1.58	1.465	1.29	1.245	1.17	1.075	1	
440	4.00	2.83	2.31	1.79	1.63	1.51	1.335	1.265	1.21	1.11	1.035	
530,433	4.125	2.915	2.38	1.845	1.685	1.56	1.375	1.305	1.245	1.145	1.065	1
600,442	4.245	3.00	2.45	1.895	1.73	1.605	1.415	1.34	1.28	1.18	1.095	1.03
611,532	4.36	3.08	2.52	1.95	1.78	1.65	1.455	1.38	1.315	1.21	1.125	1.06
620	4.47	3.16	2.58	2.00	1.825	1.69	1.49	1.415	1.35	1.24	1.155	1.085
541	4.585	3.24	2.645	2.05	1.87	1.73	1.53	1.45	1.38	1.27	1.185	1.11
622	4.69	3.315	2.71	2.10	1.915	1.77	1.565	1.485	1.415	1.30	1.21	1.135
631	4.795	3.39	2.77	2.145	1.955	1.815	1.60	1.515	1.445	1.33	1.24	1.16
444	4.90	3.465	2.83	2.19	2.00	1.85	1.635	1.55	1.48	1.36	1.265	1.185
710,550,543	5.00	3.535	2.89	2.235	2.04	1.89	1.665	1.58	1.51	1.385	1.29	1.21
640	5.10	3.605	2.94	2.28	2.08	1.925	1.70	1.61	1.54	1.415	1.315	1.235
721,633,552	5.195	3.675	3.00	2.325	2.12	1.955	1.73	1.645	1.57	1.44	1.34	1.26
642	5.29	3.74	3.055	2.365	2.16	2.00	1.765	1.675	1.595	1.47	1.365	1.285
730	5.385	3.81	3.11	2.41	2.20	2.035	1.795	1.705	1.625	1.495	1.39	1.305

SOLUTIONS TO EXERCISES

Chapter 1

1.1 Figure 1.13 shows the ray paths (neglecting the *rotations* produced by the magnetic fields of the lenses) in a conventional three-lens imaging system. Since all the lenses work with real objects (real images from the previous lens), each introduces an inversion (Fig. 1.13*a*). In a three-lens system the image is therefore inverted with respect to the specimen. In contrast to the image the diffraction pattern is not inverted by the objective lens (compare Figs. 1.13*a* and *b*). Therefore, if the only change when the diffraction pattern is produced is weakening of the intermediate lens (as envisaged in Fig. 1.13), the diffraction pattern is inverted with respect to the image (i.e., not inverted with respect to the specimen).

If, however, the number of lenses for imaging and that for diffraction are different, these relationships will be altered. In general, if *n* lenses ***after the objective*** are used to form real images (or real images of diffraction patterns), the final image will be inverted with respect to the specimen if *n* is even, while the diffraction pattern will be inverted with respect to the specimen if *n* is odd. The *rotations* produced by all the lenses must also be properly taken into account.

1.2 The largest errors in the apparent positioning of the SAD aperture are shown schematically in Figs. 1.14 and 1.15 and are combined in eq. 1.8. The typical figures given in the text indicate an error of 0.04 μm produced by spherical aberration alone at 100 kV, on the order of 10% of the 0.5 μm diameter effective aperture size (10 μm physical aperture imaged at 1/20 magnification on the specimen). The position of any smaller aperture would be in error by a significant fraction of its diameter.

Higher voltage operation alleviates the problem in general, and careful under-

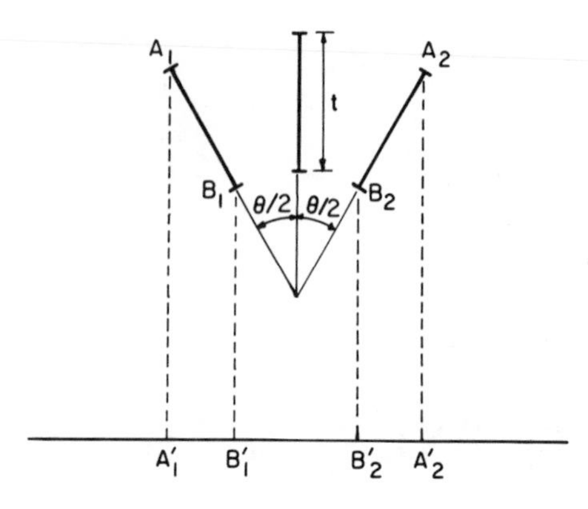

Fig. S1.3 Simple construction illustrating the parallax introduced by tilting a specimen.

focusing (negative D in eq. 1.8) may be used to remove the error for a specific value of α. Section 5.3 describes the improved area selection capability of STEM systems.

1.3 In Fig. S1.3 it may be seen that the required parallax is $A_1' - B_1' - (A_2' - B_2')$ in the specimen plane, times M when magnified into the image plane, that is,

$$p = M \times \{t \sin(\theta/2) - [-t \sin(\theta/2)]\},$$

$$p = 2Mt \sin(\theta/2).$$

1.4 The final viewing magnification is $M = 20{,}000 \times 2$. Substituting $M = 40{,}000$, $t = 5 \times 10^{-7}$ m, and $p = 4 \times 10^{-3}$ m into eq. 1.10 yields $\theta = 11.5°$.

1.5 (*a*) The structure factor is

$$F = \sum_{1}^{N} f_m e^{2\pi i(hu_m + kv_m + lw_m)},$$

summing over all the atoms in the unit cell.

(*b*) The positions $u_m v_m w_m$ of the atoms in the NaCl unit cell are as follows:

$$\text{Na:} \quad 000,\ \tfrac{1}{2}\tfrac{1}{2}0,\ \tfrac{1}{2}0\tfrac{1}{2},\ 0\tfrac{1}{2}\tfrac{1}{2},$$

$$\text{Cl:} \quad \tfrac{1}{2}\tfrac{1}{2}\tfrac{1}{2},\ 00\tfrac{1}{2},\ 0\tfrac{1}{2}0,\ \tfrac{1}{2}00,$$

$$F = f_{\text{Na}} e^{2\pi i(0)} + f_{\text{Na}} e^{2\pi i(h+k)/2} + f_{\text{Na}} e^{2\pi i(h+1)/2} + f_{\text{Na}} e^{2\pi i(k+1)/2}$$

$$+ f_{\text{Cl}} e^{2\pi i(h+k+1)/2} + f_{\text{Cl}} e^{2\pi il/2} + f_{\text{Cl}} e^{2\pi ik/2} + f_{\text{Cl}} e^{2\pi ih/2},$$

$$= [1 + e^{\pi i(h+k)} + e^{\pi i(h+1)} + e^{\pi i(k+1)}]\,[f_{\text{Na}} + f_{\text{Cl}} e^{\pi i(h+k+l)}],$$

$$= 0 \qquad \text{for } hkl \text{ mixed} \therefore \text{fcc type}$$

$$= 4(f_{\text{Na}} + f_{\text{Cl}}) - h + k + 1 \qquad \text{even} \quad \bullet$$

$$= 4(f_{\text{Na}} - f_{\text{Cl}}) - h + k + 1 \qquad \text{odd} \quad \times$$

The resulting primitive cubic reciprocal lattice is sketched in Fig. S1.5.

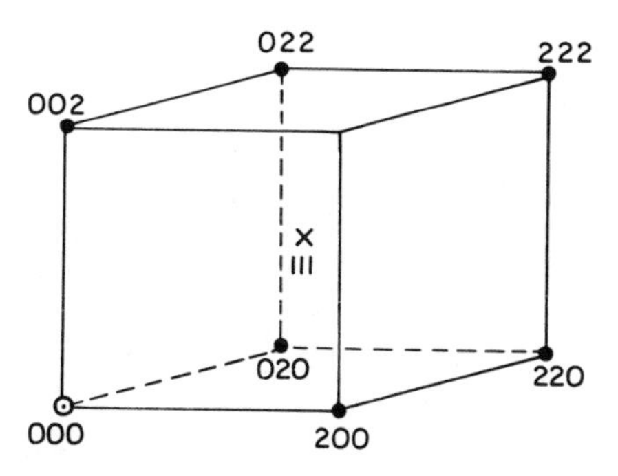

Fig. S1.5 The NaCl reciprocal lattice.

1.6 Radius of Ewald sphere $r_x = 1/\lambda$.

$$\text{X-rays:}\quad r_x = \frac{1}{\lambda_{\text{Cu}\,K_\alpha}} = 0.068\ \text{Å}^{-1}.$$

$$\text{Electrons:}\ r_e = \frac{1}{\lambda_{(100\,\text{kv})}} = 27.02\ \text{Å}^{-1}.\quad \text{(See Table 1.2)}$$

Reciprocal lattice spacing for Al 200 planes =

$$D = \frac{1}{d_{200}} = \frac{2}{4.04} \cong 0.5\ \text{Å}^{-1}.$$

Reflections will occur only where the reciprocal lattice meets the Ewald sphere. Because of the small Ewald sphere (large λ) of X-rays and thermal neutrons, radius r_x, only a very limited number of reflections will be obtained (see Fig. S1.6). Using polychromatic radiation will create a set of Ewald spheres with a continuous range of radii. This corresponds to varying λ in Bragg's law, $\lambda = 2d \sin \theta$.

For electrons the Ewald sphere, of radius r_e, is almost flat, allowing many reflections to occur at the same time (see Fig. S1.6).

1.7 Take the (001) standard projection (Fig. S1.7*a*), and rotate by 19.5° about $\bar{1}10$ so that the 114 pole moves along the equator to its center. Move all other poles by 19.5° along their latitude (small) circles to obtain the (114) projection (Fig. S1.7*b*).

The (111) twinning plane is given by the great circle 90° away from the 111 pole. Since $1\bar{1}0$ lies in this plane (Fig. S1.7*b*), it will remain unchanged by a reflection through (111).

The twinning operation can be described either by a reflection through the 111 plane or by a 180° rotation about the 111 pole. The two ways are distinguished only by the sign of the twinning matrix. These twinning matrices may be obtained by choosing three linearly independent directions whose

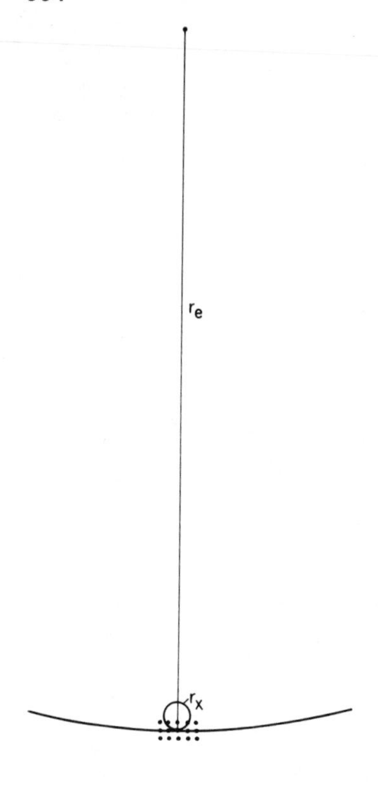

Fig. S1.6 Illustrating the differences in scale of the Ewald sphere construction for conventional X-ray and electron diffraction.

indices are known before and after twinning. For example,

$$\begin{array}{lll} \bar{1}10 \longrightarrow \bar{1}10 & & \bar{1}10 \longrightarrow \bar{1}10 \\ \bar{1}01 \longrightarrow \bar{1}01 & \text{and} & \bar{1}01 \longrightarrow 10\bar{1} \\ 111 \longrightarrow \bar{1}\bar{1}\bar{1} & & 111 \longrightarrow 111 \end{array}$$

by 180° rotation   by reflection

The transformation matrices are then

$$\frac{1}{3}\begin{pmatrix} 1 & -2 & -2 \\ -2 & 1 & -2 \\ -2 & -2 & 1 \end{pmatrix} \quad \text{and} \quad \frac{1}{3}\begin{pmatrix} -1 & 2 & 2 \\ 2 & -1 & 2 \\ 2 & 2 & -1 \end{pmatrix}$$

for reflection   for rotation.

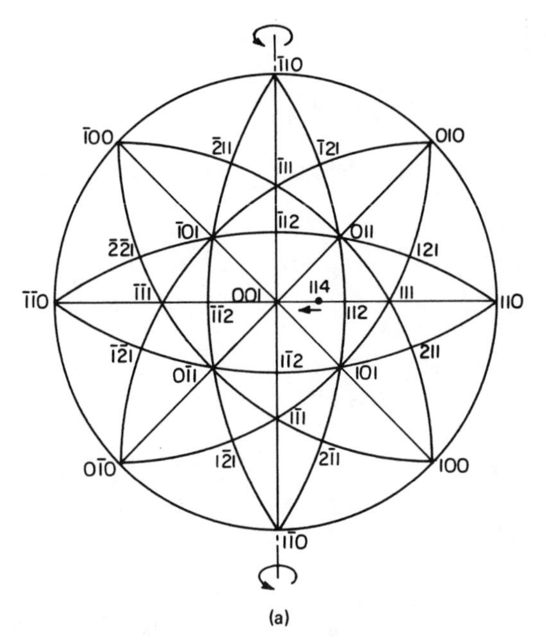

Fig. S1.7 (*a*) (001) standard stereographic projection, indicating rotation about $[1\bar{1}0]$ required to obtain the (114) projection.

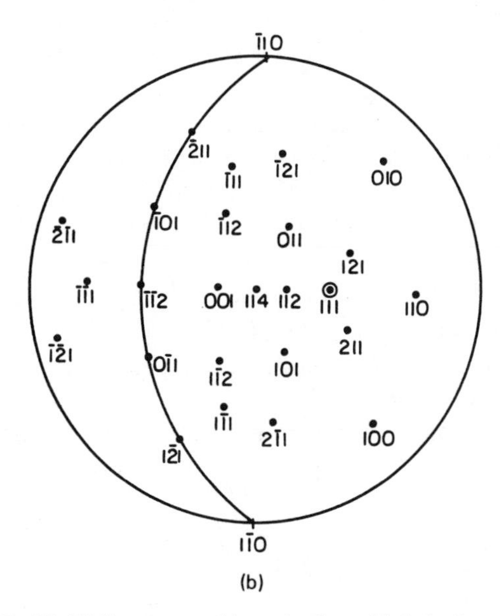

Fig. S1.7 (*b*) (114) stereographic projection with twinning on (111).

1.8 The current I through the probe is given by

$$I = B \times \frac{\pi d^2}{4} \times \pi\alpha^2,$$

where B is the brightness, d the probe diameter, and α the convergence semi-angle. Here

$$I = 10^{10} \times \frac{\pi \times (10^{-8})^2}{4} \times \pi \times (5 \times 10^{-3})^2$$

$$= 6.2 \times 10^{-11} \text{ A}$$

$$\text{Number of electrons per second} = \frac{I}{\text{charge of electron}}$$

$$= \frac{6.2 \times 10^{-11}}{1.6 \times 10^{-19}}$$

$$= 3.9 \times 10^{8} \text{ electrons/sec}$$

1.9 (*a*) Background under Ni peak = $5000 \times 2 \times 10^{-2}$ = 100 counts. Thus the Ni K_α count must exceed $5 \times \sqrt{100} = 50$. Since the cross sections for copper and nickel are, to a good approximation, the same, this corresponds to

$$\frac{50}{5000} = 1 \text{ at.\% of nickel.}$$

(*b*) When the probe diameter is increased by a factor of 10, the Cu count will increase to 500,000 (before contamination stops the experiment), and the Ni count will have to exceed $5 \times \sqrt{10{,}000} = 500$. The minimum detectable concentration of nickel will therefore improve to 0.1 at. %.

(*c*) If any increase in the background due to multiple scattering is neglected, the fourfold increase in specimen thickness will similarly improve the detection limit by a factor of 2 to 0.5 at. %.

1.10 Figure S1.10 shows sketches of the imaging errors produced by (*a*) spherical and (*b*) chromatic aberrations.

In optical microscopes, aberration-free lenses can be produced so that the resolution is limited only by diffraction at the aperture. The corresponding resolution limit is

$$\Delta r_d = \frac{0.61\lambda}{\alpha}.$$

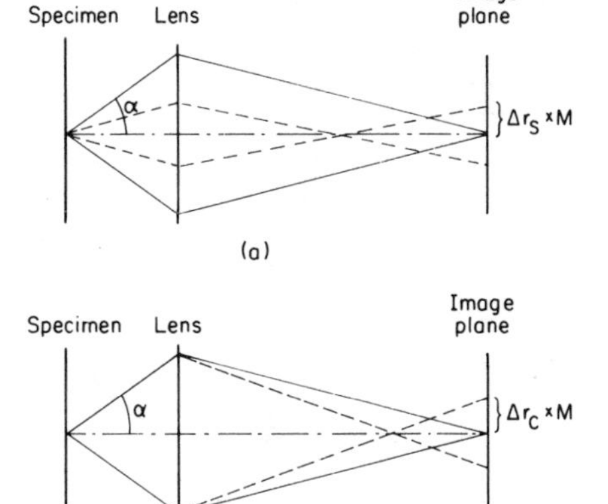

Fig. S1.10 Illustrating the errors introduced in the image by (*a*) spherical and (*b*) chromatic abberration.

In electron microscopes spherical aberration cannot be totally corrected. The resolution limit therefore depends on α: either

$$\Delta r_{\text{min}} = \Delta r_d \qquad \text{(small } \alpha\text{)}$$

or

$$\Delta r_{\text{min}} = \Delta r_s = C_s \alpha^3 \qquad \text{(large } \alpha\text{)},$$

where C_s is the spherical aberration constant, or the compromise between the two limits (optimum). At optimum aperture angle

$$\Delta r_d = \Delta r_s,$$

$$\alpha_{\text{opt}} = \lambda^{1/4} C_s^{-1/4},$$

$$\Delta r_{\text{min}} = (0.61\lambda)^{3/4} C_s^{1/4}.$$

To minimize spherical aberration, a small aperture angle α must be used.

Chromatic aberration depends on the energy spread of the electrons,

$$\Delta r_c = C_c \alpha \frac{\Delta E}{E}.$$

It can be minimized by avoiding fluctuations in the accelerating voltage or by increasing the voltage. Light elements (low Z) have stronger energy exchange with electrons, relative to elastic scattering mechanisms, than do heavy elements, that is, the energy loss ΔE will be larger and will lead to an increase in Δr_c.

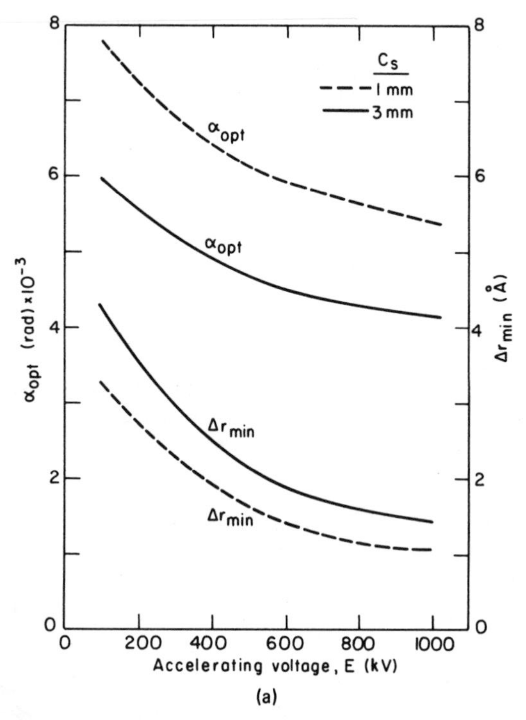

Fig. S1.11 (*a*) Optimum objective angle and ultimate resolution as a function of accelerating voltage.

1.11 (*a*) The completed table is as follows:

Voltage (kV)	λ (pm)	α_{opt} (rad.) $C_s = 1$ mm	α_{opt} (rad.) $C_s = 3$ mm	Δr_{min} (Å) $C_s = 1$ mm	Δr_{min} (Å) $C_s = 3$ mm
100	3.7	7.80×10^{-3}	5.93×10^{-3}	3.27	4.31
500	1.42	6.14×10^{-3}	4.67×10^{-3}	1.60	2.10
1000	0.87	5.43×10^{-3}	4.13×10^{-3}	1.11	1.46

The results are plotted in Fig. S1.11*a*.

(*b*) For resolution of the Na and the Cl ion positions in the crystal, it is necessary to choose an orientation with alternate rows of all Na and all Cl ions parallel to the beam. This is satisfied by the 110 orientation (Fig. S1.11*b*), which contains two sets of {111} planes. Recall the structure factors for 111 and 222 reflections (Exercise 1.5):

$$F_{111} = 4(f_{Na} - f_{Cl}),$$

$$F_{222} = 4(f_{Na} + f_{Cl}).$$

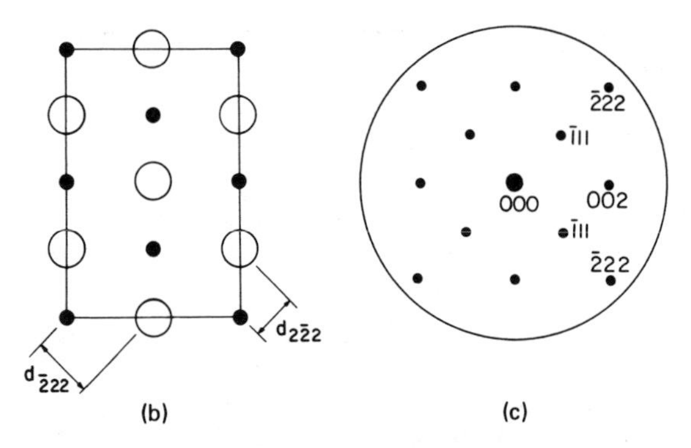

Fig. S1.11 (*b*) [110] projection of NaCl. (*c*) Schematic diffraction pattern for the orientation of (*b*) with superimposed objective aperture.

Therefore, if two sets of 111 and 222 reflections are included in the aperture (Fig. S1.11*c*), a cross grating of fringes will be obtained with different intensities for the two kinds of ions.

The spacing of {222} planes in NaCl is 1.63 Å. From Fig. S1.11*a* it may be seen that $E > 500$ kV with $C_s = 1$ mm and $E > 780$ kV with $C_s = 3$ mm is needed to obtain the necessary resolution, that is, a high voltage microscope would be essential with these values of spherical aberration.

(*c*) To find the beams scattered into the optimum aperture it is necessary to use Bragg's law,

$$\lambda = 2d \sin \theta.$$

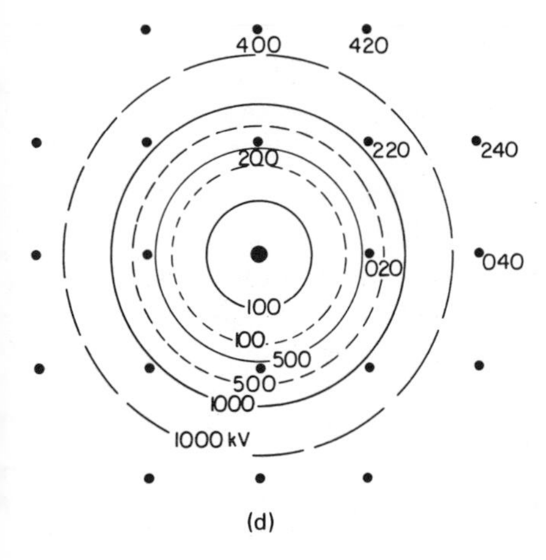

Fig. S1.11 (*d*) Effective optimum aperture size superimposed on the [001] reciprocal lattice projection of NaCl at the voltages indicated. Solid lines C_s = 3 mm, dashed C_s = 1 mm.

Since angles are small, $2 \sin \theta = 2\theta$, which here is set equal to or less than α_{opt}, that is,

$$d_{min} = \frac{\lambda}{\alpha_{opt}}.$$

The minimum resolvable spacing d_{min} corresponds to a circle of radius $1/d_{min}$ in reciprocal space. Circles for the various voltages are drawn on the 001 reciprocal lattice projection in Fig. S1.11*d*. From this figure it may be seen that the numbers of beams, in addition to the zero-order beam, included in the aperture are as follows:

Voltage (kV)	C_s	
	1 mm	3 mm
100	0	0
500	4	0
1000	8	4

Chapter 2*

2.1 Bcc cell of side $(2/a)$.

2.4 A set of self-consistent indices is shown for each pattern in Fig. S2.4, together with the corresponding beam direction $[UVW]$ for (*a*) fcc, (*b*), bcc, and (*c*) hcp. In the last case the four-index notation (see Appendix B) has been used, and the points denoted as x may be produced only by double diffraction.

2.5 (*a*) Fcc (gold, in fact).

(*b*) (i) 6.13 mm, (ii) 4.19 mm, (iii) 3.28 mm.*

(*c*) 480 mm.

(*d*) From patterns, -8%; from formulas, -4.6%.

2.6 Twinning of 001 about (111) reflects that pole to $\bar{2}\bar{2}1$ (see Fig. S2.6*a*), which is the new beam orientation. Note that $\bar{1}10$ does not change position.

The resulting diffraction pattern has superimposed spots at $\bar{2}20$ ($\bar{2}20_T$), 420 ($0\bar{2}\bar{4}_T$), and so on, as shown in Fig. S2.6*b*. No *extra* spots occur.

2.7 (*a*) The whole pattern may be indexed as $[\bar{1}\bar{1}0]$ beam direction, and the anomalously bright spots alone as [114]. Thus the high intensity spots could arise from (i) superlattice, (ii) precipitate of the same lattice constant as the matrix, (iii) twinning, or (iv) double diffraction.

Possibility (i) can be ruled out because superlattices have multiples of fundamental lattice spacings so that the diffraction pattern would show

*N.B. Solutions to problems requiring measurements: the values quoted here are from the original negatives. Figures have been reduced in printing, hence the reader should use the appropriate scale factor in each case, namely problems 2.5 and 2.9.

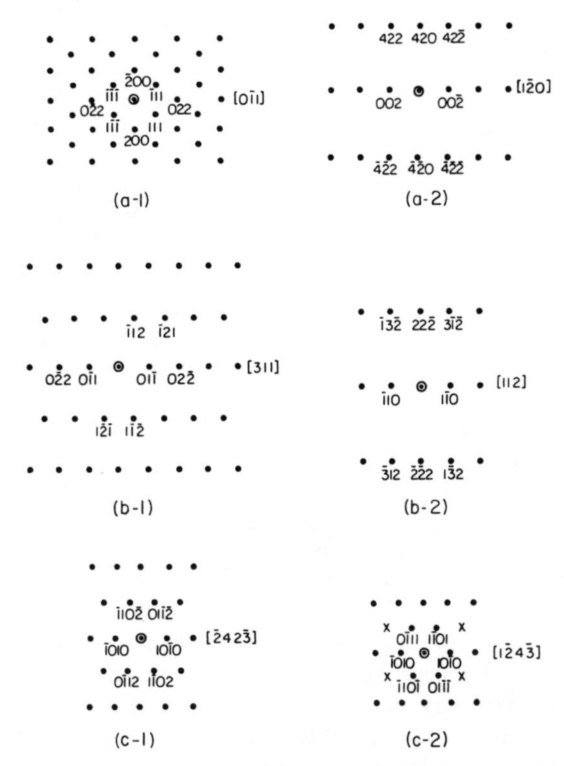

Fig. S2.4 The idealized symmetrical patterns of Fig. 2.35 indexed for (*a*) fcc, (*b*) bcc, and (*c*) hcp, together with the corresponding beam direction [*UVW*].

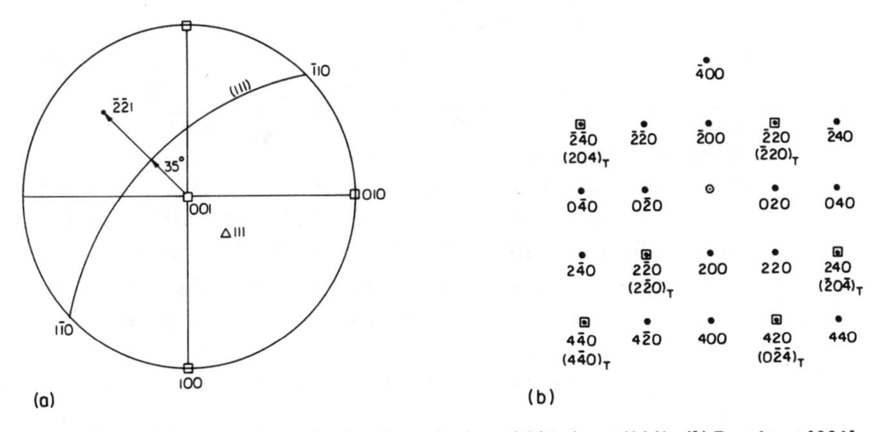

Fig. S2.6 (*a*) Stereographic projection for twinning of 001 about (111). (*b*) Resultant [001] diffraction pattern.

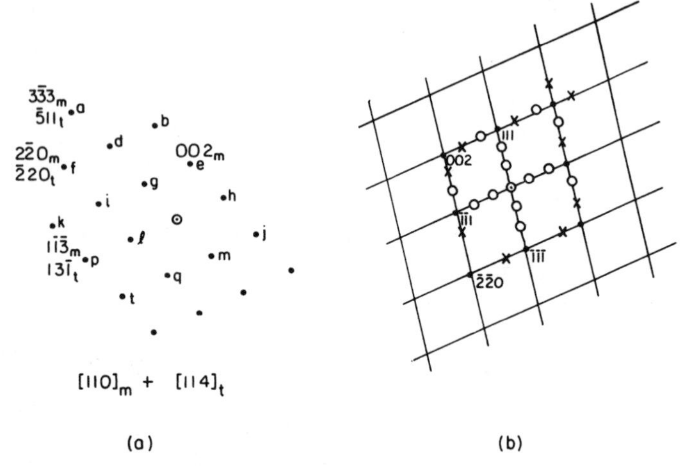

Fig. S2.7 Indexed diffraction patterns corresponding to Fig. 2.37. (*a*) The beam direction in the matrix is $[\bar{1}\bar{1}0]$ and in the (single) twin [114]. (*b*) The extra diffraction spots arise from a multiple twinned structure (X) by single diffraction and (0) by double diffraction.

these extra spots between those of the matrix. Possibility (iv) can also be ruled out because the higher intensity pattern does not have the 110 pattern symmetry. As for possibility (ii), a (114) precipitate with the same lattice parameter as the (110) matrix has a very definite orientation relationship to the matrix. If they are twin related, one has case (iii).

Figure S2.7*a* shows a few of the diffraction spots indexed according to the scheme $[\bar{1}\bar{1}0]$ matrix and [114] twin, with, for example, spot *a* as $3\bar{3}3_m$ and $\bar{5}11_t$ $(h^2 + k^2 + l^2 = 27)$.

(*b*) In the fcc system twin diffraction spots must either coincide with matrix spots (as above) or be displaced by $\frac{1}{3}\langle 111 \rangle$ (see Section 4.4). In the latter case the twin spots must of course still coincide with allowed distances from the origin; otherwise they can be produced only by double diffraction from overlaid twins. The pattern of Fig. 2.37*b* may thus be indexed, as shown schematically in Fig. S2.7*b*, as $[1\bar{1}0]_m$ with twin spots marked x for single diffraction and 0 for double diffraction.

2.8 Figure 2.38 is indeed correctly indexed. Selection of any pair produces the common zone axis [110], and the angles check in each case, for example,

$$\cos^{-1}(1\bar{1}3.2\bar{2}0) = \cos^{-1}\left(\frac{4}{\sqrt{88}}\right) = 65°.$$

2.9 (*a*) Pairs of parallel Kikuchi lines have been selected by inspection to produce a set of triangles surrounding the incident beam direction 0 (see Fig. S2.9). They may be indexed for type by measurements of their spacings,*

*See f/n p. 360.

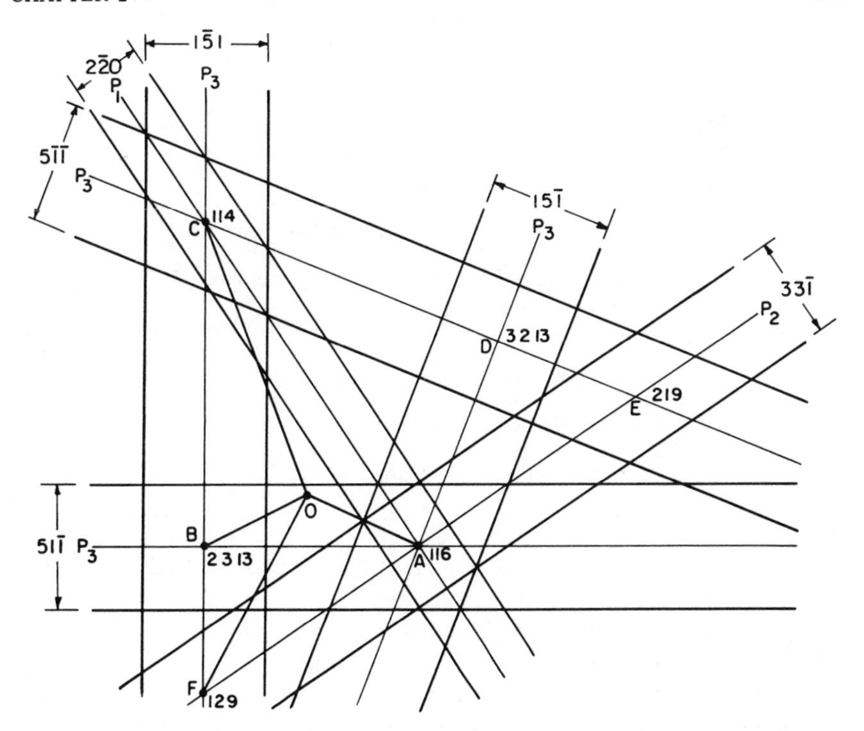

Fig. S2.9 Convenient sets of Kukuchi line pairs from Fig. 2.39, redrawn and indexed for planes of spacing P_1 and directions $A, B, \ldots$ in which the pairs of planes intersect.

and for specific indices by comparison of angles:

Measurements			Trial	Confirmation	
$p_1 = 1.63$ cm	$p_3/p_1 = 1.82$	$\widehat{p_3, p_1} = 55°$	$2\bar{2}0$	$d_{115}/d_{220} = 1.84$	$\widehat{2\bar{2}0, 51\bar{1}} = 57.02°$
$p_2 = 2.47$ cm	$p_2/p_1 = 1.52$	$\widehat{p_1, p_2} = 90°$	$33\bar{1}$	$d_{331}/d_{220} = 1.54$	$\widehat{\bar{2}\bar{2}0, 33\bar{1}} = 90°$
$p_3 = 2.95$ cm	$p_3/p_2 = 1.19$	$\widehat{p_2, p_3} = 34°$	$51\bar{1}$	$d_{115}/d_{331} = 1.19$	$\widehat{33\bar{1}, 51\bar{1}} = 32.98°$

$$\text{POLE } A = [2\bar{2}0] \times [51\bar{1}] = [116].$$

Similarly, the other poles, B, C, D, E, F, can be indexed as in Figure S2.9. The scale factor ϕ/y can be found from the known angular separation of two poles. For example,

		ϕ/y (deg cm^{-1})
$AF = 6.42$ cm	$AF = 129, 116 = 4.373°$	0.681
$FC = 11.35$ cm	$FC = 129, 114 = 7.589°$	0.669
$AC = 9.35$ cm	$AC = 114, 116 = 6.209°$	0.664

$$(\phi/y)_{av} = 0.67° \text{ cm}^{-1}$$

The angles between point O and poles A, F, and C can now be found by measuring the distances:

$$\begin{aligned} OA &= 3.02 \text{ cm} \qquad & OA &= UVW, 116 = 2.03^\circ, \\ OF &= 5.45 \text{ cm} \qquad & OF &= UVW, 129 = 3.66^\circ, \\ OC &= 7.00 \text{ cm} \qquad & OC &= UVW, 114 = 4.70^\circ. \end{aligned}$$

The following system of equations can now be set up:

$$u + v + 6w = \sqrt{u^2 + v^2 + w^2}\,\sqrt{38} \times \cos 2.03^\circ \equiv K,$$

$$u + 2v + 9w = \sqrt{u^2 + v^2 + w^2}\,\sqrt{86} \times \cos 3.66^\circ \equiv L,$$

$$u + v + 4w = \sqrt{u^2 + v^2 + w^2}\,\sqrt{18} \times \cos 4.70^\circ \equiv M.$$

By the use of Kramer's rule, the solution can be written as

$$u = \frac{\begin{vmatrix} K & 1 & 6 \\ L & 2 & 9 \\ M & 1 & 4 \end{vmatrix}}{D}, \qquad v = \frac{\begin{vmatrix} 1 & K & 6 \\ 1 & L & 9 \\ 1 & M & 4 \end{vmatrix}}{D}, \qquad w = \frac{\begin{vmatrix} 1 & 1 & K \\ 1 & 2 & L \\ 1 & 1 & M \end{vmatrix}}{D},$$

where

$$D = \begin{vmatrix} 1 & 1 & 6 \\ 1 & 2 & 9 \\ 1 & 1 & 4 \end{vmatrix},$$

$$u = 0.17\,(u^2 + v^2 + w^2)^{1/2},$$

$$v = 0.20\,(u^2 + v^2 + w^2)^{1/2},$$

$$w = 0.97\,(u^2 + v^2 + w^2)^{1/2},$$

or normalized to

$$\frac{u}{(u^2 + v^2 + w^2)^{1/2}} \equiv U = 1,$$

where

$$U = 1, \qquad V = 1.17, \qquad W = 5.75.$$

The exact orientation is [1, 1.17, 5.75].

(*b*) To calculate the absolute value of λ use

$$\lambda = 2\theta_{hkl} d_{hkl},$$

where

$$2\theta_{hkl} = p_{hkl}\frac{\phi}{y},$$

$$d_{hkl} = \frac{a}{(h^2 + k^2 + l^2)^{1/2}};$$

$$\lambda = p_{hkl}\frac{\phi}{y}\frac{a}{(h^2 + k^2 + l^2)^{1/2}}$$

$$= 1.63 \text{ cm} \times 0.67^\circ \text{ cm}^{-1} \times \frac{2\pi}{180^\circ} \times 5.428 \text{ Å} \times \frac{1}{\sqrt{8}},$$

$$\underline{\lambda = 0.0365 \text{ Å}};$$

$$\lambda = \frac{12.26}{(E/V)^{1/2}[1 + 0.9788 \times 10^{-6}(E/V)]^{1/2}} \text{ Å},$$

$$= \frac{A}{E^{1/2}(1 + BE^{1/2})} \quad (E \text{ in volts}),$$

$$E = \frac{1}{2B}\left(\sqrt{1 + \frac{4BA^2}{\lambda^2}} - 1\right),$$

$$E = 102.5 \text{ kV}.$$

2.10 Using eq. 2.30, lattice parameter 5.43×10^{-10} m, and electron wavelength 3.7×10^{-12} m yields a value of thickness $t = 240$ nm from either 220 or 440 (most easily measured in Fig. 2.34*d*).

Chapter 3

3.1 Here F is the structure factor for the reflection considered, t the specimen thickness in the direction of the beam, I_0 the incident beam intensity, I_g the diffracted beam intensity, V_c the volume of the unit cell, and s the deviation parameter (see Chapter 2, Sections 1.2 and 3.1). The equation was deduced assuming that (*a*) the scattering was weak, so that the incident beam amplitude was constant; and (*b*) there was no absorption. Neither of these assumptions is ever strictly true, but both are more valid if the crystal is very thin. Similarly, no other beam should be significantly excited. However, even in thin crystals the

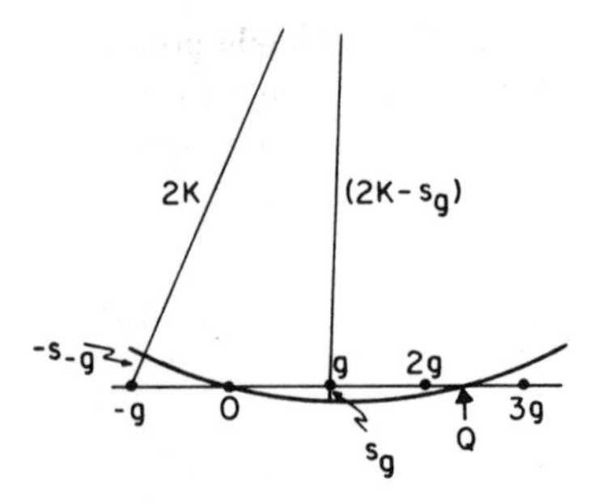

Fig. S3.2 The Ewald sphere cutting a systematic row of reflections.

equation will be incorrect for small values of s since

$$\frac{I_g}{I_0} = \frac{F^2}{V_c^2} \cdot \frac{(\pi t s)^2}{(\pi s)^2} = \left(\frac{Ft}{V_c}\right)^2,$$

that is, it increases without limit with increasing thickness!

3.2 The section through the Ewald sphere containing the incident beam direction and the systematic row is shown in Fig. S3.2. From the figure, using the intersecting chord theorem,

$$s_g(2K - s_g) = g \cdot (n - 1)g = 0g \times gQ,$$

that is,

$$s_g = \frac{(n - 1)g^2}{2K},$$

since $s_g << K$, and for $-g$

$$-s_{-g}(2K + s_{-g}) = g(n + 1)g = 0\bar{g} \times \bar{g}Q,$$

that is,

$$s_{-g} = \frac{-(n + 1)g^2}{2K},$$

where the point Q is where the sphere cuts through the row at $n\mathbf{g}$ from the origin. Now the length OQ is the sum of the distances of the components of any Kikuchi line pair from the origin, which may be seen from Fig. 3.51 to be $2.4g$.

For copper a = 3.62 Å and for 100 keV electrons $K = 1/\lambda = 27.02$ Å^{-1}; hence the values of s are as tabulated below (in units of Å^{-1}):

	g		
	111	200	220
s_g	5.9×10^{-3}	7.9×10^{-3}	1.6×10^{-2}
s_{-g}	-1.4×10^{-2}	-1.9×10^{-2}	-3.8×10^{-2}

As described more fully in Chapter 5, Section 7.1, narrow, accurately positioned weak beam images are obtained for $s \gtrsim 2 \times 10^{-2}$ Å^{-1}, that is, only for s_{-g} and the higher order reflections for the particular Kikuchi line positions considered in this case.

3.3 (*a*) Moiré fringes (whether parallel, rotation, or mixed) will rotate as the operating reflection **g** is changed. Thickness fringes will not change orientation with **g**. The two contrast effects can therefore be distinguished by changing **g**.

(*b*) Extinction contours move as the crystal is tilted. Dislocation images do not move on tilting the crystal but go out of contrast. The two contrast effects can therefore be distinguished by tilting the crystal.

3.4 Silicon (diamond cubic structure) consists of two interpenetrating, face centered cubic lattices. Allowed dislocations are as follows:

1. Perfect dislocations $\frac{1}{2}\langle 110\rangle$ (of which $\frac{1}{2}[1\bar{1}0]$, $\frac{1}{2}[10\bar{1}]$, and $\frac{1}{2}[01\bar{1}]$ have (111) as their slip plane).
2. Shockley partials $\frac{1}{6}\langle 112\rangle$ (of which $\frac{1}{6}[\bar{1}\bar{1}2]$, $\frac{1}{6}[\bar{1}2\bar{1}]$, and $\frac{1}{6}[2\bar{1}\bar{1}]$ have (111) as their slip plane).
3. Frank partials $\frac{1}{3}\langle 111\rangle$ (of which only $\frac{1}{3}[111]$ may lie in (111) and is sessile).

Tables 3.1 and 3.2 (based on fcc reflections) list many combinations of **g**'s and **b**'s for which $\mathbf{g} \cdot \mathbf{b} = 0$, for these Burgers vectors. Adding **g**'s of type 400 (and excluding 200, 420, etc., which are forbidden for diamond cubic), one may see that, for example, reflections 400, $11\bar{1}$, and $1\bar{1}1$ (all of which occur for a [011] foil orientation) distinguish between the perfect dislocations with slip plane (111). Additional reflections are required for the complete analysis.

In general, two noncolinear disappearance conditions, $\mathbf{g}_1$ and $\mathbf{g}_2$, define **b** uniquely in direction as $\mathbf{b} = \mathbf{g}_1 \times \mathbf{g}_2$.

3.5 The geometry of the situation is shown in Fig. S3.5*a*, where the foil normal **n** and fault plane normal **f** are indicated ([012] and $[1\bar{1}\bar{1}]$, respectively); these make an angle of 141° with each other in the projection plane $(\bar{1}\bar{2}1)$ of the figure. The fault is thus inclined at 39° to the surface and intersects the top (electron beam entrance) surface at *T* and the bottom at *B*. The projection $\mathbf{g}_p$ of the operating reflection, 200, in the plane of the figure is also shown. By convention the displacement **R** at the fault occurs in the lower crystal (i.e., to the left of the line *TB*), is in the directions shown for intrinsic ($\mathbf{R}_i$) and extrinsic ($\mathbf{R}_e$), and is of magnitude $f/3$. Thus the phase angles $\alpha = 2\pi\mathbf{g} \cdot \mathbf{R}$ are, respectively, $\alpha_i = +4\pi/3 \equiv -2\pi/3$ modulo 2π and $\alpha_e = -4\pi/3 \equiv +2\pi/3$.

The bright field image of a single stacking fault is symmetrical, with the initial fringe brighter than the background for $\alpha = +2\pi/3$ and darker for $\alpha = -2\pi/3$. Therefore the images expected are as sketched in Fig. S3.5*b*; see also Figs. 5.6 and 5.7.

The geometry for two parallel faults is shown in Fig. S3.5*c* with three repre-

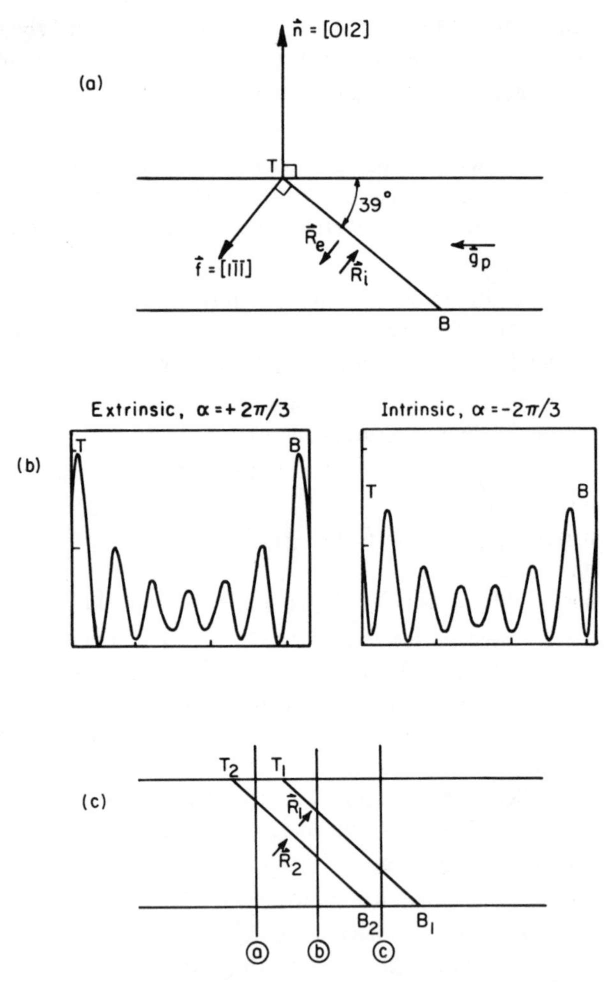

Fig. S3.5 (*a*) Defining geometry for an inclined displacement fault. (*b*) Resultant bright field images if the fault is an extrinsic or intrinsic stacking fault. (*c*) Defining geometry for an overlapping pair of faults.

sentative image "columns" indicated, one for each of the possible physical situations. If the pair is *not too close*, columns *a* and *c* are in the single-fault regions extending from T_2 to T_1 and B_2 to B_1. Under these conditions the extreme fringes will be characteristic of fault 2 at the top and fault 1 at the bottom. If the faults are dissimilar, this will lead to asymmetry of the bright field image, the doubly faulted region in the middle also having fault-type fringes. If the faults are very closely spaced (i.e., $T_2 T_1$ is smaller than a fraction of the fringe period in Fig. S3.5*b*), the result is the same as a single fault with

Table S3.6 $\mathbf{g} \cdot \mathbf{b}_p$ and α Values for $\mathbf{b}_p^1 = \frac{1}{6}[121]$, $\mathbf{b}_p^2 = \frac{1}{6}[21\bar{1}]$

	g				
	$\bar{1}11$	002	$2\bar{2}0$	$1\bar{1}3$	$\bar{1}13$
$\mathbf{g} \cdot \mathbf{b}_p^1$	+1/3	+1/3	-1/3	+1/3	+2/3
$\mathbf{g} \cdot \mathbf{b}_p^2$	-1/3	-1/3	+1/3	-1/3	-2/3
α $(=2\pi\mathbf{g} \cdot \mathbf{b}_p^1)$	$+2\pi/3$	$+2\pi/3$	$-2\pi/3$	$+2\pi/3$	$-2\pi/3$
Case	*f*	*f*	*f*	*f*	*a* (*c*)

displacement **R** equal to the sum of the separate displacements. For $\mathbf{R} = \mathbf{R}_i + \mathbf{R}_i$ this gives $\alpha = 8\pi/3 \equiv +2\pi/3$, that is, as a single extrinsic fault, while for $\mathbf{R} = \mathbf{R}_i + \mathbf{R}_e = 0$ the pair should be invisible. The closer the faults, the more complete is the lack of visibility.

3.6 Considering only the four lowest reflections of types 111, 200, 220, and 311, those for which $\mathbf{g} \cdot \mathbf{b} = 0$ for $\mathbf{b} = \frac{1}{2}[110]$ (i.e., reflections **g** for which the total dislocation is invisible) are selected, and the requisite $\mathbf{g} \cdot \mathbf{b}_p^1$ values presented in Table S3.6. The phase angle α for the stacking fault has been set equal to $2\pi\mathbf{g} \cdot \mathbf{b}_p^1$.

Under most conditions partial dislocations with $|\mathbf{g} \cdot \mathbf{b}| = \frac{1}{3}$ are invisible, whereas near the reflecting position those with $|\mathbf{g} \cdot \mathbf{b}| = \frac{2}{3}$ are visible. However, further from the reflecting position ($w > 1$) those with $\mathbf{g} \cdot \mathbf{b} = -\frac{2}{3}$ are very much less visible than those with $\mathbf{g} \cdot \mathbf{b} = +\frac{2}{3}$. Thus only cases *f*, *a*, and perhaps *c* are possible, as indicated in Table S3.6. [Reference: J. M. Silcock and W. J. Tunstall, *Phil. Mag.*, **10**, 361 (1964).]

3.7 (*a*) A superlattice dislocation may be considered as a matrix dislocation with multiple Burger's vector, that is, $\mathbf{b}_s = n\mathbf{b}_m$ (say, $n = 2$). Thus in matrix reflections $\mathbf{g}_m$ the values of $\mathbf{g}_m \cdot \mathbf{b}_s$ will be at least doubly integral, giving complicated images. In superlattice reflections, $\mathbf{g}_s = \mathbf{g}_m/n$, $\mathbf{g}_s \cdot \mathbf{b}_s$ will be singly integral, but since extinction distances are often large, complicated images will ensue. The $\mathbf{g} \cdot \mathbf{b} = 0$ disappearance condition applies to either.

(*b*) At antiphase domain boundaries the displacement **R** is a matrix lattice translation, and hence the boundaries are invisible in all matrix reflections. In superlattice reflections the phase angle α may be π (e.g., for $\mathbf{R} = \mathbf{R}_m = \mathbf{R}_s/2$, $\mathbf{g}_s = \mathbf{g}_m/2$), giving rise to characteristic fringes (see Figs. 3.17, 5.6*e*, 5.8*e*, and 5.9), although in more complicated situations α may take other values with resulting fringe patterns more like stacking fault fringes. Stacking faults will exhibit fringes in many matrix reflections as well as in superlattice reflections.

3.8 The defect strain fields assumed in row 4 of Fig. S3.8 would give rise to the contrast features in rows 1 to 3. For convenience an arbitrary sign has been

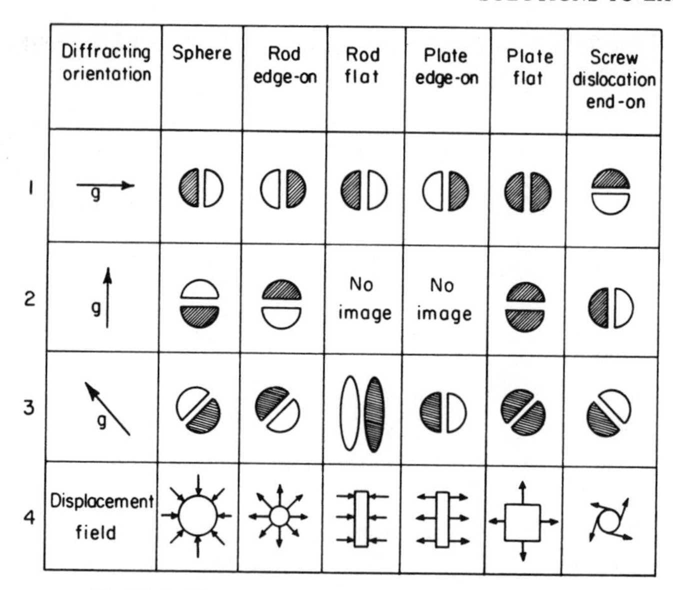

Fig. S3.8 The completed set of images from Fig. 3.55.

applied to each strain field; reversal of that sign would reverse the contrast. Such contrast is depth dependent however.

3.9 The resulting lattice image fringe patterns are shown in Fig. S3.9, together with comments on the behavior expected with change of focus (see Chapter 5, Section 9, for fuller details).

3.10 Rel-rods along ⟨111⟩ directions in reciprocal space are produced by thin plates on {111} planes in real space. Therefore take lattice images of {111} planes to observe these plates edge-on. To minimize complications due to changes with

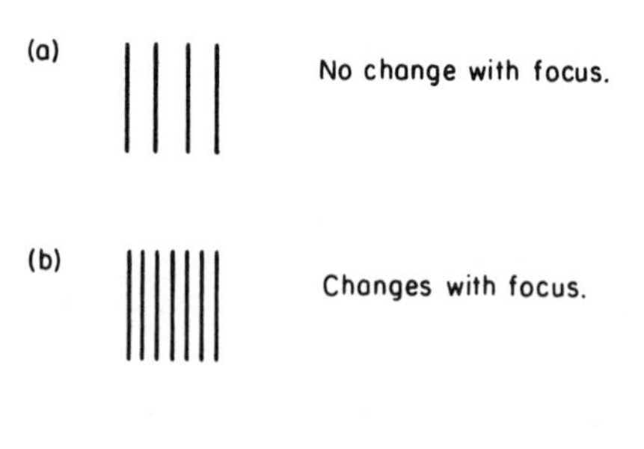

Fig. S3.9 Showing fringe patterns, with expected effect of changes in focus.

defocus (see Exercise 3.9) use two-beam tilted illumination with any g_{111} operating.

The experiment can be carried out in a specimen orientation that contains the $\langle 111 \rangle$ direction by tilting away from a symmetrical multibeam situation to a two-beam condition with g_{111} in the diffracting condition. The specimen should be as thin as possible, that is, on the order of a few hundred angstroms.

The size of the aperture must be large enough to include the transmitted and the tilted beam simultaneously. This can be determined from the Bragg angle, $\lambda/d_{111} = 2\theta = 0.016$ rad. With a focal length f of the objective lens of 3 mm the diameter of the aperture should be

$$\phi = 2\theta \times f = 47 \times 10^{-3} \text{ mm} \simeq 50\ \mu\text{m}.$$

To observe the effects of the plates on the image, monitor the fringe spacing by microdensitometer traces or optical diffraction from the micrograph.

Chapter 4

4.1 After traversing the crystal, an incident wave $\phi_0 \exp(2\pi i \boldsymbol{\chi} \cdot \mathbf{r})$ has become, by integration of eq. 4.5 for small values of t,

$$\psi = \phi_0 \exp(2\pi i \boldsymbol{\chi} \cdot \mathbf{r}) \exp\left[\frac{2\pi i m_0 e t}{h^2 \chi}(V_0 + V_1 \cos 2\pi \mathbf{g} \cdot \mathbf{r})\right].$$

Setting $A = 2\pi m_0 e/h^2 \chi$,

$$\begin{aligned}\psi &= \phi_0 \exp(2\pi i \boldsymbol{\chi} \cdot \mathbf{r} + iAV_0 t) \exp(AV_1 t \cos 2\pi \mathbf{g} \cdot \mathbf{r}) \\ &= \phi_0^1 \exp(2\pi i \boldsymbol{\chi} \cdot \mathbf{r})(1 + iAV_1 t \cos 2\pi \mathbf{g} \cdot \mathbf{r}),\end{aligned}$$

where $\phi_0^1 = \phi_0 \exp(iAV_0 t)$, and t has again been taken to be small. Hence

$$\begin{aligned}\psi &= \phi_0^1 \exp(2\pi i \boldsymbol{\chi} \cdot \mathbf{r})\{1 + \tfrac{1}{2} iAV_1 t[\exp(2\pi i \mathbf{g} \cdot \mathbf{r}) + \exp(-2\pi i \mathbf{g} \cdot \mathbf{r})]\} \\ &= \phi_0^1 \{\exp(2\pi i \boldsymbol{\chi} \cdot \mathbf{r}) + (\tfrac{1}{2} iAV_1 t) \exp[2\pi i(\boldsymbol{\chi} + \mathbf{g}) \cdot \mathbf{r}] \\ &\quad + (\tfrac{1}{2} iAV_1 t) \exp[2\pi i(\boldsymbol{\chi} - \mathbf{g}) \cdot \mathbf{r}]\},\end{aligned}$$

that is, a beam of the original amplitude in the incident direction plus two beams in diffracted directions $\boldsymbol{\chi} \pm \mathbf{g}$ of relative amplitude $AV_1 t/2$ $(= \pi m_0 e V_1 t/h^2 \chi)$.

4.2 Increasing the voltage of the electron microscope increases the values of both χ and ξ_g in Exercises 3.1 and 4.1 (but not in the simple nonrelativistic way; see eqs. 4.47 and 1.5). Therefore at first sight both phase grating and kinematical integrals may be used for *thicker* specimens. However, the kinematical formulation is less usable at higher voltages in the sense that many more beams are excited (because many more reflections are near the "flattened out" Ewald sphere).

4.3 Following solution 4.1 and again setting $A = 2\pi m_0 e/h^2\chi$ yields for the wave emitted

$$\psi = \phi_0^1 \exp(2\pi i \boldsymbol{\chi} \cdot \mathbf{r}) \exp [AV_1 t (\cos 2\pi g_x x + \cos 2\pi g_y y) + AV_2 t \cos 2\pi (g_x x + g_y y)] .$$

Expanding the second exponential, this time to second order, and selecting the required components yields only

$$\phi_{g_x 00} = \phi_0^1 \left[\frac{iAV_1 t}{2} + 0(AV_1 t)^2 \right]$$

$$\phi_{g_x g_y 0} = \phi_0^1 \left[\frac{iAV_2 t}{2} + \frac{(AV_1 t)^2}{2} + 0(AV_1 t)^3 \right] .$$

where in the latter expression the term in t^2 may be of comparable magnitude to that in t, because $V_1 >> V_2$. The term in $V_2 t$ corresponds to single scattering directly to $\mathbf{g} = (g_x, g_y, 0)$, while the term in $(V_1 t)^2$ corresponds to double scattering to $\mathbf{g} = (g_x, 0, 0) + (0, g_y, 0)$. Thus the intensities of the beams are

$$I_{g_x 00} = I_0 \frac{(AV_1 t)^2}{2},$$

$$I_{g_x g_y 0} = I_0 \left[\frac{(AV_2 t)^2}{2} + \frac{(AV_1 t)^4}{4} \right] .$$

4.4 A Bloch wave has a high potential for absorption when its wave function piles up charge at the lattice ion cores. Such a condition is satisfied by the form

$$\psi^{(1)} = A \cos \frac{\pi \chi}{d} e^{2\pi i k_z z} .$$

Strong channeling (low absorption) is allowed by the wave function which concentrates charge between ion cores:

$$\psi^{(2)} = A \sin \frac{\pi \chi}{d} e^{2\pi i k_z z} .$$

Both wave functions and the potential U experienced by their electrons are depicted schematically in Fig. S4.4.

The probability of absorption is given by the volume integral of the product of the electron density and the local absorption probability. For the given cube

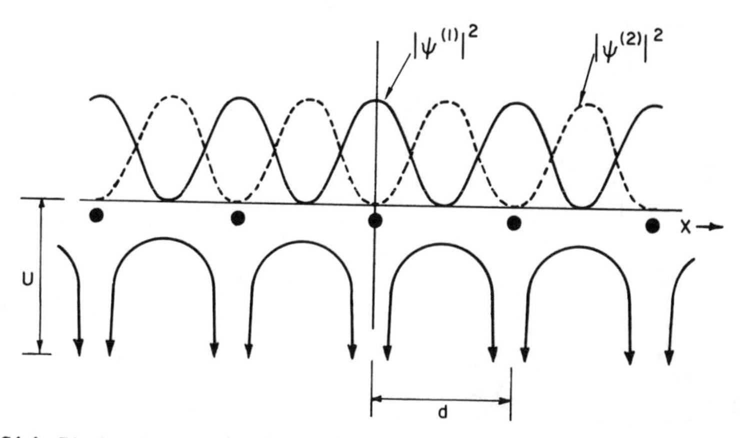

Fig. S4.4 Bloch wave electron densities and the potential U experienced by the electrons.

of side αd,

$$p^{(1)} \alpha \iiint_{\text{cube } \alpha d} A_2 \cos^2 \frac{\pi x}{d} dv = A^2 (\alpha d)^2 \left(\alpha d + \frac{d}{\pi} \sin \alpha\pi\right),$$

$$p^{(2)} \alpha \iiint_{\text{cube } \alpha d} A^2 \sin^2 \frac{\pi x}{d} dv = A^2 (\alpha d)^2 \left(\alpha d - \frac{d}{\pi} \sin \alpha\pi\right),$$

$$\text{Ratio } \frac{p^{(1)}}{p^{(2)}} = \frac{\alpha\pi + \sin \alpha\pi}{\alpha\pi - \sin \alpha\pi}.$$

(*a*) As $\alpha \to 0$, the ratio $\to \infty$, representing a strong localization of the scattering process about ion positions and hence negligible absorption in the well-transmitted beam.

(*b*) Conversely, as $\alpha \to 1$, the ratio also tends to unity; a uniform absorption probability affects both types of wave identically.

4.5 The Ewald sphere intersects the plane containing $\mathbf{g}_1$ and $\mathbf{g}_2$ in a circle as shown in Fig. S4.5*a*, where the *k*-vectors required for eq. 4.50 are also indicated. The equation may be written out in full as

$$\begin{aligned}
&[K^2 - (k^{(0)})^2] C_0 + && U_{10} C_1 + && U_{20} C_2 = 0, \\
&U_{01} C_0 + && [K^2 - (k^{(g_1)})^2] C_1 + && U_{21} C_2 = 0, \\
&U_{02} C_0 + && U_{12} C_1 + && [K^2 - (k^{(g_2)})^2] C_2 = 0,
\end{aligned}$$

Setting $\mathbf{g}_1 = 220$ and $\mathbf{g}_2 = 20\bar{2}$, and inspecting Fig. S4.5*a*, it may be seen that, for example, $\mathbf{k}_{xy}^{(0)} = (-\mathbf{g}_1 - \mathbf{g}_2)/3 = [\bar{4}\bar{2}2]/3 = \delta$, and hence that the bracketed

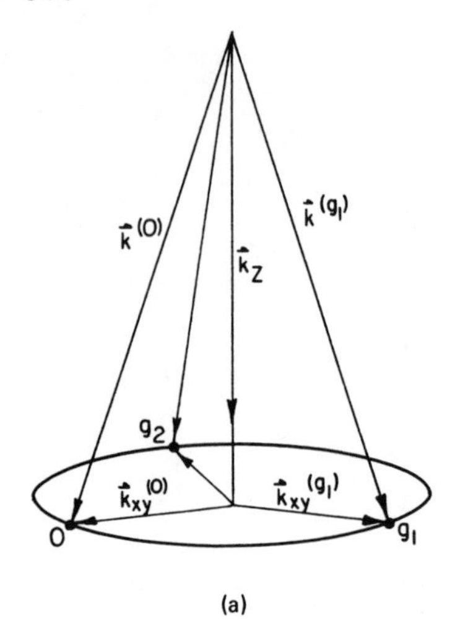

Fig. S4.5 (*a*) The Ewald sphere construction for Exercise 4.5.

terms all reduce to $X = K^2 - k_z^2 - \delta^2$. Also, for the centrosymmetric (fcc) crystal $U_{10} = U_{01}$, $U_{20} = U_{02}$, $U_{21} = U_{12}$, and for the present geometry all are equal to $U = U_{220}$. Hence the condition for a nontrivial solution for the C's reduces to

$$\begin{vmatrix} X & U & U \\ U & X & U \\ U & U & X \end{vmatrix} = 0,$$

that is, $X = U$ (twice), $-2U$.

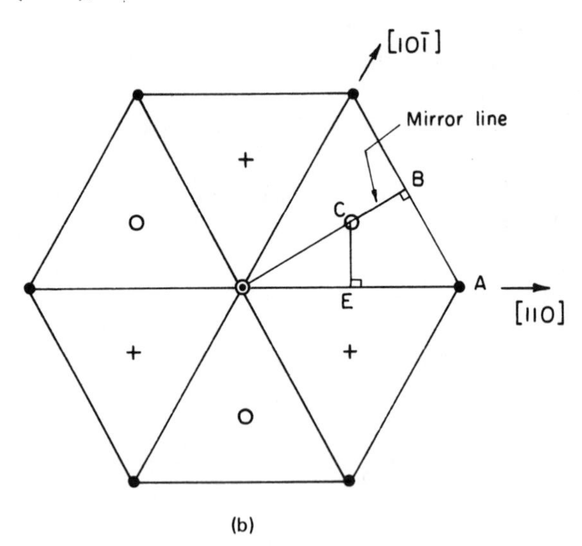

Fig. S4.5 (*b*) Projection of the fcc structure on (111).

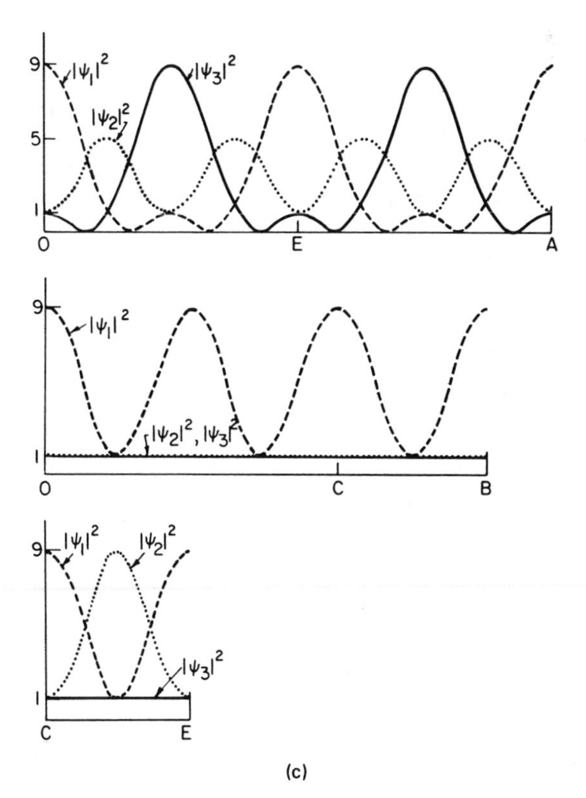

Fig. S4.5 (*c*) The relative Bloch wave electron densities along the indicated lines in Fig. S4.5*b*.

Hence, since $K^2 - k_z^2 \simeq 2k(K - k_z) = -2k\gamma$,

$$\gamma = \frac{2U - \delta^2}{2k}, \quad \frac{-(U - \delta^2)}{2k} \text{ (twice)}.$$

Backsubstitution yields the full solution (neglecting the δ^2 offset common to all) as

$$\gamma^{(1)} = \frac{+U}{k}, \quad C_0 = C_1 = C_2,$$

$$\gamma^{(2)} = \frac{-U}{2k}, \quad C_0 = C_1 = -C_2,$$

$$\gamma^{(3)} = \frac{-U}{2k}, \quad C_0 = -C_1 = C_2.$$

The electron densities are given by $D = |\psi|^2$, where

$$\psi = C_0 \exp(2\pi i \mathbf{k}_{xy}^0 \cdot \mathbf{r}) + C_1 \exp(2\pi i \mathbf{k}_{xy}^{(g_1)} \cdot \mathbf{r}) + C_2 \exp(2\pi i \mathbf{k}_{xy}^{(g_1)} \cdot \mathbf{r}).$$

At the atom sites in the reference crystal lattice plane (denoted in Fig. S4.5*b*) these relative densities are $D^{(1)} = 9, D^{(2)} = D^{(3)} = 1$. The same densities are found at positions where there are atoms in the close-packed planes above (denoted as 0) and below (+) the reference plane (e.g., $\mathbf{r} = [21\bar{1}]/3$, point C in Fig. S4.5*b*). Density plots are drawn in Fig. S4.5*c* along convenient lines in the typical triangle OAB, the total distribution being produced by the symmetry operations of the mirror line OB and sixfold axis at O.

The corresponding two-beam solutions are

$$\gamma^{(1)} = \frac{+U}{2k}, \qquad C_0 = C_1,$$

$$\gamma^{(2)} = \frac{-U}{2k}, \qquad C_0 = -C_1,$$

with the electron densities as shown in Figs. 4.12 and S4.4, the interatom distances being equivalent to OA in Fig. S4.5*b*.

4.6 The z-components of the wave vectors of the Bloch waves generated on [100] incidence, at the exact 022 reflecting condition, differ by an amount Δk, given by

$$\Delta k = \frac{1}{\xi_{022}} = \frac{1}{76\ \mathrm{nm}} = 1.3 \times 10^7\ \mathrm{m}^{-1},$$

as indicated in Fig. S4.6; tie-line ABC is normal to the top surface. On exit from the (210) face two wave points, D and E, are excited, their angular splitting being $\Delta\theta$, where

$$\Delta\theta = \frac{CD}{k}.$$

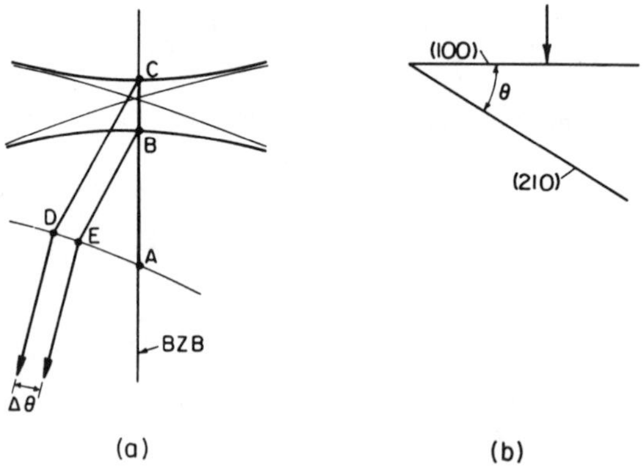

(a) (b)

Fig. S4.6 (*a*) Dispersion surface construction and (*b*) the corresponding crystal geometry pertinent to Exercise 4.6.

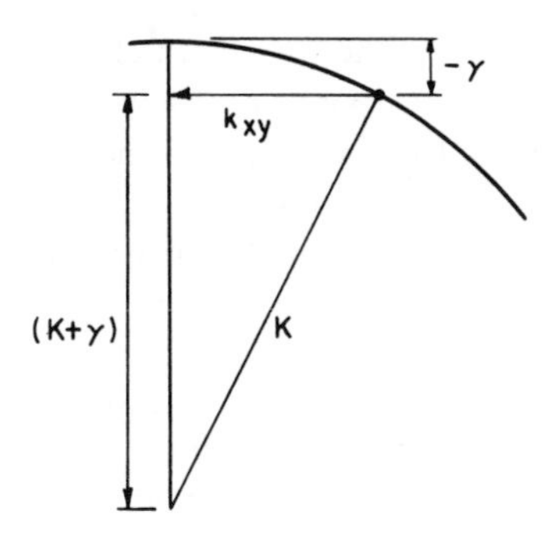

Fig. S4.7 "Free electron" wave relations.

Now $DE = \Delta k \tan\theta$, $\theta = \cos^{-1}(100 \cdot 210) = 26.6°$, and for 100 keV electrons $k = 2.7 \times 10^{11}\ \text{m}^{-1}$. Hence

$$\Delta\theta = \frac{\Delta k \tan\theta}{k} = 2.4 \times 10^{-5}\ \text{rad}.$$

These two beams give rise to interference fringes perpendicular to direction DE with period $1/DE = 1/\Delta k \tan\theta = \xi_g/\tan\theta$. The latter is the same formula as is used for calculating thickness fringe spacings. In this geometrical sense thickness fringes may therefore alternatively be thought of as interference fringes between the emissions from points D and E.

4.7 As discussed in Section 4.2, the dispersion surface $[\gamma(k_{xy})]$ diagrams and conventional $E(k)$ diagrams are similar because they both relate energy and **k**-vector in a particular direction in the x–y plane. Likewise they are strictly three dimensional (the term "dispersion *surface*" emphasizing this point). To confirm that γ and E are equivalent consider Fig. S4.7, which is drawn for the "free electron" situation. Here

$$(k + \gamma)^2 + k_{xy}^2 = k^2,$$

or, since $\gamma << k$,

$$-\gamma = \frac{k_{xy}^2}{2k},$$

which compares with $E = h^2k^2/2m$ for the $E(k)$ diagram. Also note here the unfortunate difference in convention adopted in the two approaches in respect to the scale of reciprocal space–in solid state physics usually $2\pi/d$, and in X-ray and electron diffraction usually $1/d$.

Chapter 5

5.1 The small contrast features in Fig. 5.39*a* show the characteristic "black-white" contrast parallel to **g**, which was discussed in Section 3.3 (See also problem 3.8). This contrast occurs for small defects with significant strain fields **R** when they are near either surface and the orientation of the crystal is very near the exact reflecting condition. The black-white vector **1** is along +**g** for some defects and −**g** for others, indicating that they lie in more than one of the depth bands shown in Fig. 5.16. The variation in size and strength of the image may be an indication of the true size of the defect, but may also arise from this same depth dependence.

The "black-on-white" and "white-on-black" spots on either side of the thickness fringes in Fig. 5.39*b* are caused by defects with very small strain fields (there is some suggestion of black-white contrast in some of them). The principal contrast mechanism in this case is structure factor contrast. Defects of different structure from the matrix (small voids here) act to a good approximation (at the reflecting position) to make the specimen locally thinner (or thicker if of the same structure and more strongly scattering). Thus, where contrast is varying rapidly with thickness [$t = (n + 1/4)\xi_g$ for small integral values of n; see Fig. 4.6], the defects will be visible as brighter spots on the thin side of a dark thickness fringe and as darker spots on the thick side, as in the figure. The reverse would be true if the small defect were of greater scattering power.

The larger contrast features are dislocation loops.

5.2 From Fig. S5.2

$$s_{+2}(2k - s_{+2}) = g \cdot 2g,$$

that is,

$$s_{+2} = s_{-1} = -\frac{g^2}{k}.$$

Equation 4.61 may be written as

$$[\mathrm{A} - \{\gamma\}]\mathbf{C}^{(i)} = 0.$$

Then also

$$\begin{bmatrix} s_{-1} - \gamma & U_1/2K & U_2/2K & U_3/2K \\ U_1/2K & -\gamma & U_1/2K & U_2/2K \\ U_2/2K & U_1/2K & s_{+1} - \gamma & U_1/2K \\ U_3/2K & U_2/2K & U_1/2K & s_{+2} - \gamma \end{bmatrix} \begin{bmatrix} C_{-1} \\ C_0 \\ C_{+1} \\ C_{+2} \end{bmatrix} = 0.$$

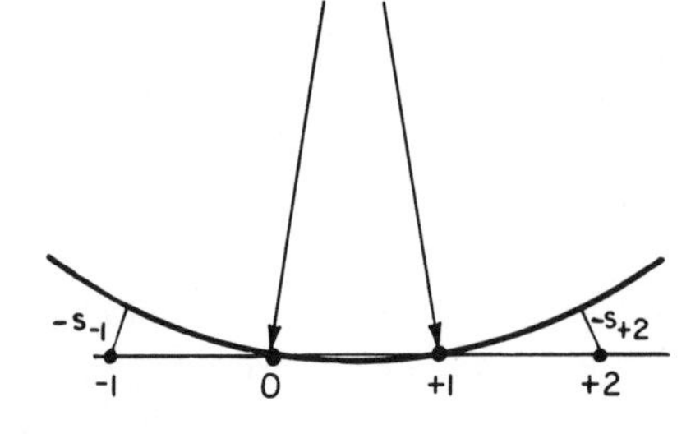

Fig. S5.2 The Ewald sphere construction for the four-beam symmetrical systematic case.

Now $s_{-1} = s_{+2} = -g^2/K$ and $s_{+1} = 0$, to produce a very symmetrical matrix. Hence the solutions must have symmetry.

1. $C_{-1} = C_{+2} = 2KD$, $C_0 = C_{+1} = 2KE$, symmetric.

$$D(-2g^2 - 2K\gamma + U_3) + E(U_1 + U_2) = 0, \tag{1a}$$

$$D(U_1 + U_2) = E(U_1 - 2K\gamma) = 0, \tag{1b}$$

that is, for nontrivial solution determinant (coefficients) = 0

$$(U_1 - 2K\gamma)(-2g^2 - 2K\gamma + U_3) + (U_1 + U_2)^2 = 0,$$

$$4K^2\gamma^2 - \gamma(-2g^2 + U_3 + U_1)(2K) - (U_1 + U_2)^2 + U_1 U_3 - 2g^2 U_1 = 0,$$

$$\gamma = \frac{(-2g^2 + U_1 + U_3) \pm \begin{bmatrix} 4g^4 + U_3^2 + U_1^2 - 4g^2 U_1 - 4g^2 U_3 + 2U_1 U_3 \\ + 4(U_1 + U_2)^2 - 4U_1 U_3 + 8g^2 U_1 \end{bmatrix}^{1/2}}{4K},$$

$$4K\gamma = (U_1 + U_3 - 2g^2) \pm [(2g^2 + U_1 - U_3)^2 + 4(U_1 + U_2)^2]^{1/2}. \tag{2}$$

2. $C_{-1} = -C_{+2} = 2KF$, $C_0 = -C_{+1} = 2KG$, antisymmetric.

$$F(-2g^2 - 2K\gamma - U_3) + G(U_1 - U_2) = 0, \tag{3a}$$

$$F(U_1 - U_2) + G(-U_1 - 2K\gamma) = 0, \tag{3b}$$

$$4K^2\gamma^2 - \gamma(-2g^2 - U_3 - U_1)(2K) - (U_1 - U_2)^2 + U_1 U_3 + 2U_1 g^2 = 0,$$

$$4K\gamma = (-U_1 - U_3 - 2g^2) \pm \begin{bmatrix} 4g^4 + U_3^2 + U_1^2 + 4g^2 U_3 + 4g^2 U_1 \\ + 2U_1 U_3 + 4(U_1 - U_2)^2 - 4U_1 U_3 - 8U_1 g^2 \end{bmatrix}^{1/2},$$

$$= (-U_1 - U_3 - 2g^2) \pm [(2g^2 - U_1 + U_3)^2 + 4(U_1 - U_2)^2]^{1/2}. \tag{4}$$

The $\Delta\gamma$ of interest is that between the two largest values of γ, and obviously the + sign from each is likely to be what is required; one solution from each of the symmetric (eq. 2) and antisymmetric (eq. 4) sets (by analogy with the two-beam solution)

$$\begin{aligned} 4K\,\Delta\gamma_{++} = 2U_1 + 2U_3 &+ [(2g^2 + U_1 - U_3)^2 + 4(U_1 + U_2)^2]^{1/2} \\ &- [(2g^2 - U_1 + U_3)^2 + 4(U_1 - U_2)^2]^{1/2}, \end{aligned}$$

which is essentially eq. 5.21.

A check should be made that this is indeed the most important γ by confirming that the values of C_0 for these two solutions are much larger than those for the − sign solutions. For example, in the symmetric case $E/D = C_2/C_0$ must be small. Substituting for γ from eq. 2 into eq. 1b gives

$$\frac{E}{D} = \frac{U_1 + U_2}{2K\gamma - U_1}.$$

Obviously the solution γ_+^s produces the smaller value of E/D.

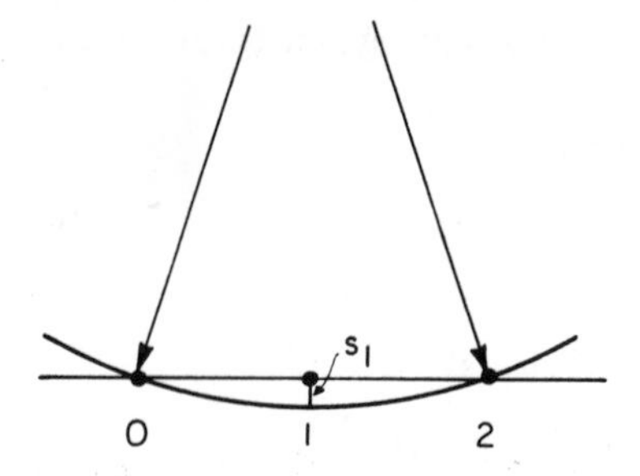

Fig. S5.3 The Ewald sphere construction for the three-beam systematic case.

Similarly, γ_+^A produces the smaller value of G/F.

5.3 From Fig. S5.3 $s_0 = s_2 = 0$, $s_1 = g^2/2K$. In this case eq. 4.61 becomes

$$\begin{bmatrix} -\gamma & U_1/2K & U_2/2K \\ U_1/2K & g^2/2K - \gamma & U_1/2K \\ U_2/2K & U_1/2K & -\gamma \end{bmatrix} \begin{bmatrix} C_0 \\ C_1 \\ C_2 \end{bmatrix} = 0.$$

The solutions must have symmetry.

1. $C_0 = C_2 = 2KM$, $C_1 = 2KN$, symmetric.

In other words,

$$(U_2 - 2K\gamma) M + U_1 N = 0,$$

$$2U_1 M + (g^2 - 2K\gamma) N = 0,$$

and for a nontrivial solution

$$(U_2 - 2K\gamma)(g^2 - 2K\gamma) - 2U_1^2 = 0$$

or

$$\gamma_s = \frac{(g^2 + U_2) \pm \sqrt{(g^2 + U_2)^2 - 4(U_2 g^2 - 2U_1^2)}}{4K}.$$

2. $C_0 = -C_2 = 2KP$, $C_1 = 0$, antisymmetric.

In other words,

$$P(U_2 + 2K\gamma) = 0,$$

and for a nontrivial solution

$$\gamma_A = \frac{-U_2}{2K}.$$

Branches 2 and 3 coincide when $\gamma_s^- = \gamma_A$, that is,

$$(g^2 + U_2) - \sqrt{(g^2 + U_2)^2 - 4(U_2 g^2 - 2U_1^2)} = -2U_2,$$

which reduces to eq. 5.26, namely,

$$U_2^2 + g^2 U_2 = U_1^2 .$$

Noting that

$$U_g = \frac{K}{\xi_g} \qquad \text{(eq. 4.57)},$$

$$U = \frac{2me\,V}{n} \qquad \text{(eq. 4.46)},$$

and

$$m = m_0 \left(1 + \frac{E_0}{m_0 c^2}\right) \qquad \text{(eq. 4.47)},$$

substituting, and solving for E_0 yields

$$E_0(2g) = \frac{2.26 \times 10^{-18} (\xi_g^{100})^2 (\xi_{2g}^{100})\, g^2}{(\xi_{2g}^{100})^2 - (\xi_g^{100})^2} - 512,$$

where the 100 kV values of extinction distances ξ_g^{100} and ξ_{2g}^{100} are in nanometers, g is in reciprocal meters (m^{-1}), and E_0 is in kilovolts.

5.4 For atomic fractions x of gold and $1 - x$ of copper it may be assumed that potentials U are averaged and hence that

$$\frac{1}{\xi_g} = \frac{x}{\xi_g(\text{Au})} + \frac{1 - x}{\xi_g(\text{Cu})}.$$

Differentiation of eq. 5.27 to find its rate of change yields a very complicated analytical expression; it is sufficiently accurate to use a graphical method instead. Substitution of suitable values yields the following results for 222 and 400 reflections, both of which are in the 100 kV to 1 MV range:

	Cu_2Au $x = 0.33$	Cu_3Au $x = 0.25$	Cu_4Au $x = 0.20$
V_c (222)	520	561	586
V_c (400)	841	897	935

Interpolating ±5 kV in V_c corresponds to $\pm 10^{-2}$ and $\pm 6 \times 10^{-3}$ in x, respectively, for 222 and 400 reflections.

5.5 Simplification of eq. 5.28 to the two-beam situation yields (remembering that $\beta_0' = 0$)

$$\frac{d\epsilon^{(1)}}{dz} = 2\pi i \,\{\epsilon^{(1)} C_g^{(1)*} C_g^{(1)} + \epsilon^{(2)} C_g^{(1)*} C_g^{(2)} \exp(2\pi i\, \Delta\gamma z)\}\, \beta_g',$$

$$\frac{d\epsilon^{(2)}}{dz} = 2\pi i \, \{\epsilon^{(1)} C_g^{(2)*} C_g^{(1)} \exp(2\pi i \, \Delta\gamma z) + \epsilon^{(2)} C_g^{(2)*} C_g^{(2)}\} \, \beta'_g.$$

The two-beam values for $C_g^{(j)}$ are, from eq. 4.39,

$$C_g^{(1)} = \cos\left(\frac{\beta}{2}\right), \qquad C_g^{(2)} = -\sin\left(\frac{\beta}{2}\right).$$

Hence the expressions become

$$\frac{d\epsilon^{(1)}}{dz} = 2\pi i \beta'_g \left\{\epsilon^{(1)} \cos^2 \frac{\beta}{2} - \tfrac{1}{2} \sin\beta \, \epsilon^{(2)} \exp(2\pi i \, \Delta\gamma z)\right\}.$$

$$\frac{d\epsilon^{(2)}}{dz} = 2\pi i \beta'_g \left\{-\tfrac{1}{2} \sin\beta \, \epsilon^{(1)} \exp(2\pi i \, \Delta\gamma z) + \epsilon^{(2)} \sin^2 \frac{\beta}{2}\right\}.$$

Making the substitutions

$$\epsilon^{(1)} \longrightarrow E^{(1)} \exp\left(2\pi i \mathbf{g} \cdot \mathbf{R} \sin^2 \frac{\beta}{2}\right),$$

$$\epsilon^{(2)} \longrightarrow E^{(2)} \exp\left(2\pi i \mathbf{g} \cdot \mathbf{R} \sin^2 \frac{\beta}{2}\right).$$

produces the final expressions

$$\frac{dE^{(1)}}{dz} = -\pi i \beta'_g E^{(2)} \sin\beta \exp(2\pi i \, \Delta\gamma z) \exp(-\pi i \cos\beta),$$

$$\frac{dE^{(2)}}{dz} = -\pi i \beta'_g E^{(1)} \sin\beta \exp(2\pi i \, \Delta\gamma z) \exp(+\pi i \cos\beta),$$

neither of which includes an intraband term.

5.6 For completeness most parts of Fig. 5.34 may be written in tabular form as follows:

	(1)	(2)	(3)	(4)
a	$R_g(z)$	$R_g(z)$	$-R_g(-z)$	
b			$R_g(-z)$	
c	$R_g(z)$	$R_g(z)$	$R_g(-z)$	$R_g(-z)$
d			$-R_g(-z)$	

where R_g is the component of the defect displacement field $\mathbf{R}$ in the direction of $\mathbf{g}$ (i.e., the component which yields $\mathbf{g} \cdot \mathbf{R}$), and the origin coordinates have been taken at the center of the specimen in the indicated column.

The bright field condition of eq. 5.35 then follows immediately by comparing $a(1)$ and $a(3)$ in the table. The dark field condition of eq. 5.36 follows by comparing $c(1)$ and $c(3)$ at the exact reflecting position ($\phi = \theta_g$).

The comparison of weak beam images becomes possible for different orientations ϕ:

Figure 5.34

ϕ	$c(1)$	$c(2) = d(3) = d(4)$	$c(3) = c(4)$
θ_g	(g, g)	$(-g, -g)$	(g, g)
0	$(g, 0)$	$(-g, -2g)$	$(g, 2g)$
$-\theta_g$	$(g, -g)$	$(-g, -3g)$	$(g, 3g)$

From comparison of $c(1)$ and $c(3)$ for $\phi = -\theta_g$ it may be seen that $(g, -g)$ and $(g, 3g)$ images are related in a way that is analogous to Ball's strong dark field criterion. In particular, the background intensity in perfect crystal regions, $\mathbf{R} = 0$ (and hence thickness fringes), will be identical.

5.7. (*a*)

$$\chi(g) = \pi\lambda g^2\left(\frac{C_s\lambda^2 g^2}{2} + D\right)$$

and

$$D = D_{\text{Sch}} = -(C_s\lambda)^{1/2}.$$

Substitution of $C_s = 5$ mm, $\lambda = 1.42$ pm, and $g_{200} = (2/0.352) \times 10^9\ m^{-1}$ yields the value of χ_{Sch} (200) = + 11.3 rad.

(*b*) Conditions for resolution in the axial mode are usually defined by the position of the first zero in the phase contrast function at Scherzer defocus (point A in Fig. 5.36*d*, although for lattice images it may be better to consider the real part–point E–which is marginally different). Setting $\chi = 0$ and $D = D_{\text{Sch}}$ gives the result

$$\delta_0 = \frac{1}{g_0} = \frac{C_s^{1/4}\lambda^{3/4}}{\sqrt{2}} = 0.7 C_s^{1/4}\lambda^{3/4} = 0.24 \text{ nm}.$$

[The corresponding value for the zero in the real part ($\chi = \pi/2$) is $0.64 C_s^{1/4}\lambda^{3/4}$ = 0.22 nm.] This resolution limit is greater than the 200 spacing of 0.176 nm, and thus the planes will not be imaged.

Even if resolution according to the formula were possible, it would be necessary to check that the phase factor was correct as well, for contrast to be obtained.

5.8 Fourier images occur at values of defocus D for which the phase factor χ (eq. 5.39) differs by $2n\pi$, that is,

$$\Delta D = \frac{2}{\lambda g^2} \quad \text{or} \quad \frac{2d^2}{\lambda}.$$

(*a*) For $Nb_{12}O_{29}$ in [010] projection there are two possible periodicities, $d = a$ and $d = c$:

$$\Delta D_a = 11.8\ \mu\text{m}, \quad \Delta D_c = 6.0\ \mu\text{m}.$$

Note that the defocus increments for the *a*-spacing are about twice as large as those for the *c*-spacing.

(*b*) For gold the spacing concerned is the 200 one, 0.204 nm:

$$\Delta D_{200} = 58.6 \text{ nm}.$$

Periodic defocusing situations like those above do not occur for symmetrical situations; in principle, the 400 fringes in gold (from the two dark field beams, $\mathbf{g}_1 = \bar{2}00$ and $\mathbf{g}_2 = 200$) would be present at all values of focus (but below the resolution limit).

INDEX